Algorithms for Intelligent Systems

This book series publishes research on the analysis and development of algorithms for intelligent systems with their applications to various real world problems. It covers research related to autonomous agents, multi-agent systems, behavioral modeling, reinforcement learning, game theory, mechanism design, machine learning, meta-heuristic search, optimization, planning and scheduling, artificial neural networks, evolutionary computation, swarm intelligence and other algorithms for intelligent systems.

The book series includes recent advancements, modification and applications of the artificial neural networks, evolutionary computation, swarm intelligence, artificial immune systems, fuzzy system, autonomous and multi agent systems, machine learning and other intelligent systems related areas. The material will be beneficial for the graduate students, post-graduate students as well as the researchers who want a broader view of advances in algorithms for intelligent systems. The contents will also be useful to the researchers from other fields who have no knowledge of the power of intelligent systems, e.g. the researchers in the field of bioinformatics, biochemists, mechanical and chemical engineers, economists, musicians and medical practitioners.

The series publishes monographs, edited volumes, advanced textbooks and selected proceedings.

More information about this series at http://www.springer.com/series/16171

Mohammad Shorif Uddin · Avdhesh Sharma ·
Kusum Lata Agarwal · Mukesh Saraswat
Editors

Intelligent Energy Management Technologies

ICAEM 2019

Springer

Editors
Mohammad Shorif Uddin
Department of Computer Science
and Engineering
Jahangirnagar University
Dhaka, Bangladesh

Kusum Lata Agarwal
Department of Electrical Engineering
Jodhpur Institute of Engineering
and Technology
Jodhpur, India

Avdhesh Sharma
Department of Electrical Engineering
MBM Engineering College
Jodhpur, India

Mukesh Saraswat
Jaypee Institute of Information Technology
Noida, India

ISSN 2524-7565 ISSN 2524-7573 (electronic)
Algorithms for Intelligent Systems
ISBN 978-981-15-8819-8 ISBN 978-981-15-8820-4 (eBook)
https://doi.org/10.1007/978-981-15-8820-4

This Springer imprint is published by the registered company Springer Nature Singapore Pte Ltd.
The registered company address is: 152 Beach Road, #21-01/04 Gateway East, Singapore 189721, Singapore

Preface

Twenty-first century is the transition from fossil-fuel-based energy to renewable energy, and after Montreal and Kyoto Protocol, the world is inclined toward renewable energy sources. Intelligent energy management technologies are now appearing very cost-effective compared to the projected high cost of fossil fuels. In this context, numerous intelligent control schemes are proposed for developing smart energy management systems. With this background, the idea of the International Conference on Advances in Energy Management System (ICAEM 2019) is conceived to develop a platform for networking; disseminating; and exchanging challenges, ideas, concepts, and results among the researchers from academia and industry.

This book contains the good-quality research papers as the proceedings of this International Conference on Advances in Energy Management System (ICAEM 2019). ICAEM 2019 has been jointly organized by the Rajasthan Technical University (RTU), Kota, and Jodhpur Institute of Engineering and Technology (JIET), Jodhpur, Rajasthan, India. It was held on December 20–21, 2019 at JIET, Jodhpur, Rajasthan, India. The conference focused on the application of artificial intelligence, soft computing, optimization, machine learning, intelligent software, data science, data security, and big data analytics on advanced energy management systems.

We have tried our best to enrich the quality of the ICAEM 2019 through a stringent and careful peer-review process. ICAEM 2019 received 132 papers on four conference tracks. Among these 65 papers were selected through the peer-review process for presentation during the conference. However, the final proceedings contain only 38 papers after careful editorial reviewing.

The book presents the intelligent computing research results in reliable power systems, power quality, smart grids, energy management, conversion techniques, energy economics, etc. We believe that this book serves as a reference material for advanced research in the field of energy management.

Dhaka, Bangladesh — Mohammad Shorif Uddin
Jodhpur, India — Avdhesh Sharma
Jodhpur, India — Kusum Lata Agarwal
Noida, India — Mukesh Saraswat

Contents

1 **Solar PV-Based Electric Vehicle Charging System with Power Backup** 1
Utkarsh Gupta, D. K. Yadav, and Dheeraj Panchauli

2 **A Survey of Energy Theft Detection Approaches in Smart Meters** 9
Divam Lehri and Arjun Choudhary

3 **Renewable Energy Conversion: Sustainable Energy Development and Efficiency Enhancement of Solar Panels: A Review** 25
Bhavik J. Pandya, Megha C. Karia, and Kamlesh B. Sangani

4 **Evaluation of Thermal Degradation Behavior of Cardboard Waste** 45
Samit Kumar Singh and Sadanand A. Namjoshi

5 **Wavelength Optimization in Gigabit Passive Optical Network by Proposed Quad Play Architecture** 53
Md. Hayder Ali and Mohammad Hanif Ali

6 **A Priority-Based Deficit Weighted Round Robin Queuing for Dynamic Bandwidth Allocation Algorithm in Gigabit Passive Optical Network** 65
Md. Hayder Ali and Mohammad Hanif Ali

7 **Protection of Six-Phase Transmission Line Using Demeyer Wavelet Transform** 73
Gaurav Kapoor

8 **Analysis of Voltage Sag and Swell Problems Using Fuzzy Logic for Power Quality Progress in Reliable Power System** 85
Ankit Tandon and Amit Singhal

9 Vehicle Detection and Its Speed Measurement 95
Morium Akter, Jannatul Ferdous, Mahmuda Najnin Eva, Sumaita Binte Shorif, Sk. Fahmida Islam, and Mohammad Shorif Uddin

10 Fuel Cells as Naval Prime Movers: Feasibility, Advances and Implications 101
Akshat Mathur and Sushma Dave

11 Overview of Eco-Friendly Construction Materials 111
Nidhi Sharma and Aashish Kumar Jha

12 Protection of Nine-Phase Transmission Line Using Biorthogonal-2.2 Wavelet Transform 119
Gaurav Kapoor

13 An Intelligent Hybrid Model for Forecasting of Heart and Diabetes Diseases with SMO and ANN 133
Shalini, Pawan Kumar Saini, and Yatendra Mohan Sharma

14 Power Quality Assessment of Solar PV Standalone System Using Various DC-DC Converters 139
Surbhi Shringi, Santosh Kumar Sharma, and Utkarsh Gupta

15 Coordinated Control of UPFC-Based Damping Controller with PID for Power System 149
Amit Singhal and Ankit Tandon

16 Computational Neuroscience and Its Applications: A Review 159
Aisha Jangid, Laxmi Chaudhary, and Komal Sharma

17 Optimization of Band Notch Characteristic in Ultra-Wideband Microstrip Patch Antenna for Wireless Power Transfer 171
Ashish Mathur, Geetika Mathur, and Abhijit kulshrestha

18 Experimental Investigation for Energy-Conscious Welding Based on Artificial Neural Network 181
Sudeep Kumar Singh, Suvam Sourav Swain, Amit Kumar, Prashanjeet Patra, Nitesh Kumar, and A. M. Mohanty

19 Detection of GSM Signal Using Energy Detection and Matched Filter-Based Techniques 191
Bablu Kumar Singh and Sanjay Bhandari

20 Simulation of Performance Characteristics of Different PV Materials 199
Harish Kumar Khyani and Jayashri Vajpai

21 Stabilization Analysis of Clay Soil by Using Waste Tile Particles 209
Saraswati Chand Dhariwal and Rajat Mangal

22 **Renewable Energy as Biofuel from Mirabilis Jalapa Seed Oil** 219
Akleshwar Mathur and Harish Kumar Khyani

23 **A Comparative Study of Various Approaches to Lossy Image Compression Process** 227
Nitesh Agarwal and Rajendra Purohit

24 **Practical Design Considerations of DC/DC Converter Used in MPPT for Solar PV Systems** 251
Kusum Lata Agarwal and Dhanraj

25 **Machine Learning-Based Solar Energy Forecasting: Implications on Grid and Power Market** 267
Chandershekhar Singh and Ajay Kushwaha

26 **Performance Comparison of Minimum Shift Keying and Gaussian Minimum Shift Keying in Additive White Gaussian Noise Environment** 277
Bablu Kumar Singh and Supriya Vyas

27 **MATLAB-Based Comparative Analysis of Alternative PV Models** 285
Khamma Kanwar and Jayashri Vajpai

28 **Nuclear Fusion: Energy of Future** 295
Jayesh Nehiwal, Harish Kumar Khyani, Shrawan Ram Patel, and Chandershekhar Singh

29 **Medical Image Processing Using Soft Computing Techniques and Mathematical Morphology** 303
Pratik Bhansali and Sandip Mehta

30 **Analysis of Hotspot Development in Power Transformer and Its Life Estimation** 319
Vinit Mehta and Jayashri Vajpai

31 **Photovoltaic Module Cleaning Prediction Using Deep Neural Networks** 335
Kapil Panwar, Aditya Jindal, and Kusum Lata Agarwal

32 **Optimal Controller Design for DC–DC Buck Converter** 343
Shubham Sharma and Kusum Lata Agarwal

33 **Comparative Analysis of Different Maximum Power Point Techniques for Solar Photovoltaic Systems** 357
Shyam Lal Vishnoi and Kusumlata Agarwal

34 **MATLAB/Simulink-Based Tracking of Maximum Power Point in a Generalized Photovoltaic Module by Using DC-DC Boost Converter** 381
Yogesh Joshi and Vinit Mehta

35 Design and Simulation of MPPT-Operated DC-DC Flyback Converter Used for Solar Photovoltaic System 397
Ranjana Choudhary and Shrawan Ram Patel

36 To Improve Power Transfer Capacity Using TCSC FACTS Controller 409
Kishore Singh Gehlot and Shrawan Ram

37 Stockwell Transform and Hilbert Transform Based Hybrid Algorithm for Recognition of Power Quality Disturbances 425
Ramesh Aseri and Ashwani Kumar Joshi

38 Detection and Classification of Transmission Line Faults Using Combined Features of Stockwell Transform, Hilbert Transform, and Wigner Distribution Function 449
Tanmay Bhati and Harish Kumar Khyani

Author Index 469

About the Editors

Dr. Mohammad Shorif Uddin is a Professor of Computer Science and Engineering at Jahangirnagar University, Bangladesh. He completed his Bachelor of Science in Electrical and Electronic Engineering at BUET in 1991, Master of Technology Education at Shiga University, Japan, in 1999, Doctor of Engineering in Information Science at Kyoto Institute of Technology, Japan, in 2002, and an M.B.A. at Jahangirnagar University in 2013. He undertook postdoctoral researches at Bioinformatics Institute, Singapore, Toyota Technological Institute, Japan, Kyoto Institute of Technology, Japan, Chiba University, Japan, Bonn University, Germany, and Institute of Automation, Chinese Academy of Sciences, China. His research is motivated by applications in the fields of artificial intelligence, imaging informatics, and computer vision. He is the Editor-in-Chief of ULAB Journal of Science and Engineering and an Associate Editor of IEEE Access and has served as General Chair or Co-Chair of various conferences, including the IJCCI 2018 and 2019, EICT 2017, and IWCI 2016. He holds two patents for his scientific inventions, is a senior member of several academic and professional associations, and has published extensively in reputed international journals and conference proceedings.

Dr. Avdhesh Sharma is a Professor of Electrical Engineering at M.B.M. Engineering College, India. He completed his Ph.D. in Electrical Engineering at IIT, Delhi, M.Tech. in Computer Science & Data Processing at IIT, Kharagpur, M.Sc. at AMU, and B.Sc. in Electrical Engineering at D.E.I. Engineering College, India. He has over 33+ years of experience in teaching and research. He has published around 75 papers in international journal and conference proceedings. He guided a good number of PG and Ph.D. students. He is a Fellow of the Institute of Engineers, India.

Dr. Kusum Lata Agarwal is a Professor and Head of Electrical Engineering at JIET, Jodhpur, India. She completed her Ph.D., M.E., and B.E. degrees at M.B.M. Engineering College, India. She has over 18 years of experience in teaching and research. She published around 30 papers in journal and conference proceedings.

She supervised a good number of PG theses. She played a key role in organizing 4 national and international conferences and also achieved notable research grants. She is in the reviewer panel of many journals and conferences.

Dr. Mukesh Saraswat is an Associate Professor of Computer Science & Engineering at Jaypee Institute of Information Technology, India. He completed his B.Tech. in CSE at MJP Rohilkhand University, M.Tech. in CSE at UPTU, and Ph.D. in CSE at IIITM, Gwalior, India. He has more than 15 years of teaching and research experience. He has guided more than 72 UG & PG theses and presently guiding 4 Ph.D. students. He is a member of several academic and professional bodies and has published more than 35 papers in journals and conference proceedings in the area of imaging, pattern recognition, and soft computing.

Chapter 1
Solar PV-Based Electric Vehicle Charging System with Power Backup

Utkarsh Gupta, D. K. Yadav, and Dheeraj Panchauli

1 Introduction

With the rapid increase in technologies and popularity of EV, a need develops for the improved charging infrastructure for their successful propulsion [1].

The charging could basically be powered by electricity generated from either conventional or non-conventional source of energy. But to make the concept of EVs completely environment-friendly a charging station powered by a renewable source of energy is considered to be the best [2].

Out of many types of renewable sources present on earth like tidal, geothermal, solar, wind energy, etc., the use of solar PV array for the EV charging station is preferred the most due to its easy availability, ease of installation, and less maintenance due to the absence of moving parts [3, 4].

The solar-powered charging system with power backup provides various advantages to the infrastructure where it is installed as it provides free fuel for the EVs throughout its lifetime, 20–25 years approximately, after a single investment and also eliminates the need of the power backup sources like diesel generator, inverters, etc. as it can supply the load by the EV batteries under emergency conditions [5].

U. Gupta (✉) · D. K. Yadav · D. Panchauli
Rajasthan Technical University, Kota 324010, India
e-mail: Utkarshgupta3095@gmail.com

D. K. Yadav
e-mail: dkyadav@rtu.ac.in

D. Panchauli
e-mail: dheerajpanchauli777@gmail.com

M. Shorif Uddin et al. (eds.), *Intelligent Energy Management Technologies*, Algorithms for Intelligent Systems,
https://doi.org/10.1007/978-981-15-8820-4_1

2 System Description

See Fig. 1.

2.1 PV Array

A solar PV array of 1 Soltech 1 STH-215-P is used with 10 series modules and 40 parallel strings. The PV and IV characteristics of the PV array are as follows (Figs. 2).

Characteristics of PV array
See Fig. 3.

2.2 Boost Converter

A boost converter is a DC–DC converter that steps up voltage (while stepping down the current) from its input to output. It is a category of SMPS (switch mode power supply) containing at least two semiconductors (a diode and a transistor) and at least one energy storage element, a capacitor, an inductor, or two in combination (Fig. 4).

2.3 Single-Phase Full Bridge Inverter for R-L Load Inverter

A single-phase square wave type voltage source inverter produces square-shaped output voltage for a single-phase load. Such inverters have very simple control logic and the power switches need to operate at much lower frequencies compared to switches in some other types of inverters (Fig. 5).

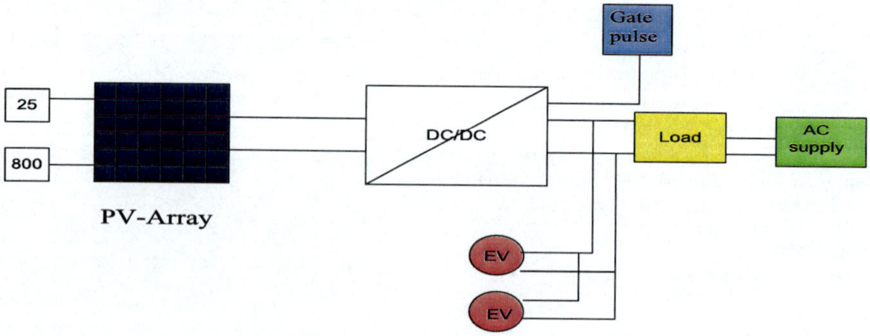

Fig.1 Block diagram of the system

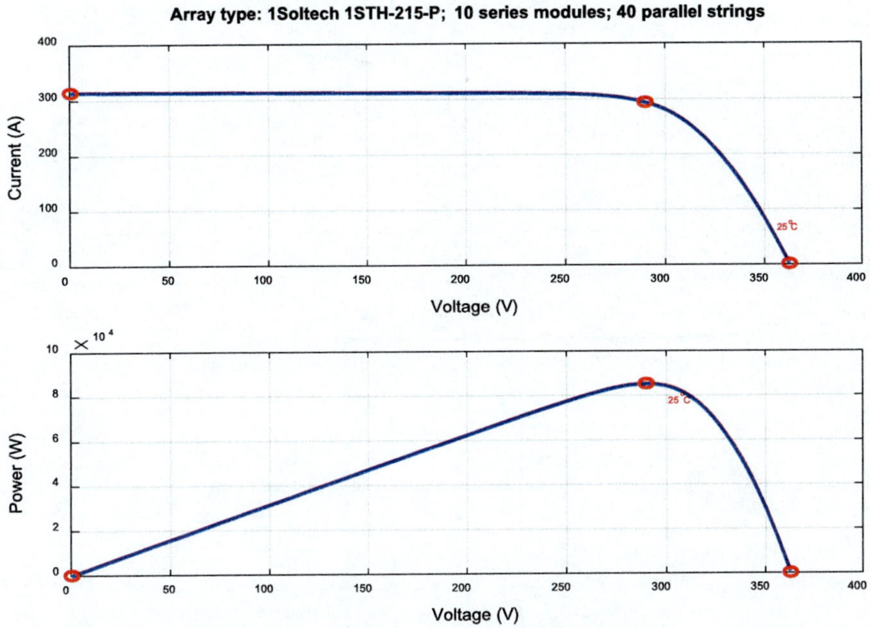

Fig. 2 The PV and IV characteristics of the PV array are as follows

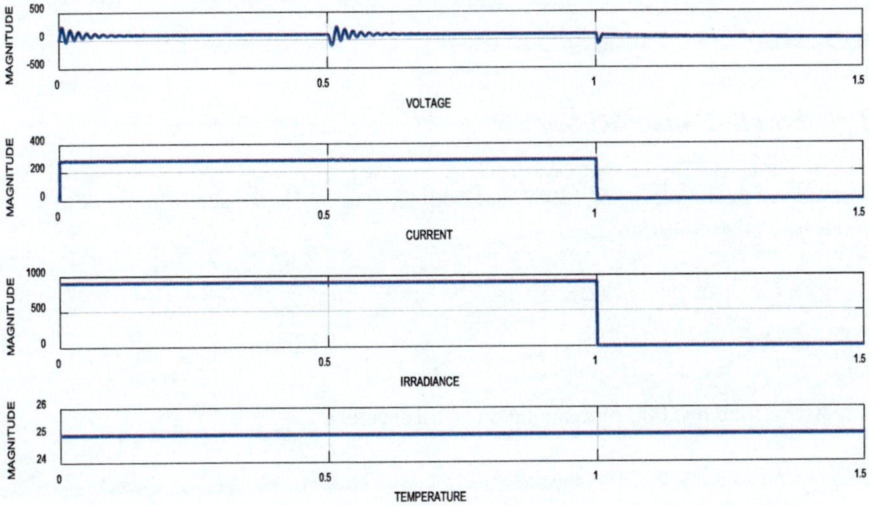

Fig. 3 The output waveform of the PV array used in this system

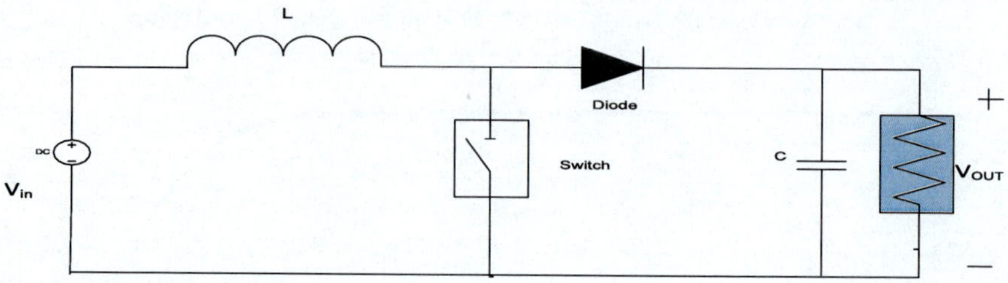

Fig. 4 Basic diagram of a boost converter

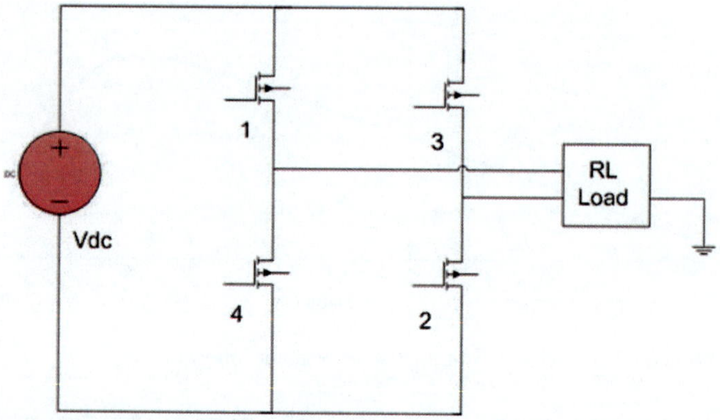

Fig. 5 Basic diagram of a single-phase inverter

2.4 *Single-Phase AC Supply*

A supply of peak voltage of 180 V and frequency of 50 Hz is used to supply the load under normal conditions.

2.5 *Load*

A resistive load of 5000 ohms is taken in this system.

3 Working of the System

The working of this system is explained basically in three modes.

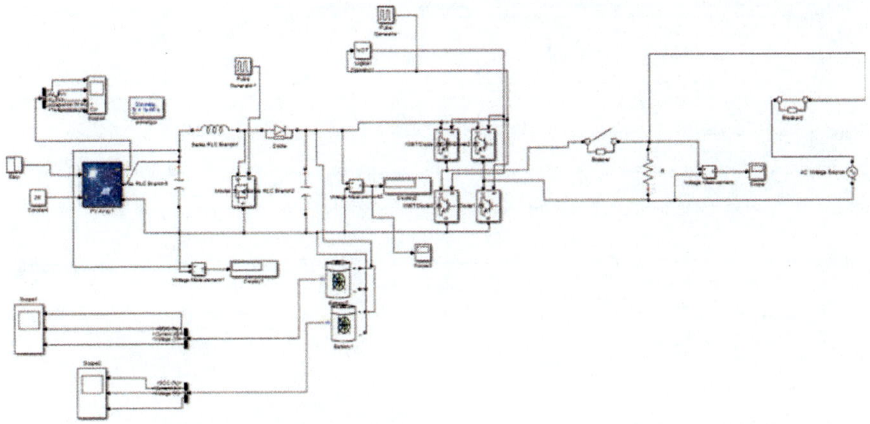

Fig. 6 Simulation model of the system in Matlab

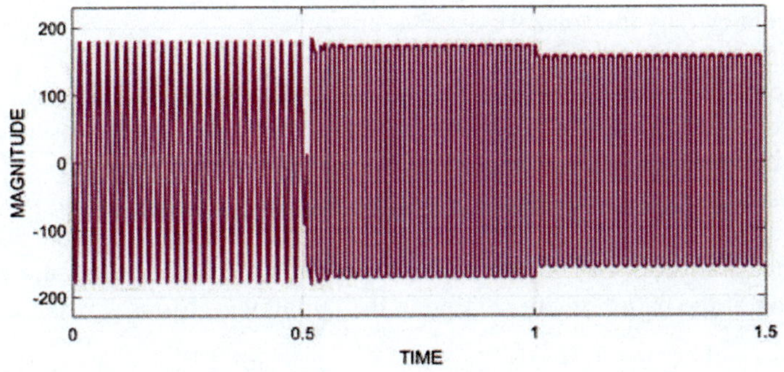

Fig. 7 Output waveforms across the load

MODE-I In this mode, both the grid and the SPV are ON where the grid supplies the load and the SPV charges the batteries of electric vehicles.
MODE-II In this mode, the grid is absent and the SPV maintains the supply across the load through a single-phase inverter and also charges the batteries of the EVs.
MODE-III In this mode, both the PV and the grid are absent and the batteries of the EVs supply the load and their discharging could be seen (Figs. 6, 7 and 8).

4 Simulation Model and Results

In Mode 1, i.e., from a time period of 0–0.5, the output waveforms across the load show sinusoidal waveforms as in this period of time the load is supplied by the grid

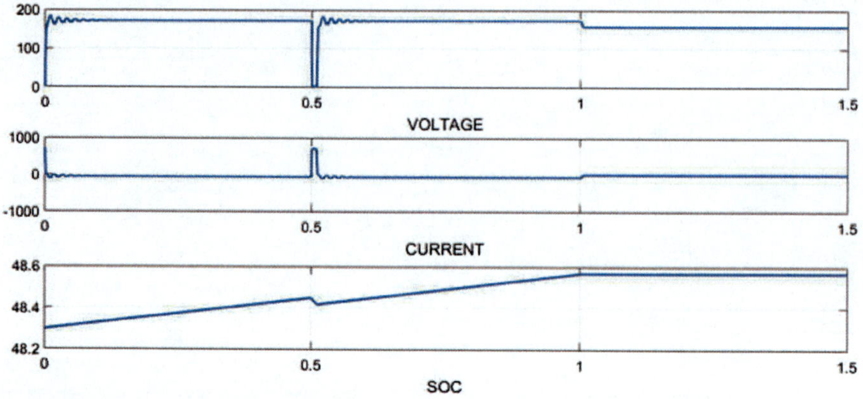

Fig.8 Output waveforms across the battery

and the output waveforms across the battery show an increasing value of SOC as it is getting charged by the solar PV array at this time.

In Mode 2, i.e., from time period of 0.5–1, the output waveforms across the load can be seen as a square wave as it is supplied by the solar PV inverter via a single-phase inverter in the absence of grid and the output waveforms across the battery show the increasing SOC as it is getting charged by the solar PV array.

In Mode 3, i.e., from time period of 1–1.5, the output waveforms across the load show the square wave of 144 V approximately which is supplied by the battery via a single-phase inverter due to the absence of both the sources (solar and grid) and discharging of the batteries can be seen in the output waveforms across the battery.

5 Conclusion

The charging station developed in this system proves to be a very economical system as with eliminating the requirements of expensive power backup sources. It also provides free fuel to the EVs throughout its life. The system could be further expanded and could be made more useful by applying some control techniques.

References

1. Suman N, Yadav DK, Jangid T (2018) Modeling and analysis of photovoltaic system with improved inverter technique. In: 2018 9th international conference on computing, communication and networking technologies (ICCCNT). https://doi.org/10.1109/icccnt.2018.8494168
2. Koduri N, Kumar S, Uday kumar RY (2014) On-board vehicle-to-Grid (V2G) integrator for power transaction in the smart grid environment. In: 2014 IEEE international conference on computational intelligence and computing research. https://doi.org/10.1109/iccic.2014.7238404

3. Chen J, Zhang Y, Su W (2015) An anonymous authentication scheme for plug-in electric vehicles joining to charging/discharging station in vehicle-to-Grid (V2G) networks. China Commun 12(3):9–19. https://doi.org/10.1109/cc.2015.7084359
4. Yang P, PengT, Wang H, Han H, Yang J, Wang H (2017) A single-phase current-source bidirectional converter for V2G application. In: 2017 IEEE 3rd international future energy electronics conference and ECCE Asia (IFEEC 2017 - ECCE Asia). https://doi.org/10.1109/ifeec.2017.7992125
5. Li Y, Yu G, Liu J, Deng F (2017) Design of V2G auxiliary service system based on 5G technology. In: 2017 IEEE conference on energy internet and energy system integration (EI2). https://doi.org/10.1109/ei2.2017.8245513
6. Han H, Lv Z, Huang D, Li Q (2017) Research on charge and discharge power tracking control for V2G system. In: 2017 IEEE 2nd information technology, networking, electronic and automation control conference (ITNEC). https://doi.org/10.1109/itnec.2017.8284928

Chapter 2
A Survey of Energy Theft Detection Approaches in Smart Meters

Divam Lehri and Arjun Choudhary

1 Introduction

Since last decade numerous efforts have been made by the governments and distribution companies to counter electricity theft but it still remains a challenge. The entire loss suffered by the power sector is known as Transmission and Distribution Losses (TD Losses) which comprises an aggregate of Technical Losses (TL) and Non-Technical Losses (NTL). TD losses represent the difference between the electricity generated and the electricity consumed. Technical losses are those losses which are internal to the system such as energy dissipation by the electrical equipments used in distribution lines, transformers, transmission lines, and iron losses in transformers. On the other hand, NTL constitutes losses arising due to defective meters, errors in billing, flaws in supply, unmetered connections, and malicious activities by the consumer such as tampering of meter. Table 1 provides an overview of different types of electricity losses caused by different components of power sector.

The easiest way to determine the amount of non-technical losses (NTL) is by merely calculating the technical losses (TL) in the system and subtracting it from total losses (TD).

We can evaluate it as follows:

$$\text{NTL} = \text{Total Energy Losses(TD)} - \text{TL} \tag{1}$$

$$\text{Total Energy Losses} = \text{Energy Supplied} - \text{Bills paid} \tag{2}$$

D. Lehri (✉) · A. Choudhary
Sardar Patel University of Police, Security and Criminal Justice, Jodhpur, India
e-mail: lehridivam@gmail.com

A. Choudhary
e-mail: a.choudhary@policeuniversity.ac.in

M. Shorif Uddin et al. (eds.), *Intelligent Energy Management Technologies*, Algorithms for Intelligent Systems,
https://doi.org/10.1007/978-981-15-8820-4_2

Table 1 Classification of methods of electricity theft

Elements	Methods of theft
Meters	Bypassing the meter Deliberately damaging the meter seals or removing of the meter
Wires/Cables	Illegal tapping to bare wires or underground cables
Transformers	Illegal tapping of transformer terminals and junction boxes of overhead lines
Billing irregularities	Errors made by meter readers
Unpaid bills	Unpaid bills by individuals or institutions

Some of the losses such as TL are unavoidable. The energy theft in India is majorly due to unmetered usage of electricity. The concept of Transmission and Distribution (TD) losses has been extended further to Aggregate Technical and Commercial losses (AT&C).

$$\text{AT \& C Losses} = \{1 - (\text{BE} \times \text{CE})\} \times 100 \tag{3}$$

$$\text{T \& D Losses} = \{1 - (\text{BE})\} \times 100 \tag{4}$$

where

$$\text{Billing efficiency (BE)} = \text{Total unit Billed/Total unit Inputs} \tag{5}$$

$$\text{Collection efficiency (CE)} = \text{Revenue collected/Amount Billed} \tag{6}$$

TD loss is the difference in input energy and energy billed. There is no account for the losses arising due to low collection. AT&C loss is the difference in input energy and energy for which revenue has been collected. Simply stated AT& C Loss can be aggregated as

$$\text{AT \& C Losses} = \text{TL} + \text{CL} \tag{7}$$

Statistics on electricity losses in India shows that around 10–12% of AT&C losses amount to technical reasons, while remaining 18–20% comprises commercial reasons [1] known as commercial losses (CL). According to U.S. Energy Information Administration (EIA) [2] in the countries with low rate of theft and optimal technical efficiency TD losses generally span between 6 and 8%. Figure 1 shows graphical representation of AT &C loss percentage of different Indian states.

In such scenarios adoption of smart meters by the government of India could prove as a game changer to curb electricity theft. There is also a grave need to develop a common framework in the country where governments, manufacturers, research institutions, Distribution system operators (DSOs) and academia work with mutual

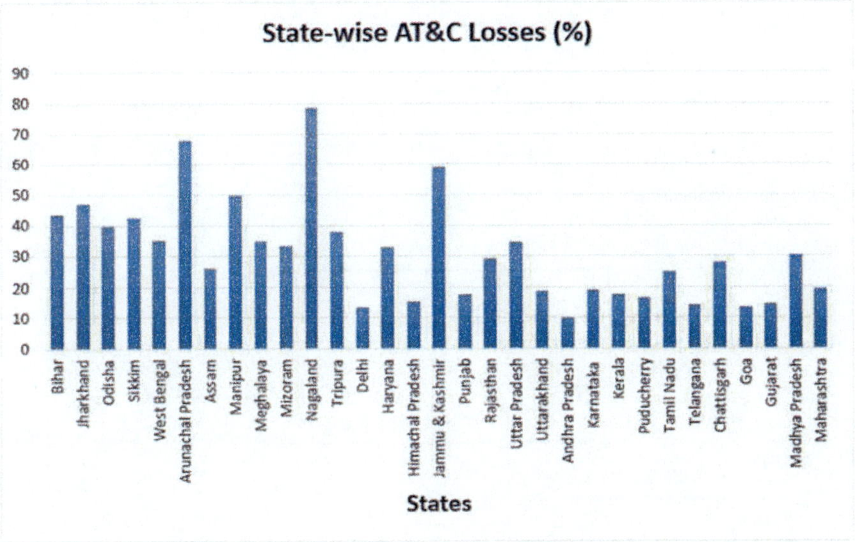

Fig. 1 A visual representation of state-wise AT&C losses (%) for the period Apr'14–Mar'15, According to catalog available on Open Government Data Platform (OGD), India [3]

cooperation to ensure resiliency, privacy and security of smart grid. A paradigm of one such framework is SEGRID Project [4] for European Digital Grid. While smart meters may not be completely theft resistant but they are capable of minimizing the number of theft cases due to their immunity against traditional electricity theft methods as well as the real-time monitoring of data between the utility companies and the consumers.

Due to the complex architecture and large attack surface of Advanced Metering Infrastructure (AMI), Smart Meters are vulnerable to tampering thereby requiring effective theft prevention and detection techniques. In this paper, we present a survey of available energy theft detection techniques.

2 Meter Tampering Methodologies

There are various mechanisms through which an adversary can tamper Smart Meters. Methods of meter tampering can be divided into four classes:

- Current related tampering methods.
- Voltage related tampering methods.
- Mechanical tampering methods.
- Tampering by hacking and altering the memory.

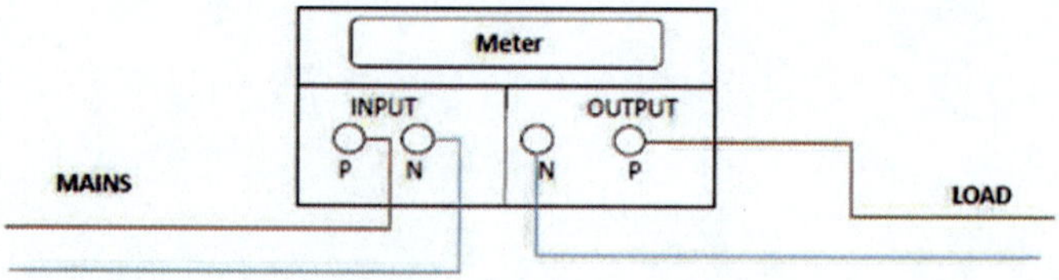

Fig. 2 Actual connection

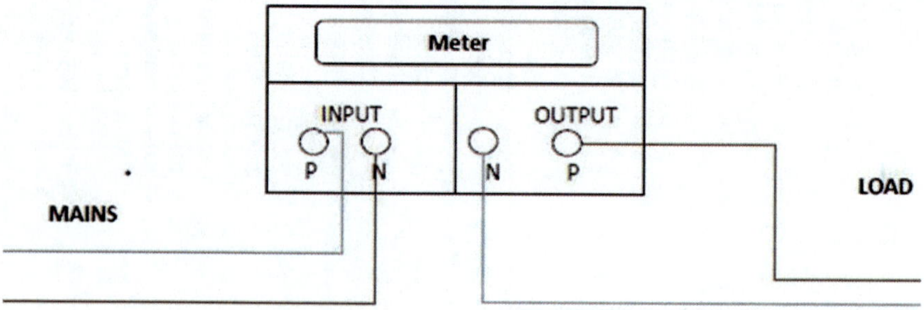

Fig. 3 Swapping of phase and neutral lines

A summary of mechanisms that are generally used to tamper smart meters is presented in this section.

2.1 *Swapping of Phase and Neutral Lines*

In this method of tampering the adversary interchanges the phase and neutral lines. This swapping of phase and neutral lines reverses the energy flow thereby effecting the billing calculation (Figs. 2 and 3).

2.2 *Double Feeding*

Double feeding as the name suggests is a meter bypassing technique where an additional feeder is connected to the meter in such a manner that meter gets bypassed and the energy consumption is not accounted for. Under such scenario the consumption for the load affixed to the supplementary feeder won't be recorded by the meter even if the connection is legitimate. This type of tampering is generally done to connect any heavy electric appliance so that it's consumption remain unnoticed (Fig. 4).

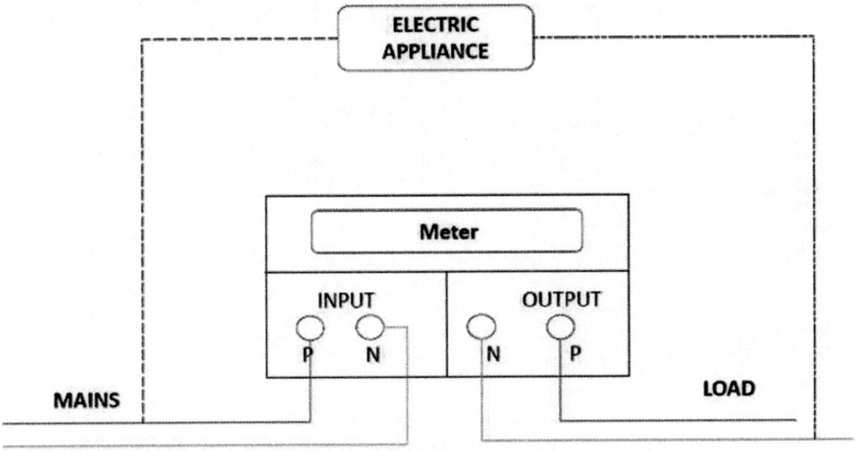

Fig. 4 Double feeding

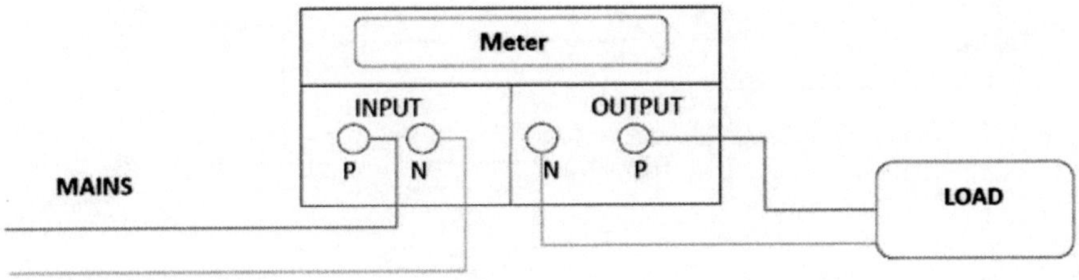

Fig. 5 Actual condition

2.3 *Neutral Missing*

In this method of meter tampering, the neutral line is completely cut-off from the meter thereby resulting in zero input voltage. Hence the power computed by the meter is zero (Since $P = V * I$ and for given condition $V = 0$, therefore $P = 0$) This tampering method is also referred to as single wire operation [5] (Figs. 5 and 6).

2.4 *Neutral Disturbance*

In this method of tampering some noise (High-Frequency voltage signals) is added to the neutral line of the meter by connecting it through diode/variable resistance/capacitor. The neutral of the meter gets deviated from its original point and becomes unbalanced leading to less voltage recording by the meter and therefore less energy consumption is recorded by the meter (Fig. 7).

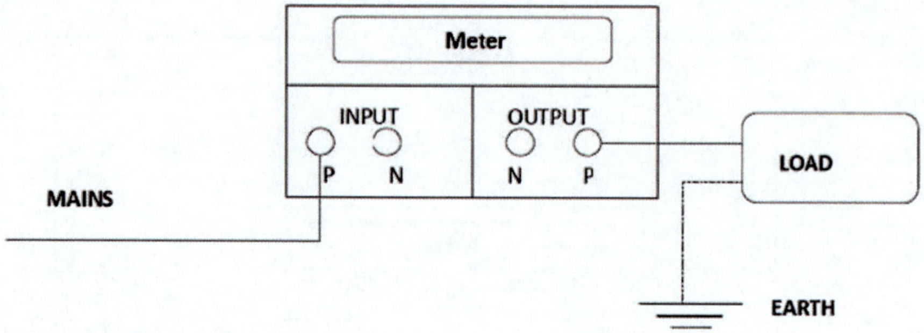

Fig. 6 Neutral missing

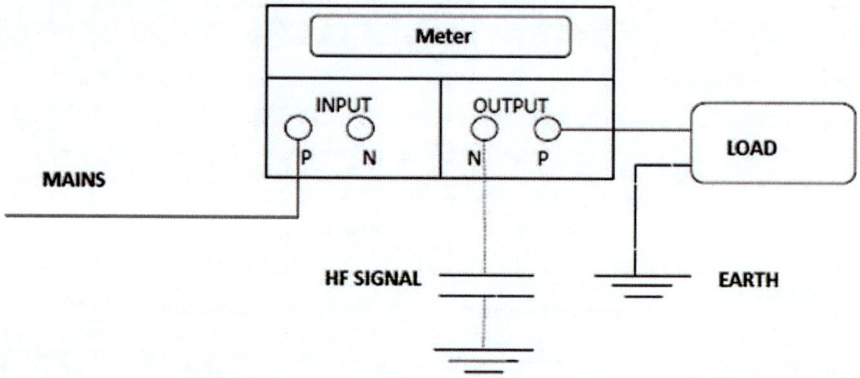

Fig. 7 Neutral disturbance

2.5 Current Reversal by Connecting Input and Output in Reverse

In this tampering event, the adversary connect the phase and neutral wires to the wrong inputs. This causes the current to change direction from it's original path in which it was intended to flow. The intention of this kind of tampering is to dupe the billing computation by reversing the route of current flow [6] (Fig. 8).

2.6 Partial Earth Fault Condition

It is a tampering method in which the load is connected to the earth due to which the return current going back to the meter is reduced. This generates a difference in the current stream flowing through the neutral wire and phase wire leading to current in

Fig. 8 Current reversal

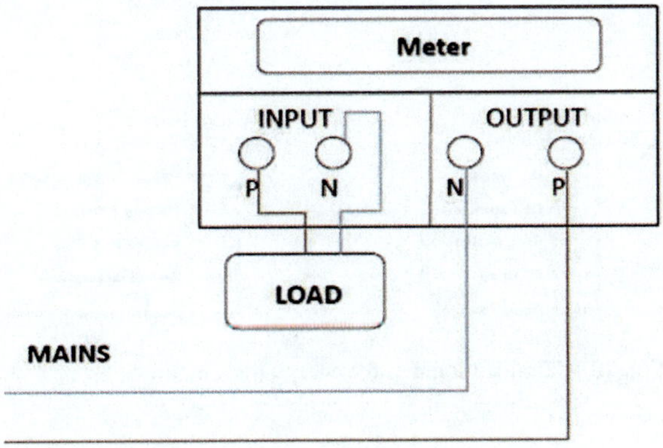

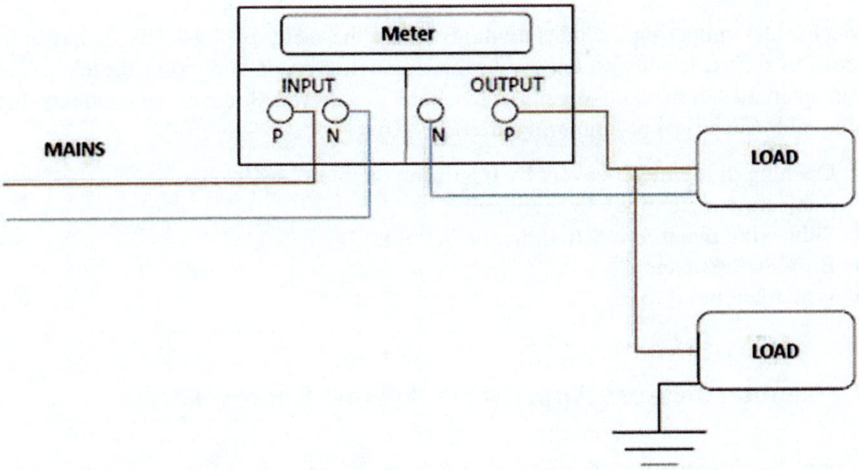

Fig. 9 Partial fault connection

neutral wire become less than the current in the phase wire. Under normal conditions, the current in the phase wire and the neutral wire is equal (Fig. 9).

2.7 *Tampering Using High Frequency/Voltage*

In this type of meter tampering a remote-controlled device is placed in close proximity to the meter. The device is capable of generating high-electrostatic discharge. The discharge so generated causes a spark in the meter thereby thwarting the meter from recording the electricity consumption. Such method of theft is a concern as it does not leave any trace or evidence.

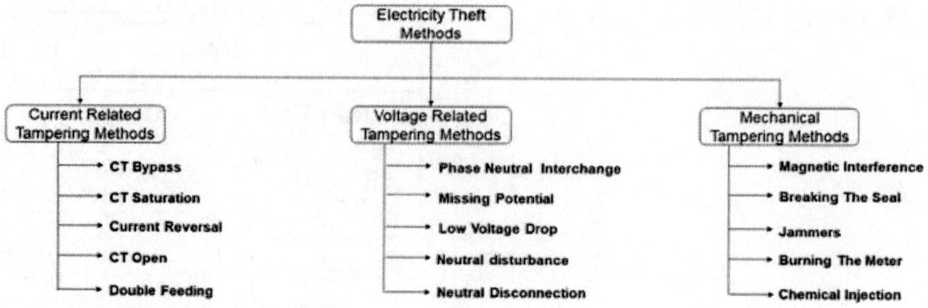

Fig. 10 Classification tree for energy theft methods

2.8 Mechanical Tampering

Mechanical tampering includes methods where the meter is physically damaged in order to record less or no energy usage. In such type of tampering the electrical characteristics of the components of the meter are altered. Some of the conducts that amount to such type of tampering are (Fig. 10)

- Opening of the meter covers by fracturing the meter seals.
- Subjecting meter body to chemicals.
- Subjecting meter to external magnetic field.
- Burning the meter.
- Using jammer devices.

3 Countermeasure Approaches Against Energy Theft

Along with the adoption of new technologies such as smart grid, a new era of attacks are expected to emerge. The government and the utilities are now becoming aware of these scenarios and are taking steps toward mitigating next generation of threats. Rapid developments in the AMI have captured the attention of research organizations and scholars from academia all around the sphere and a range of approaches have been proposed to curb the menace of electricity theft. In this section, we will provide a survey of the available approaches for energy theft detection.

3.1 Game Theory Based Detection Technique

In this technique, the stealing of electricity is represented as a play-off sandwiched amidst the adversaries involved in electricity thief and the distribution utility. It is a model projected on the concept of game theory where the main objective of the adversary is to whip a predetermined amount of electricity and at the same time minimizing the possibility of being identified, whereas the electricity utility desires to augment the chance of detection of adversary and the level of operative price it will sustain in administering this anomaly recognition operation [7]. However, it still remains a challenge to construct a potential game plan and all players that include regulators, thieves, and distributors.

Moreover, game theory is based on assumption that the number of players participating in the game is finite. In country like India which is one of the largest in terms of population, equipping smart meter in every household simply means a drastic increase in number of players which makes game theory difficult to implement.

3.2 *Supervised Learning Approach*

In this approach load profiles for each customer is developed based upon the historical data which is used as a classifier dataset. A pre-selection is made on the subset of smart meters which are straightforwardly confirmed by the technicians within a specified region and time. This process is carried out by the utility company and is referred to as campaign [8]. Information such as consumption, profile, and external information along with other parameters is used to design the profile. In general, a classification problem of fraud identification is formulated which employs supervised learning approach over the historical dataset of fraud cases that occurred in the past [8]. The main criterion for evaluation is the (Odds Ratio) OR. OR may be computed between the falsified clients against all the clients not incorporated in any campaign, called as ORPG or between falsified clients and the non-falsified clients known as ORPN. The ratios obtained from the campaign are mentioned in Table 1 in [8] and are based on some of the characteristics obtained from the campaign. Based on probability a fraud score is computed for each customer according to which the customer can be classified as Fraudulent, Non-fraudulent, and Absent. However, this methodology has performance challenges in scenarios where rate of campaigns is excessively high or the size of campaigns is on a large scale.

3.3 *Linear Error Correction Block Codes*

Linear error-correcting block codes have a linear dependency between the bits of input message and the parity bits. In other words, the resultant of sum of any two codewords is also a codeword. At the receiving end, these bits are utilized to detect and correct errors in the transmission. A computation of the total amount of power in distinct combinations of the cables is computed repetitively and then these readings are utilized to detect and correct errors in the meter readings [6]. In this approach, the concept of syndrome decoding is applied where a generator matrix (G) is used by sender to generate the codeword and decoding matrix, also called parity check matrix (H) is used by the receiver to detect and correct the errors. If $G.H = 0$, then the received codeword is correct. In case G.H $\neq 0$, we can determine the error using the position of non zero bit and correct it. Additional meters, called check meters are used to detect and correct single-bit errors in meter readings. It is assumed that there are M check meters, which are capable of computing the sum of energies of desired cable combinations [6]. However, this Linear block code detection mechanism is prone to magnetic interference and can only detect that there is an error but could not identify the actual meter on which the error exists.

3.4 Dynamic Programming Algorithm Based on Probabilistic Detection

A novel algorithm based on dynamic programming which utilizes the tree structure of the distribution network has been proposed. It makes use of Feeder Remote Terminal Unit (FRTU), which is capable of measuring analog and digital signals and transmitting energy usage data back to the control unit wirelessly. The power consumption in the downstream network commencing from FRTU can be tracked down by means of the information gathered from FRTU. If the power consumption varies notably compared to the aggregate of the readings of smart meters in the downstream network than it is concluded that at least one of the meters has been compromised [9]. The algorithm aims to install FRTUs in minimal quantity in power distribution networks due to cybersecurity concerns. Moreover the algorithm is optimized to increase efficiency by utilizing solution pruning techniques. The main parameters that the algorithm uses to determine theft are the Attacking Probability and the Anomaly Score which are defined in [9]. Based on the anomaly score it is decided whether a meter is anomalous or not. The proposed algorithm works under the supposition that the adversary can only attack the smart meter equipped in her/his own apartment. It may not provide a fruitful solution in situations where adversary uses advanced techniques such as remotely attacking meter in Neighborhood Area Network (NAN). Another probabilistic approach has been proposed in [10] which provides an estimate of Technical and Non-Technical Losses in a segregated manner.

3.5 Temperature-Dependent Predictive Model

"Temperature-Dependent Technical Loss Model (TDTLM)" is the advancement of the "Constant Resistance Technical Loss Model (CRTLM)" [11] by making the resistance temperature dependent. To estimate NTL, TDTLM utilizes the property that there is a linear dependency between the resistance of material and its temperature in. The power consumption values along with other instantaneous measurements are aggregated and sent back to the utility repeatedly after a fixed interval of 30 min for calculation of NTL. Based on the threshold value of NTL cases are classified as theft and non-theft. To train the predictive model data from the first two days (no theft) is utilized. Whenever NTL estimate exceeds the threshold value, it is suspected that a power theft has occurred in the user group.

3.6 Current Bypass Anti-Tampering Algorithm

A single-chip solution has been developed where an anti-tampering algorithm has been implemented on an "ARM Coretext-M3 (STM32L152VB)" microcontroller. It is a low power microcontroller operating at 32 MHz using "ADE7953 (Single Phase Smart Meter)" and "ADE7878 (Three Phase Smart Meter)" Analog Devices [12]. The unbalance current difference (Irr) is calculated by extracting data values from IRMSA (I_a) and IRMSB (I_b) registers of ADE7953, where current in phase line is denoted as I_a and current in neutral line is denoted as I_b. The unbalance current difference (Irr) is expressed in Eq. (1) of [12] as

$$|\mathrm{Irr}| = I_a - I_b \div (I_a + I_b) \tag{8}$$

Verification of current bypass tampering event by the smart meter is done by comparing the calculated Irr value in (4) against the pre-defined threshold values which are 2.5% in case of three-phase meter and 1% in case of single-phase meter [12]. In case of any uneven event, an interrupt is sent to the MCU. The MCU verifies the tampering event by examining the status bits in the registers of ADE7953 and ADE7878.

3.7 Microprocessor-Based Theft Control System

The theft control system based on "ARM-Cortex M3 processor" has been implemented to prevent the energy meter from tampering attacks such as disconnections of phase/neutral lines, entire meter bypassing, and meter tampering. This approach utilizes the current difference in phase line and neutral line to detect tampering event. Two current transformers, one in each phase line and neutral line are inducted. In case of any disconnection of either of the lines from the meter, it would result in a significant drop in current measured by the each current transformer [13]. This difference is computed by the microcontroller by measuring current through ADC. In case of any irregular difference an SMS is sent to the electricity utility by the microcontroller. This functionality can be integrated into existing meters in addition to manufacturing it in new meters. The module uses GSM network for communication which is already well established in India.

3.8 AMIDS Framework

Advanced Metering Infrastructure Intrusion Detection System (AMIDS) is a framework developed using an amalgamation of a variety of approaches for detection and reporting of energy theft in smart meters. An attack graph-based data fusion algorithm is used by AMIDS to merge artefacts of on-going attacks from numerous sources [14]. The attack graph so composed is a directed graph based on state which consists of different stages from initial to final. To achieve information fusion online, the attack graph is considered as a Hidden Markov Model (HMM). AMIDS makes use of both a supervised learning methodology that can compute individual application usage and an unsupervised methodology that learns by clustering load events. AMIDS takes into account numerous information sources to collect adequate amount of artefacts regarding an on-going attack prior to identifying an activity as a malicious energy theft.

3.9 Model Based on Harmonic Generators

A model has been proposed in [15] which uses harmonic sensors to determine the uncertainty in smart meter readings. The proposed model consists of harmonic sensor, ICS, energy meter, circuit breaker, and communication system. The External Control Station (ECS) situated at the utility company receives instant values from end user side. The non-technical loss is calculated by ECS and in case of loss being more than 5% it would break the supply to the meter by indicating the control system to disconnect the customer. The core of the model is the harmonic sensor which compute the uncertainty in meter reading based on Total Harmonic Distortion (THD). In [16] this approach based on the harmonic generators is extended by placing two smart meters along with harmonic sensors and generators at either ends. One meter is placed at the consumer end and other one at the utility end which makes it possible to keep track of the generated power as well as the consumed power.

The difference in generated and consumed power is calculated to determine the theft.

3.10 MIDAS Framework

A novel framework MIDAS [17] is developed which is the integration of several techniques such as statistical analysis, data mining, and neural network. This framework is different from other approaches as it classifies not only suspected users but also classifies users without technical losses. Data mining is performed for fault and theft sensing and to analyze load profiles of individual customers. The neural network is trained with multiple methods. Different neural network topologies are developed and at end of training the Root Mean Square (RMS) value for each model is computed. The model with minimum RMS value is presented as final neural network.

3.11 Measuring Voltage Drop Between Smart Meters

This approach utilizes the magnitude of the voltage drop between two smart meters to detect and decrease illegal consumption. The concept is to grab the voltage and power data from the meter. It functions on pre-condition that there should be more than two consumers involved in the powering of transformer, because detection of unauthorized spending is computed by comparing the drop of voltage of each measuring point to the transformer. In case if a drop of voltage occurs than it is deduced that the consumer is having unauthorized connection to the meter [18].

3.12 Energy Lens

Energy Lens system intelligently integrates electricity meter data with sensors on smartphones. It is expected that users using energy lens possess an android phone with the capability to sample microphone audio and WiFi signal strength. During initial phase users are required to turn on electricity appliances which they want to get recognized by the energy lens. Users wait for some time for its power consumption to reach a steady state and then turn it off. Based on this data acquired, it is identified that when, where, and by whom the activity is performed upon the execution of algorithm on the server [19]. However, energy lens faces several challenges such as acquisition of ground truth statistics for building up the precision, the effect of phone's direction and privacy of the customer.

3.13 FNFD ("Fast NTL Fraud Detrection and Verification")

FNFD is a mathematical method constructed on the notion of Recursive Least Square (RLS) to represent adversary behavior. Using FNFD the NTL fraud, in Smart Grid is detected in real time. FNFD is capable of verifying a fraud even with one single measurement, given that the historical data supplied to it is accurate [20]. FNFD employs linear functions to simulate the behavior of adversary. The main advantage of FNFD as compared to other schemes is that it requires much less data and supports NTL fraud verification, a unique feature that is not available in other schemes and it is much faster than the other similar frameworks.

4 Proposed Work

All the existing works available on energy theft detection in smart meters are dealing only with types of thefts where by some means either phase or neutral wires were swapped or removed which led to significant change in the voltage or current values or due to billing irregularities. Our work extends the existing approaches to a new threat scenario where theft detection in smart meter occurs due to tampering of the hardware chip of the meter itself. Our approach is mainly concerned with the chip-level tampering of smart meters. An adversary could Reverse engineer the meter and obtain the low-level assembly instructions of the meter. Further disassembling of the smart meter could be done, thereby attempting to read the firmware directly from the chip. Obtaining of the low-level assembly instructions would reveal the hard-coded cryptographic keys among other sensitive information that can be used in later attacks. Moreover, by exploiting the low-level assembly code the adversary could alter the consumption readings. The common methods of exploiting the hardware chip include:

- Logical Analyzer.
- Circuit Bending.
- JTAG Method.
- Hacking Over UART.

The logical analyzer is an instrument which sniffs the signals when placed on different test points on the circuit board thereby revealing potential information that could be interpreted into something useful, adding or removing circuit components such that the functionality of the circuit is affected, also known as Circuit Bending and using Joint Test Action Group (JTAG) method to read full memory hex dump.

We will begin with exploring the embedded hardware of the smart meter, examining individual components present on the circuit board. To get a better understanding of the working of each component we will probe the datasheets associated with each component. Extending our approach further we will examine the interconnections between different components using multi-meter. This will provide us insight of how data and signal transmission is taking place on the device. Now we will hunt for debug ports present on the device. JTAG and UART are the most common debug ports and we can easily identify them by monitoring the voltage levels using a multi-meter or with the help of oscilloscope. Once debug ports are identified we will start interacting with the device by making connections between the debug ports and any USB bridge. USB Bridge will provide us with the capability to interact with the device through console and finally we will begin the process of extracting data\firmware from the device. We will modify the data dump that we acquired and rewrite it to the device such that we can manipulate the device.

In continuation to this paper, we will be showcasing this kind of meter tampering using these mentioned method along with the experimental results. We will also propose mitigation measures for such type of chip-level energy theft approach such as assembly code obfuscation.

5 Conclusion

Curbing the energy theft menace is a huge concern for the governments and utilities. The scope of tampering comprises straightforward approaches like controlling live or neutral wires to more grave means like retrieving device firmware. Appliances like smart meters are part of critical infrastructure and any compromise to it would be causing chaos in the power sector and huge loss of revenue to the government. Most of the critical infrastructure devices are procured from global sources and may come pre-installed with hardware backdoors. Adversary can also intrude through the weakest point in the supply chain and compromise the device by installing hardware backdoor. This shows that attackers are now moving down the stack from application layer attacks to embedded hardware of the devices. The tools required to carry out physical attacks are also proliferating and becoming inexpensive. Such scenarios call for importance to hardware-level security which is not usually considered as important as application-level security. Organizations need to reshape their security approach from the viewpoint of attackers and conduct red team assessments to enhance the security of the assets. In recent years, the advancement of smart grid and adoption of smart meters has called for proposals from industry, universities, and governments to tackle the vulnerabilities existing in the AMI. In this paper, we have classified various ways of energy theft and detection techniques along with their challenges. However, it still remains a fresh topic and has a lot of room to be worked upon in the future.

References

1. Sharma T, Pandey KK, Punia DK, Rao J (2016) Of pilferers and poachers: Com-bating electricity theft in India. Energy Res Soc Sci 11:40–52
2. U.S. Energy Information Administration—Eia—Independent Statistics and Analysis. https://www.eia.gov/todayinenergy/detail.php?id=23452
3. Aggregate Technical & Commercial (AT&C) Losses in power sector. saumitra.cea@nic.in. https://data.gov.in/catalog/aggregate-technical-commercial-atc-losses-power-sector. Accessed Dec 17
4. Securing the smart grid of tomorrow. https://segrid.eu/
5. Warudkar D, Chandel P, Salwe BA (2014) Anti-tamper features in energy meters. Int J Electr, Electron Data Commun 2(5), May-2014. ISSN 2320-2084
6. Mesbah W (2016). Detection and correction of tampering attempts of smart electricity meters. In: PES innovative smart grid technologies conference Europe (ISGT-Europe), October, pp 1–6. IEEE
7. Cárdenas AA, Amin S, Schwartz G, Dong R, Sastry S (2012) A game theory model for electricity theft detection and privacy-aware control in AMI systems. In: Communication, control, and computing (Allerton), 2012 50th annual Allerton conference on, October, pp 1830–1837. IEEE
8. Coma-Puig B, Carmona J, Gavalda R, Alcoverro S, Martin V (2016) Fraud detection in energy consumption: a supervised approach. In: Data science and advanced analytics (DSAA), 2016 IEEE international conference on, October, pp 120–129. IEEE

9. Zhou Y, Chen X, Zomaya AY, Wang L, Hu S (2015) A dynamic programming algorithm for leveraging probabilistic detection of energy theft in smart home. IEEE Trans Emerg Topics Comput 3(4):502–513
10. Neto EAA, Coelho J (2013) Probabilistic methodology for technical and non-technical losses estimation in distribution system. Electric Power Syst Res 97:93–99
11. Sahoo S, Nikovski D, Muso T, Tsuru K (2015) Electricity theft detection using smart meter data. In Innovative smart grid technologies conference (ISGT), 2015 IEEE power & energy society, February, pp 1–5. IEEE
12. Tangsunantham N, Ngamchuen S, Nontaboot V, Thepphaeng S, Pirak C (2013) Experimental performance analysis of current bypass anti-tampering in smart energy meters. In: Telecommunication networks and applications conference (ATNAC), 2013 Australasian, November, pp 124–129. IEEE
13. Dineshkumar K, Ramanathan P, Ramasamy S (2015) Development of ARM processor based electricity theft control system using GSM network. In: Circuit, power and computing technologies (ICCPCT), 2015 international conference on, March, pp 1–6. IEEE
14. McLaughlin S, Holbert B, Zonouz S, Berthier R (2012) AMIDS: a multi-sensor energy theft detection framework for advanced metering infrastructures. In: Smart grid communications (SmartGridComm), 2012 IEEE third international conference on, November, pp 354–359. IEEE
15. Depuru SSSR, Wang L, Devabhaktuni V (2011) Electricity theft: overview, issues, prevention and a smart meter based approach to control theft. Energy Policy 39(2):1007–1015
16. Prasad J, Samikannu R (2017) Overview, issues and prevention of energy theft in smart grids and virtual power plants in Indian context. Energy Policy 110:365–374
17. Guerrero JI, Monedero Í, Biscarri F, Biscarri J, Millán R, León C (2014) Detection of non-technical losses: the project MIDAS. Advances in secure computing, Internet services, and applications, pp 140–164
18. Bula I, Hoxha V, Shala M, Hajrizi E (2016) Minimizing non-technical losses with point-to-point measurement of voltage drop between "SMART" meters. IFAC-PapersOnLine 49(29):206–211
19. Saha M, Thakur S, Singh A, Agarwal Y (2014) EnergyLens: combining smartphones with electricity meter for accurate activity detection and user annotation. In: Proceedings of the 5th international conference on Future energy systems, June, pp 289–300. ACM
20. Han W, Xiao Y (2016) FNFD: a fast scheme to detect and verify non-technical loss fraud in smart grid. In: Proceedings of the 2016 ACM international on workshop on traffic measurements for cybersecurity, May, pp 24–34. ACM

Chapter 3
Renewable Energy Conversion: Sustainable Energy Development and Efficiency Enhancement of Solar Panels: A Review

Bhavik J. Pandya, Megha C. Karia, and Kamlesh B. Sangani

1 Introduction

In the present time, Solar PV is rapidly growing to become a top source of electric power generation. Worldwide installed capacity is the third-largest source of renewable resource after hydro as well as wind. It should be noticed that the progress in the development in the area of solar thermal is quite slower, but it indicates that it is in just the beginning period and it may be a major source for electricity generation later in century. Solar power is the most abundant form of energy in our mother earth and it is a mandatory source for its inhabitants. Now this paper presents an overview on generation of electricity from solar energy utilization. This can be achieved most simply by exploiting the heat contained in the sun's radiation.

1.1 Present Scenario in Energy Generation

The power generation capacity of installed power plants in India as on 29.02.2012 is 1,90,593 MW (share of renewable energy sources is 22,253 MW). The gross electricity generated in the year of 2011–12 in India (up to February 2012) includes

B. J. Pandya (✉) · M. C. Karia
V.V.P. Engineering College, Rajkot, India
e-mail: bhavikpandya.education@gmail.com

M. C. Karia
e-mail: megha.karia.me@vvpedulink.ac.in

K. B. Sangani
Super Specialist Technocrats LLP, Rajkot, India
e-mail: kamleshsanganisir@gmail.com

M. Shorif Uddin et al. (eds.), *Intelligent Energy Management Technologies*,
Algorithms for Intelligent Systems,
https://doi.org/10.1007/978-981-15-8820-4_3

bring from Bhutan was 798.9 Billion Units. The energy requirement, availability, and shortage in India during April 2011–February 2012 are given below:

Year	Energy requirement (MU)	Energy availability (MU)	Deficit (MU)	Deficit (%)
2011–12 (till Feb 2012)	853,324	782,124	71,200	8.3%

1.2 *History of Electricity Generation from Solar Energy Resources:*

Solar thermal power exploitation, the solar heat utilization can be found by going back in history to Archimedes but it must be included that the generation of power by the utilization of solar energy is recent the use of the sun as a source of heat, can be traced back at least to Archimedes but its application as a means to generate power is more recent. The utilization of solar heat is started from nineteenth century; parabolic reflectors were used to concentrate the heat of sun and due to this heat steam produce which used to drive steam engine. From the beginning of twentieth century, solar energy was applied to get power in engine for work like pump water for irrigation in the agricultural sector; this technique was used in Egypt. The first power plant operated by solar thermal power was established in 1960s in Italy. There is another method for generating electricity from solar with the help of an electronic device which is known as photovoltaic or PV cell. Another method of transformation of sunlight to electricity discovered, by the research of scientist named Antoine-Ce'sar Becquerel. The first person who observed the effect of photovoltaic conversion to the voltage when sun rays supplies to electrode was Becquerel. After this, a gold-coated selenium solar cell was invented by Charles Fritts. This first cell was inefficient but it gave thought for large scale electricity generation by solar. In 1941, the first silicon solar cell was invented by Russell Ohl after a number of experiments. After the 13 years an American scientist Gerald Pearson, Calvin Fuller, and Daryl Chapin developed a cell which has conversion efficiency of 6%.

1.3 *Potential of Solar Energy*

We all know that sun has energy which produces by nuclear reactions at its center. The electromagnetic radiation helps it to reach till earth's surface. This radiation is passed from 7% ultraviolet zone, 36% visible radiation zone, and 56% infrared zone of electromagnetic spectrum.

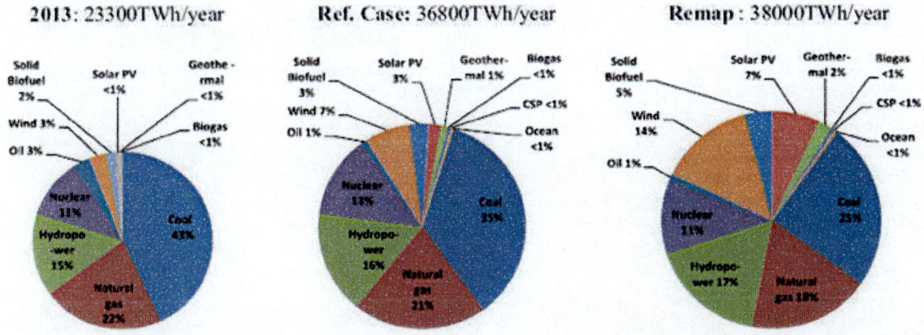

Fig. 1 Global power generation in 2013 and 2030 [1]

1.4 *Solar Energy and the Earth*

Solar constant at the distance of the earth from the sun	1367 W/m^2
Total solar energy reaching the earth in a year	3,400,000 EJ
Total solar flux reaching earth	1.083108 GW
Average solar energy density at earth's surface	170 W/m^2

Source World Energy Council

1.5 *Current Situation of Solar Power Plant in India*

In January 2016, a new international body for solar energy utilization, **International solar alliance (ISA)** has been initiated by the hardly efforts of Indian Prime minister Narendra Modi and with the support of French president Francois Hollande and also established it's headquarter at Gwal-Pahari, Gurgaon. The development and utilization of solar energy and products works by it for countries between Tropic of cancer and Tropic of Capricorn is the main focus of ISA. More than 120 countries agreed and sign the alliance at the summit held at Paris named Paris COP21 climate change (Figs. 1 and 2).

In India utilization of solar PV is at peak so some statistical data is shown below (Table 1).

Installed Solar PV till 31st March 2019

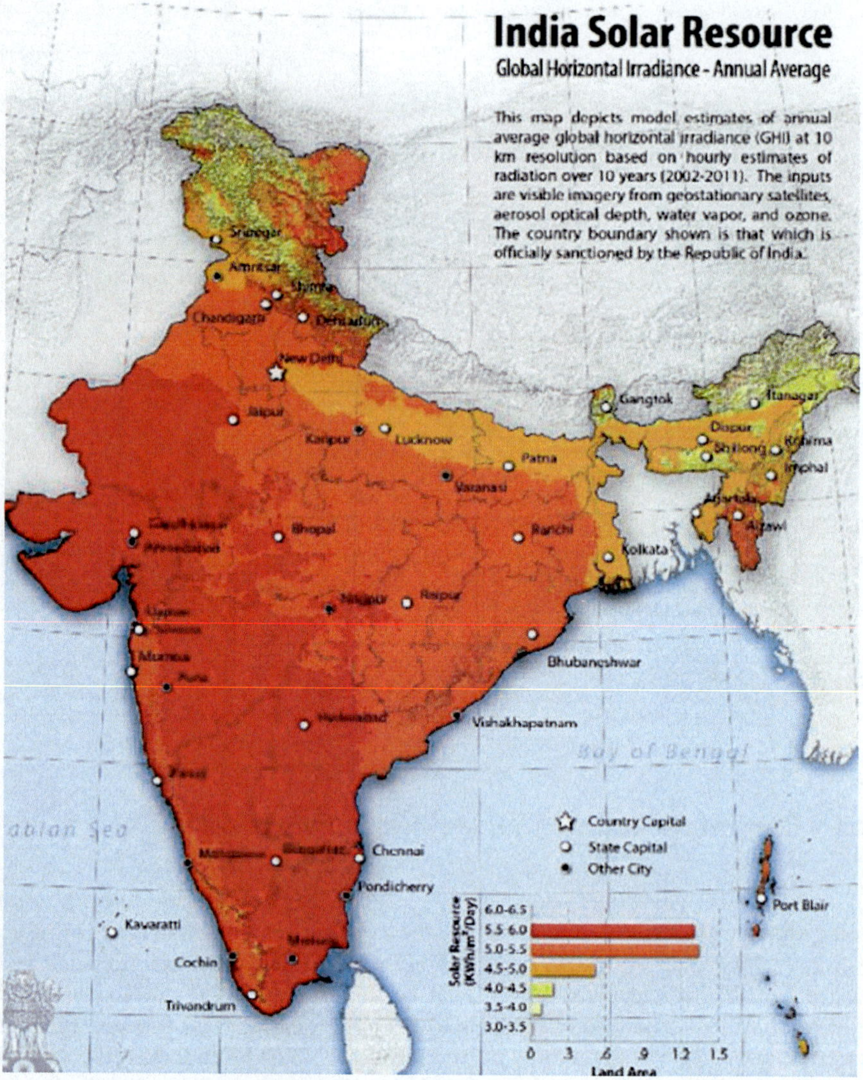

Fig. 2 Global horizontal irradiance (*Source* NREL)

2 Classification of Solar Power Plant

There are certain methods for conversion of energy of sun to the electrical power.

The first option is the direct conversion to electricity mainly by the following methods:

Table 1 Installed solar PV till 31st March 2019 (*Source* Wikipedia)

Year	Cumulative capacity (MW)
2010	161
2011	461
2012	1205
2013	2319
2014	2632
2015	3744
2016	6763
2017	12,289
2018	21,651
2019	28,181

1. Photovoltaic.
2. Photogalvanic.
3. Photoemissive.
4. Photomagnetic.

The direct conversion is practically not adopted for commercial purpose for large power demand due to high cost of cell.

Thermal energy obtained by solar energy can also be directly converted into electricity by the following methods:

1. Thermo electric.
2. Thermionic.
3. Ferro electric.
4. Magneto hydrodynamic.
5. Electro gas dynamic.

Above-mentioned methods have generally very low rate of conversion of solar energy to electricity, so this power system is not cost effective.

There is yet another very important way for electrical conversion from solar this is thermodynamic way, in which solar heat is converted into thermal energy and this thermal energy will be converted to shaft work through heat engine based on the principle of either Rankine cycle, Stirling cycle, or Brayton cycle and shaft work (mechanical energy) into electricity using alternator.

2.1 Concentrating Solar Power (CSP)

In this system, solar rays are concentrated by mirrors and lenses and then it reflects area under sunlight to small section beam, and thus by the help of this concentrated solar beam we can utilize this solar energy. Here lenses capture large area and utilize its heat in small area so we can get efficient work output. Here sun rays are focused

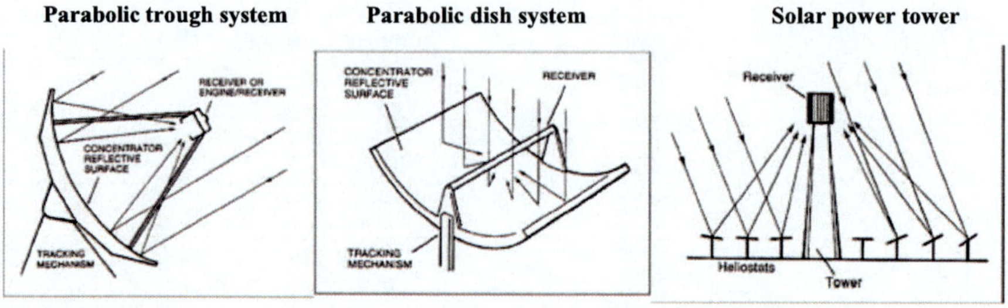

Fig. 3 [2]

on working medium, then working medium gets heat by concentrated solar rays, thus power generation takes place. It is mandatory to mention that for the purpose of focusing sun rays various tracking techniques are used. Parabolic trough, solar power tower, sterling dish, linear Fresnel reflector, and many more technologies exist in CSP and some in developing stage also (Fig. 3).

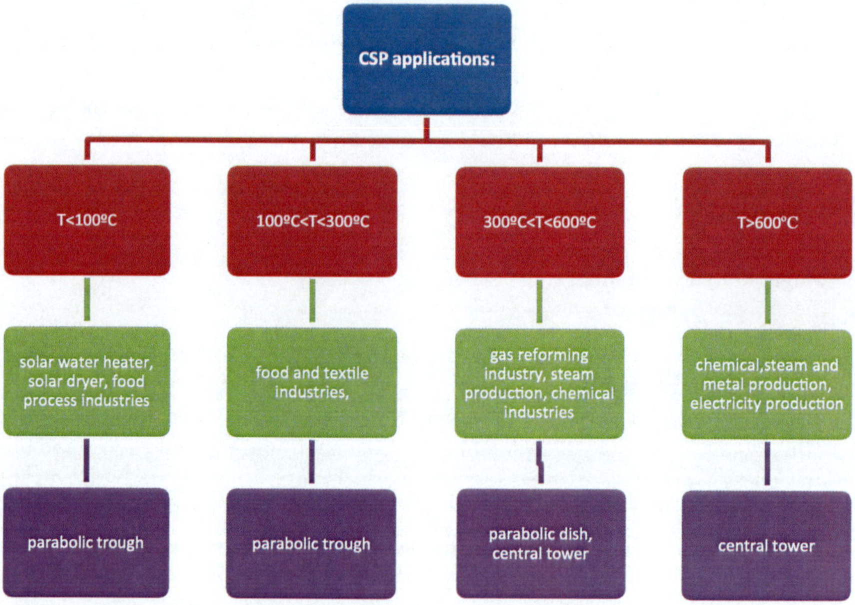

Layout 1 Application of concentrated solar power system application

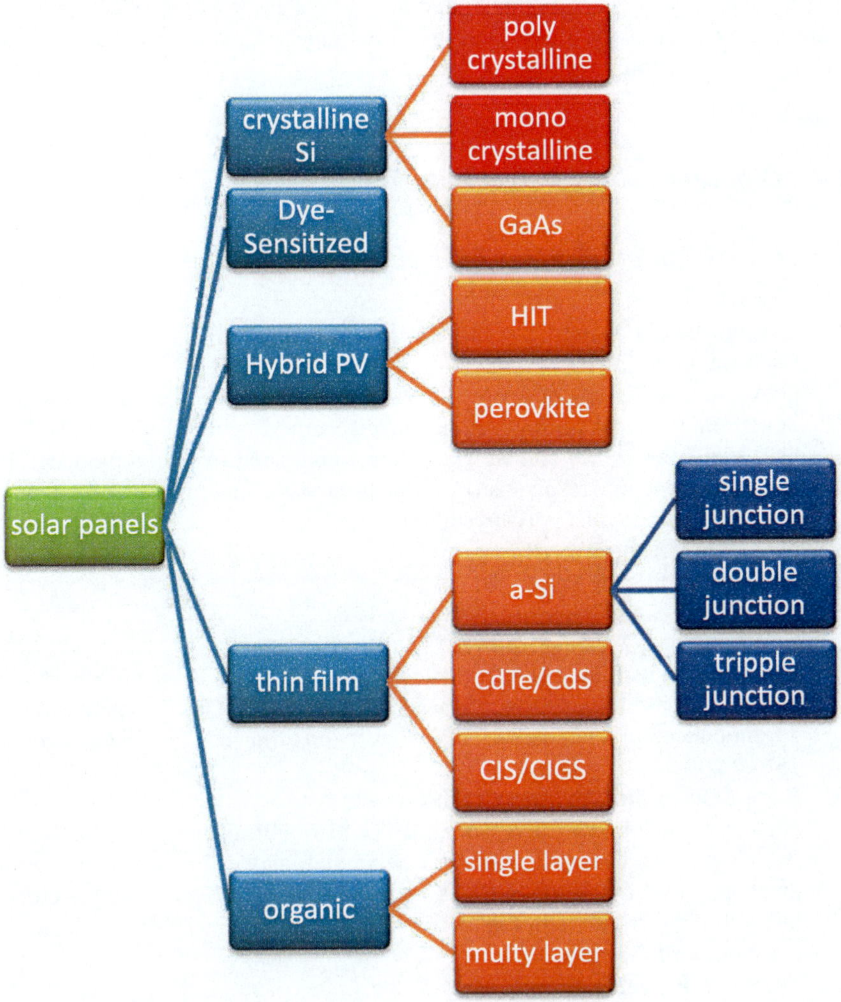

Layout 2 Different solar PV technologies

2.2 Concentrated Solar Power System Application

2.3 Solar Photovoltaic Cell Power Generation

The photovoltaic cell is solid-state device composed of thin layers of semiconductor materials which produce an electric current when light falls on it. It contains number of cell filled by photovoltaic material. Mono-crystalline Si, Poly-crystalline Si, amorphous Si, Cu-indium sulfide, and Cd-telluride are materials which are currently used for PV. Generally Poly-crystalline Si and Mono-crystalline Si are extensively used in India. Silicon gives optimum result so it is extremely suitable for PV applications among all the materials available in the earth [18].

2.4 Different Solar PV Technologies

2.5 Classification of Solar PV Cells

I. **Crystalline Si PV cell**

A. **Gallium Arsenide (GaAs)**
GaAs solar cell is part of a compound semiconductor which are a combination of gallium (Ga) and arsenic (As) having the same structure as silicon [7]. They have higher efficiency and weight is lower compared to silicon produced PV cells. Here it should be mandatory to mention that economically GaAs is much costly then mono- and poly-crystalline Si cell.

B. **Mono-crystalline PV cell**

C. **Poly-crystalline PV cell**
Note: B & C we will discuss in detail in later.

D. **Emitter Wrap-Through cell (EWT) cells**
EWT utilizes its full cross-sectional area so it can focus large amount of solar radiation. Reason behind this is it has modified design of back content cell with a front collection junctions [8]. The conversion efficiency of EWT is between 15 and 20%.

E. **Edge-defined film-fed growth (EFG) cells**
For the reduction of cost of wafers EFG Si ribbon is used. The kerf losses (which generated because of wafer sawing and ingot) elimination and cutting down of poly-Si consumption is a major benefit of EFG. As per the current research data, researchers found 16% conversion efficiency of two materials when reduced shunting.

II. **Thin-film PV cell**
Thin-film solar cells generally made by very thin layers (<10 μm) settled on the metals, polymers, or glass, by sputtering process [6]. The detailed classification of thin-film PV is given below.

A. **Amorphous silicon (a-Si) solar cells**
Amorphous silicon solar cells have a disordered structure form of Si and have 40 times higher light absorption rate compared to the mono-Si cells. They are majorly utilized and most developed thin-film solar cells. At laboratory level, efficiency of single junction a-Si cells may reach up to 12.2%.

B. **Cadmium Telluride (CdTe) or CdS cells**
Cadmium Telluride thin-film PV solar cells consist mainly two types of raw materials which are cadmium and tellurium, where cadmium is a by-product of zinc mining whereas tellurium is found out by copper processing. Both of

them have lower production cost and higher cell efficiency (higher than15%) compare to the a-Si cells. Here highest 21% conversion efficiency is achieve by a leading company named "First solar" in august 2014.

C. **Copper-Indium-Selenide (CIS) and Copper-Indium-Gallium-Diselenide (CIGS) cells**
CIS cells are competing with other thin-film cells due to their high conversion efficiency 21% reported at laboratory level. Here interfacial, stable back contact, grain boundry, and junction activation are the main problems with CIGS. Bareness of indium is also one of the major problem in CIGS.

III. **Organic PV cell**
It is developing and latest technology for alternate material of solar cell. Disposable, flexible, light weight, less production cost, semitransparent, and low material required are a characteristics of it [6, 9, 10] (Table 2).

Table 2 Laboratory-based efficiency of terrestrial PV cell and sub-module measured at STC [11]

Classification	Efficiency %	Area Cm^2	Voc (Open circuit voltage)V	Jsc (Current density) MA/cm^2	Fill factor %	Description
Silicon						
Si (crystalline)	25.6 ± 0.5	143.7 (da)	0.740	41.8	82.7	Panasonic HIT, rear Junction
Si (multi crystalline)	21.25 ± 0.4	242.74 (t)	0.6678	39.80	80.0	Trina solar
Si (thin transfer sub-module)	21.2 ± 0.4	239.7 (ap)	0.687	38.50	80.3	Solexel
Si (thin-film minimodule)	10.5 ± 0.3	94.0 (ap)	0.492	29.7	72.1	CSG solar (<2 μm on glass; 20 cells)
III-V cells						
GaAs (thin-film)	28.8 ± 0.9	0.9927 (ap)	1.122	29.68	86.5	Alta devices
GaAs (multicrystalline)	18.4 ± 0.5	4.011 (t)	0.994	23.2	79.7	RTI,Ge substrate
InP (crystalline)	22.1 ± 0.7	4.02 (t)	0.878	29.5	85.4	Spire, epitaxial
Thin-film chalcogenide						
CIGS (minimodule)	18.7 ± 0.6	15.892 (da)	0.701[a]	35.29	75.6	Solibro, 4 serial cells
CdTe (cell)	21.0 ± 0.4	1.0623 (ap)	0.8759	30.25	79.4	First solar on glass

(continued)

Table 2 (continued)

Classification	Efficiency %	Area Cm^2	Voc (Open circuit voltage)V	Jsc (Current density) MA/cm^2	Fill factor %	Description
Amorphous/microcrystalline Si						
Si (amorphous)	10.2 ± 0.3	1.001 (da)	0.896	16.36	69.8	AIST
Si (microcrystalline	11.8 ± 0.3	1.044 (da)	0.548	29.39	73.1	AIST
Organic						
Organic thin-film	11.0 ± 0.3	0.993 (da)	0.793	19.40	71.4	Toshiba
Organic (minimodule)	9.7 ± 0.3	26.14 (da)	0.806	16.47	73.2	Toshiba (8 series cells)
Perovskite						
Perovskite thin-film	15.6 ± 0.6	1.020 (da)	1.074	19.29	75.1	NIMS

3 Comparison of Various Solar PV Cell

3.1 Laboratory-Based Efficiency of Terrestrial PV Cell and Sub-Module Measured at STC

3.2 Laboratory-Based Efficiency of Terrestrial PV Modules Measured at STC

From the above laboratory experimental data, we have seen the performance of various PV cells but generally in India only two types of PV cells are used, i.e., (1) Mono-crystalline Si solar PV cell and (2) Poly-crystalline Si solar PV cell (Table 3 and Fig. 4).

Mono-Crystalline Solar PV Cell

Mono-Si photovoltaic panels are the most efficient panels. Each cell is made by only one Si crystal. Its efficiency is also much higher than the Poly-Si and thin-film. As per economic point, it is expensive than both Poly-Si and thin-film [1] the method used for production of mono-crystalline Si is known as Czochralski process. In Czochralski method is done in cylindrical boules. After this boules are cut into thin "pseudo-square" shapes [5]. To manufacture this PV cell, diffusion, chemical etching, antireflection coating, edge isolation, and development of metal this process

Table 3 Laboratory-based efficiency of terrestrial PV modules measured at STC [11]

Classification	Efficiency %	Area (cm^2)	V_{oc} V	J_{sc} A	Fill factor	Test center	Description
Si (crystalline)	22.9 ± 0.6	778 (da)	5.60	3.97	80.3	Sandia (9/96)	UNSW/Gochermann
Si (large crystalline)	22.8 ± 0.6	15,738.9 (ap)	69.36	6.459	80.0	NREL (6/15)	Sun power (96 serial cells)
Si (multicrystalline)	19.2 ± 0.4	15,126.5 (ap)	77.93	4.726	78.93	FhG-ISE (6/15)	Trina solar (120 serial cells)
GaAs (thin-film)	24.1 ± 1.0	858.5 (ap)	10.89	2.255	84.2	NREL (11/12)	Alta devices
CdTe (thin-film)	18.6 ± 0.6	7038.8 (ap)	110.6	1.533	74.2	NREL (4/15)	First solar, monolithic
CIGS (Cd free)	17.5 ± 0.5	808 (da)	47.6	0.408	72.8	AIST (6/14)	Solar frontier (70 cells)
CIGS (thin-film)	15.7 ± 0.5	9703 (ap)	28.24	7.254	72.5	NREL (11/10)	Miasole
a-Si/nc-Si (tandem)	12.3 ± 0.3	14,322 (t)	280.1	0.902	69.9	ESTI (9/14)	TEL solar, Trubbach Labs
Organic	8.7 ± 0.3	802 (da)	17.47	0.569	70.4	AIST (5/14)	Toshiba

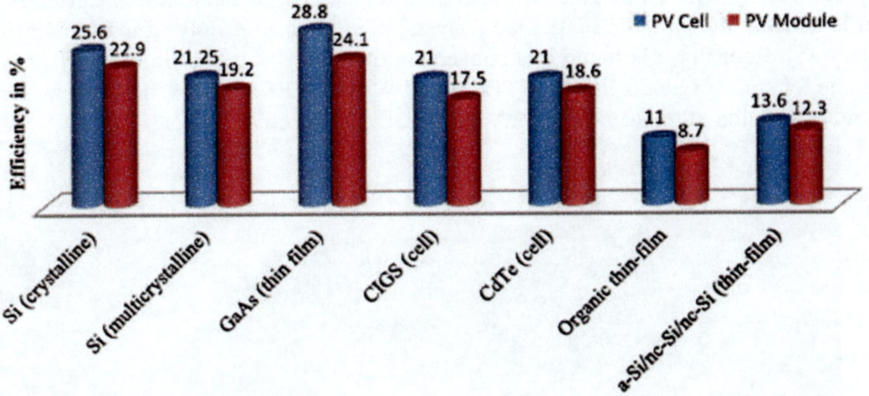

Fig. 4 Comparison between laboratories-based PV cell and module efficiencies

done on pseudo-square shape wafers [4]. Generally, mono-crystalline Si has 15–20% infield conversion efficiency [3, 5]. Lower cost silicon tri-Si and poly-Si are used as an alternative of it because of its higher cost.

Poly-Crystalline Solar PV Cell

Here number of small silicon crystal is used as raw material. To manufacture this, large amount of molten Si is solidified and converted into orient crystals. By producing these crystals in same direction, it makes cast square shape ingots of it, which will then cut to blocks to wafers [6]. Poly-Si has lower efficiency than mono-Si but its cost is also low so generally it is used for module manufacturing. Currently, it is noticed that on large scale filed conversion efficiency of Poly-Si is from 13 to 16% [5]. Recently it is found that conversion efficiency of mono-Si and poly-Si is same for some products like SUNTECH 250 W. SUNTECH 250 W has a fill factor and conversion efficiency of 77.52% and 15.38%, respectively [7].

4 Comparison of Mono-Crystalline and Poly-Crystalline Solar Cell by AHP

AHP means Analytical Hierarchy Process.

In this process, we consider some parameters like mechanical properties, electrical, environment, customer satisfaction, and economic criteria. Here we consider above-mentioned parameters analyses various kind of solar panels. According to that we do comparative analysis of that solar panels and come to conclude best brand for solar panels as per AHP. Here the evaluation is done for 200 W capacity solar panels of different manufacturer companies and find best among all [12].

Environment	P1	P2	P3	P4	P5	P6
Area	1.40	1.27	1.46	1.27	1.48	1.16
Material	Poly	Mono	Mono	Mono	Poly	Mono
Electrical characteristics						
PTC power rating (W)	175	180	179	184.8	177.5	185.9
STC power per unit of area (W/m^2)	142.1	156.7	136.5	156.7	135.1	172.3
Peak efficiency (%)	14.21	15.67	14.2	15.67	13.5	17.2
Power tolerances (%)	−9/+9	0/+3	−3/+32	0/+5	−3/+3	0/+10
Number of cells	50	72	54	72	54	96
Imp (A)	8.16	5.42	7.89	5.17	7.60	5.59
Vmp (V)	24.50	36.90	25.38	38.70	26.30	55.80
Isc (A)	8.70	5.80	8.24	5.50	8.22	3.83
Voc (V)	30.80	45.60	33.53	45.90	33.30	68.70
NOCT (°C)	–	45	–	45	–	–
Temp. coefficient of power (%K)	−0.50	−0.40	−0.50	−0.38	−0.45	−0.29
Temp. coefficient of voltage (V/K)	−0.5	−0.4	−0.5	−0.38	−0.45	−0.17
Series fuse rating (A)	15	10	15	15	15	15
Maximum system voltage (V)	600	1000	600	10,000	600	600
Lower energy density(W/m^2)	11.52	11.92	11.82	13.28	11.26	14.89
Mechanical characteristics						
Frame color	Bronze	Clear	Black	Clear	Clear	Black

(continued)

(continued)

Environment	P1	P2	P3	P4	P5	P6
Weight (kg)	15.4	14.5	35	15.5	18	15
Financial properties						
Price ($)	300	300	499	300	319	600
Cost per Watt ($)	1.05	1.05	1.75	1.05	1.05	2.10
Customer satisfaction						
Service support	3	1	6	5	2	4
Spare part	5	2	6	4	1	3
Reliability	5	2	6	4	3	1

The above table which shows AHP analysis by considering environmental, mechanical, electrical, customer satisfaction, and financial criteria. After studying this table we came to conclude that solar panel brand P6 is the best suitable. Here it must be mention that these results are for specific cases and for a particular location, it may be changed as per change in locations [12].

4.1 Solar Tracking System

The solar trackers are generally used for minimizing the shading effect and for performance enhancement of solar PV.

The solar tracker can be classified as (i) fixed-rack tracker; (ii) single-axis tracker; and (iii) double-axis tracker.

(1) Fixed-rack tracker: As the name suggests, it cannot change its angle as per sun's movement. It generally remains fixed at particular angle. At the same time its cost is least among all three because it does not require any additional device. It is durable as well as less maintenance required.

(2) Single-axis tracker: Single-axis tracker can just track one daily motion of sun either north-south or east-west. It can rotate in only single axis of rotation either horizontal or vertical. It can generate higher amount of electricity compare to fixed tracker but it also requires extra device for the tracking purpose. Due to this extra device its cost is higher.

(3) Double-axis tracker: It can track in both north-south as well as east-west motion of sun. As name indicates it has two-axis rotation in horizontal and vertical direction. So it has highest electricity production capacity but its cost is also highest among the three.

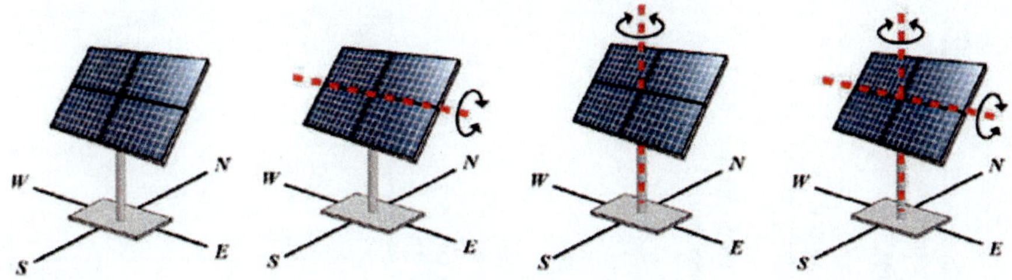

1. Fixed rack Tracker [26] 2. Single axis tracker 3. Double axis tracker

4.2 Novel and Emerging Solar Cell Technology Concept

In the above study, we just discussed few commercially available traditional solar cells. Carbon nanotubes, superlattice technology, quantum wells, hot carrier (HC), and Quantum wires/dots are newly invented materials [13, 14]. Improve efficiency and enhance performance is the main purpose for development of new technology. Here nanomaterials and nanoscale components control band gap and improve efficiency and performance.

A semiconductor which can absorb spectrum of particular size for the purpose of match with solar spectrum is called Quantum dots. It has also one additional benefit that it can generate multiple electron-hole pairs per photon. As mentioned above, for efficiency improvement it can adjust band gap [14].

Generally conversion efficiency of carbon nanotubes varies from 3 to 4%. It is generally coated with CdS—Cadmium sulfide and CdTe—cadmium telluride for the purpose of trap photons and completely absorb it. It is noticed that it has good mechanical and electrical properties.

4.3 Waste Solar Panels Recycling Technology

The recycling of end-of-life (EOL) panel is also one of the important parameters which we generally neglect. There is a lack of knowledge and awareness regarding this [15]. The table shown below is a comparison of various solar panels recycling technology.

Key technology	Method	Result-Advantages	Disadvantages	Reference
Component repair	Two kinds of methods to dismantle and analyze faults in the junction box	Can increase the output power of old solar panels	Can be used only if the external junction box and outer layer are aging	[16]
Module separation	Organic solvent method Compared to three methods: artificial disassembly, use of organic solvent to dissolve components, and heat treatment Artificial disassembly, crushing, cryogenic breaking, and electrostatic separation Organic solvent-assisted ultrasonic method Heat treatment and chemical etching methods	Mechanical pressure is critical to suppress the swelling of the EVA. The silicon panel was recovered successfully with no damage The heat treatment process was the optimal solution Obtained a mixture of different types of materials significantly shortens the dissolution time of EVA in organic solvents Module separation	Organic liquid waste produced Did not involve the resource recovery of Silicon Not sufficient for single component separation; still in the laboratory research stage Organic liquid waste produced Non-purified silicon wafers	[17] [18] [19] [20]
Recycling of silicon	Cement-based thermal insulation system and chemical method	Recycling of silicon	Organic liquid waste produced	[21]

(continued)

(continued)

Key technology	Method	Result-Advantages	Disadvantages	Reference
Recycling of rare metals	Grinding and hydrometallurgy	Recycling of indium and gallium	High price of chemicals	[22]
	Physical and chemical methods: blasting, mechanical processing and dissolution of the semiconductor, and precipitation, plating, and ion exchange	Recycling of cadmium telluride; net cost estimated at \$0.04–\$0.06/W	High price of chemicals	[23]
	Recycling (CdTe and CIS) by wet mechanical treatment such as grinding and flotation, or dry mechanical processing methods such as vacuum blasting	Recycling (CdTe and CIS)	Process is complicated	[24]
	Treated multiple types of solar panels together. Two physical methods were used: panels were broken and then heat-treated or broken with a hammer	Mix of recycled poly-silicon, amorphous silicon, and CdTe solar panels; glass directly recycle	All types of mixtures, hard to separate	[25]

5 Conclusion

After exhaustive review based on available literature presented by renowned researchers in field of solar power generation using photo volatile technology is discussed briefly here. We have observed various characteristics of solar PV cell and effect of variables in solar PV cells.

Generally electrical performance of PV cells and modules are measured by manufacturer at standard temperature condition but in actual power plant modules operate at different climatic conditions. Hence there is variation in performance, conversion efficiency, and generated power are diverse from experimental results within laboratory and real field environment. A noticeable deviation observed for experimental result between solar PV cell and module.

Here we have mentioned laboratory-based performance analysis of different PV cells like silicon crystalline, thin-film, organic, etc. and laboratory-based performance analysis of different PV modules; done by various researchers. After comparing it, it is observed that there is less power generated in PV module.

In India, poly-crystalline and mono-crystalline PV panels are majorly used so here we have discussed a comparative AHP (Analytical Hierarchy Process) analysis performed by honorable researchers and made summary of it as per different conditions. At the end of AHP analysis, we came to the conclusion that mono-crystalline panel of brand P6 is best suitable according to financial, mechanical, environmental, electrical, and customer satisfaction criteria.

Here we also discussed novel solar cell technologies and some methods for recycling waste solar panels.

References

1. Manishkumar, Arun kumar, Performance assessment and degradation analysis of solar photovoltaictechnologies: a review
2. Sharma A (2011) Comprehensive study of solar power in India and world. India. Renew Sustain Energy Rev 15: 1767–1776
3. Park S, Pandey A, Tyagi V, Tyagi S (2014) Energy and exergy analysis of typical renewable energy systems. Renew Sustain Energy Rev 30:105–123
4. Maehlum MA (2013) Which solar panel type Is best? Mono vs. polycrystalline vs. thin–film. Energy Inf
5. Tyagi V, Rahim NA, Rahim N, Jeyraj A, Selvaraj L. Progress in solar PV technology: research and achievement. Renew Sustain Energy Rev
6. El Chaar L, El Zein N (2011) Review of photovoltaic technologies. Renew Sustain Energy Rev 15:2165–2175
7. Akinyele D, Rayudu R, Nair N (2015) Global progress in photovoltaic technologies and the scenario of development of solar panel plant and module performance estimation– Application in Nigeria. Renew Sustain Energy Rev 48:112–139
8. Gee JM, Schubert WK, Basore PA (1993) Emitter wrap-through solar cell. In: Photovoltaic specialists conference, conference record of the twenty third IEEE, pp 265–270. IEEE

9. Pandey A, Tyagi V, Jeyraj A, Selvaraj L, Rahim N, Tyagi S (2016a) Recent advances insolar photovoltaic systems for emerging trends and advanced applications. Renew Sustain Energy Rev 53:859–884
10. Goetzberger A, Hebling C, Schock H-W (2003) Photovoltaic materials, history, status and outlook. Mater Sci Eng: R: Rep 40:1–46
11. Green MA, Emery K, Hishikawa Y, Warta W, Dunlop ED (2016) Solar cell efficiency tables (version 47). Progress Photovolt: Res Appl 24
12. The selection of the best solar panel for the photovoltaic systemdesign by using AHP FigenBaloa, LütfüŞağbanşuab, * Energy Procedia 100 (2016) 50–53
13. Pandey A, Tyagi V, Jeyraj A, Selvaraj L, Rahim N, Tyagi S (2016b) Recent advances in solar photovoltaic systems for emerging trends and advanced applications. Renew Sustain Energy Rev 53:859–884
14. Jha AR (2008) MEMS and nanotechnology-based sensors and devices for communications, medical and aerospace applications. CRC Press
15. Xu Y, Li J, Tan Q, Lauren Peters A (2018) Global status of recycling waste solar panels: a review. ChinaWaste Manag
16. Lin W, Chen E, Sun YL (2011) Analysis of old photovoltaic component junction box disassembling mode. Sol Energy 7:26–29
17. Doi T, Tsuda I, Unagida H, Murata A, Sakuta K, Kurokawa K (2001) Experimental study on PV module recycling with organic solvent method. Sol Energy Mater Sol Panels 67:397–403
18. Dong L (2009) Research on waste crystalline silicon solar panels resource recovery
19. Kim Y, Lee J (2012) Dissolution of ethylene vinyl acetate in crystalline silicon PV modules using ultrasonic irradiation and organic solvent. Sol Energy Mater Sol Panels 98:317–322
20. Klugmann-Radziemska E, Ostrowski P (2009) Chemical treatment of crystalline silicon solar panels as a method of recovering pure silicon from photovoltaic modules. Renew Energy 2009:1–9
21. Fernandez LJ, Ferrer R, Aponte DF, Fernandez P (2011) Recycling silicon solar cell waste in cement-based systems. Sol Energ Mat Sol C 95:1701–1706
22. Guangdong Xiandao Rare Material Co. Ltd (2011) Recovery of copper indium gallium selenide thin-film solar panel, involves crushing solar panel, soaking in sulfuric acid, filtering, extracting, separating, stripping extraction liquid, adding reducing agent to raffinate and filtering (Chn) CN 103184338-A[P] 2011-12-29
23. Berger W, Simon FG, Weimann K, Alsema EA (2010) A novel approach for the recycling of thin film photovoltaic modules. Resour Conserv Recy 54:711–718
24. Granataa G, Pagnanelli F, Moscardini E, Havlik T, Toro L (2014) Recycling of photovoltaic panels by physical operations. Sol Energy Mater Sol Panels 123:239–248
25. Kang H, Hong T∗, Jung S, Lee M. Techno-economic performance analysis of the smart solar photovoltaic blinds considering the photovoltaic panel type and the solar tracking method
26. Hafez AZ, Yousef AM, Harag NM (2018) Solar tracking systems: technologies and trackers drive types –A review. Renew Sustain Energy Rev 91:754–782. https://doi.org/10.1016/j.rser.2018.03.094
27. Heslop S, MacGill I (2014) Comparative analysis of the variability of fixed and tracking photovoltaic systems. Sol Energy 107:351–364. https://doi.org/10.1016/j.solener.2014.05.015
28. König D, Casalenuovo K, Takeda Y, Conibeer G, Guillemoles J, Patterson R et al (2010) Hot carrier solar cells: principles, materials and design. Phys E: Low-Dimens Syst Nanostruct 42:2862–2866

Chapter 4
Evaluation of Thermal Degradation Behavior of Cardboard Waste

Samit Kumar Singh and Sadanand A. Namjoshi

1 Introduction

The waste-to-energy is really a challenging task and to accomplish this task, the pyrolysis is one of the best options. The composition of MSW varies because of socio-economical status of the society, economic development, demographic division, and waste collection potential also for waste-to-energy options. The extensive rises in the MSW generation having severe impact on environment but its disposal associated with social and economic problems.

The pyrolysis having gas, char, and liquid yield products; it means three types of fuels are available. The significance of pyrolysis is also due to the decomposition of polymers into fuels which are not easily degradable and contaminate atmosphere heavily when thrown as a waste at open places. The design of pyrolyzer is based on kinetics studies which correlate the weight degradation with respect to time and temperature; it also represents space or in other words, volume requirement of fuel in the form of pyrolytic products with respect to time temperature. The quantities of yield products are also depends on available space otherwise temperature gradient will exist and conduction losses increase.

Garcia et al. [1] focused kinetics of pyrolysis at different heating rate between 1.5 and 200 °C min^{-1}. After observing the collected results, a theoretical study was carried out in connection with the effect of heating rate and the transmission of heat on the kinetic parameters which can be obtained using the experimental data.

Fontet al. [2] investigated the kinetics of the almond shells impregnated and non-impregnated with $CoCl_2$, polyethylene lignin, and MSW with the help of an analytical

S. K. Singh (✉) · S. A. Namjoshi
Madhav University, Pindwara (Sirohi) 307026, Rajasthan, India
e-mail: samit.meet@gmail.com

S. A. Namjoshi
e-mail: namjoshisadanand@yahoo.co.in

M. Shorif Uddin et al. (eds.), *Intelligent Energy Management Technologies*, Algorithms for Intelligent Systems,
https://doi.org/10.1007/978-981-15-8820-4_4

pyro probe 1000 or pyroprobe 100, thermogravimetric fluidized sand bed reactor was used to understand the thermal decomposition of heterogeneous materials.

Wu et al. [3] reported kinetic study of painted printed and writing paper by TGA in nitrogen atmosphere for the temperature range is from 450–900 K at 1, 2, and 5 K/min heating rates with two stages of decomposition.

Park et al. [4] suggested a new method for estimation of kinetics constants of low-density polyethylene (LDPE), linear low-density polyethylene (LLDPE), and high-density polyethylene (HDPE), at heating rates of 100 °C/min and 500 °C/min. The numerous analytic methods mentioned were used for comparing with the kinetic parameters estimated from the dynamic method.

Sorum et al. [5] investigated in-depth knowledge of pyrolysis process and estimation of chemical kinetics important constituents of MSW. TGA estimates the kinetic constants which are executed at 100 °C/min heating rate in N2 gas atmosphere. The degradation of hemicellulose, cellulose, and lignin was shown by three autonomous parallel reactions the cellulosic fraction of MSW.

David et al. [6] explained pyrolysis process depending upon the numbers of TGA experimental work on cardboard. The research work includes the determination of kinetics parameters for cardboard. These research works also focus pyrolysis process with the support of numbers of TGA experimental works also compare the results of graphical method with numerical results which are in good level of satisfaction.

Molto et al. [7] focused on kinetic studies for the decomposition of used cotton fabrics and for explaining the behavior of decomposition all the runs were performed and have been proposed and tested. In the pyrolysis of the used cotton fabric waste, the model comprises two parallel reactions during decomposition.

Miranda et al. [8] studied thermal degradation behavior of textile waste by TGA at different heating rates also decomposition was studied by semi-batch pyrolysis process.

The DTG curves show three peaks; two out of which were hemicellulose and cellulose and one might be due to decomposition synthetic polymer. The major research in present work focuses on thermal degradation behavior of cardboard waste at 5, 10, and 15 °C/min heating rate and also ash obtained during decomposition of cardboard analyzed using Scanning Electron Microscope (SEM).

2 Experimental Methodology

2.1 Material

The experiments were conducted on cardboard collected from waste packing boxes which was sheared into small pieces of 1–2 mm by using scissors. The size of the particle used in TGA having micro-weight and sample amount limitation of container. Table 1 indicates proximate, ultimate result, and HHV of cardboard.

Table 1 Proximate analysis, ultimate analysis, and higher heating value cardboard

Fixed carbon (% wt)	Volatile matter (% wt)	Ash (% wt)	Carbon (% wt)	Hydrogen (% wt)	Nitrogen (% wt)	Sulfur (% wt)	Oxygen (% wt)	HHV MJ/kg
12.6	74.9	12.4	42.13	5.3	0.18	0.54	39.51	15.35

2.2 *Thermogravimetric Analysis (TGA)*

The engineering design of modern pyrolysis reactors and gasifiers requires a deep knowledge of kinetics studies for different MSW constituents at different temperature ranges and heating rates. The TGA and DTG were performed in TG/DTG system. Samples' temperature range varies from room temperature (RT) to 600 °C with 5, 10, and 15 °C/min heating rate in nitrogen atmosphere and flow rate of gas was 50 ml/min. The sample size of cardboard was taken as 6 mg. Samples were placed in platinum container and reference used was α alumina.

3 Kinetics

It is necessary to understand the reaction mechanism of pyrolysis for obtaining the reactions with respect to its kinetic parameters. The heated deterioration of MSW constituents is explained to be proceeding through many complicated reactions [5, 6, 8].

A method was formulated to calculate activation energy from single dynamic TGA data employing following model [10, 11].

Here F062 is heating rate, E is activation energy A is pre-exponential or frequency factor, R is universal gas constant (8.314 kJ mol^{-1} K^{-1}).

This equation has no exact solution; hence, different approximations have been done by Flynn wall Ozawa method [12].

For cardboard decomposition, in case TGA, the Flynn wall Ozawa [12] method used for activation energy for three heating rates are used.

4 Results and Discussion

Figure 1 represents Thermogravimetric curves of cardboard at 5, 10, and 15 ^{0}C/min heating rates, respectively, while Fig. 2 represents the corresponding Differential Thermogravimetric curves for same heating rates.

The decomposition is observed in three stages as shown in Figs. 1 and 2. In first stage, moisture present in sample gets removed and hemicellulose gets decomposed. The degradation rate in first stage is slow due to low temperature range and mass

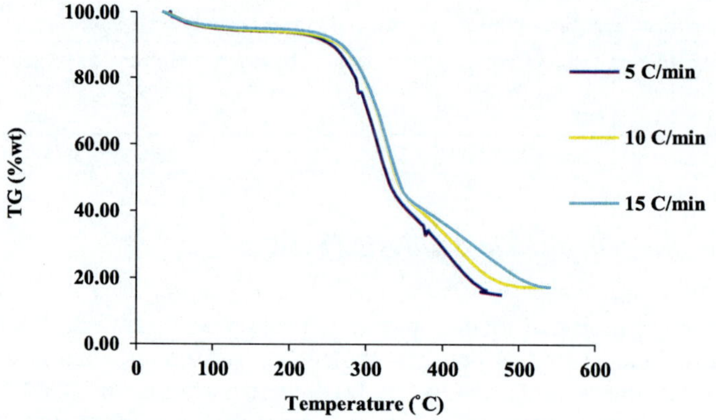

Fig. 1 TG curves

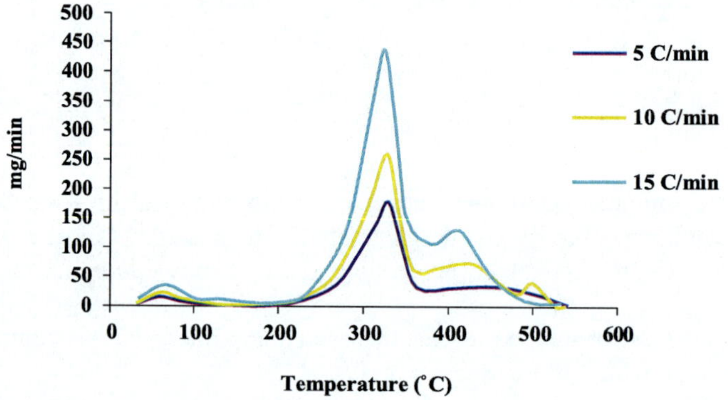

Fig. 2 DTG curves

transfer resistance [1, 5, 6, 8]. The degradation of cellulose and lignin occurs in second and third stage, respectively. The similar results are reported by Sorum et al. [5], David et al. [6], Miranda et al. [8] and which substantiates the present work.

Figure 1 represents same TG curves pattern for heating rates 5, 10, and 15 °C/min; however, rates and kinetics parameters are differing. During second stage, degradation occurred is faster because cellulose gets decomposed compared to first and third stage where hemicellulose and lignin are decomposed, respectively, as heating rate increases which is represented by TG curves in Fig. 1 [5]. The loss in weight escalates in all three phases at a quicker amount as heating rate rises which is quite obvious that heating rate enhances the values of kinetics parameters too. In connection to three stages of decomposition on the TG curves correspondingly, three peaks are noticed on the DTG curves. DTG curves show as heating rates expand there will be a lateral shift toward greater temperatures and which may be the united effects of the heat transfer at various heating rates and resulting in delayed the decomposition as stated by Sorum et al. [5]. Figure 1 reports that rise in heating rate gives the greater value of activation energy, which can be attributed to the formation of active molecules also enhances in molecular collision at high heating rates [5, 8]. It also shows that the activation energy value is maximum in second stage in comparison to other two stages in all cases all heating rates. During third phase of decomposition, activation energy value gets lowered because of low deterioration rate. It is observed that amount of decaying rises with the rise in heating rate and a declination is observed in total decaying. It is because of the inadequate decaying time for concluding reaction with rising heating rates [8]. In case of all cellulosic fractions, the major weight loss occurs between the temperatures ranges of 250–400 °C as also reported by Sorum et al. [5].

Figure 2 represents the DTG curves which are nothing but decomposition rate with respect to temperature in association with 5, 10, and 15 °C/min heating rates. Results show that rise in heating rate leads to rise in weight loss and corresponding maximum temperature also increases linearly due to delay in decaying which penetrates extra heat into the specimen for shorter while [5]. In case of cardboard, the three peaks are commonly observed and the temperature range in which maximum peak can be observed varies from 175 to 250 °C, 300 to 425 °C, and 450 to 525 °C which corresponding to hemicellulose, cellulose, and lignin, respectively [5, 6].

The description of less marked shoulder on the DTG curves for hemicellulose in case of cardboard is due to the minor content of hemicellulose and/or catalytic effects. Catalytic effects generated by inorganic species such as ash and in the case of cardboard, residues from sulfate process causes the decaying of cellulose to occur at reduced temperature [5].

Tables 2 and 3 present the values of decomposition with temperature range and kinetic constants for different stages of decomposition, respectively.

Table 2 % Weight loss in stages of decomposition corresponding to temperature range

Heating Rate (°C/min)	Stage I		Stage II		Stage III		% Residual remains
	% Loss in weight	Range of temp °C	% Loss in weight	Range of temp °C	% Loss in weight	Range of temp °C	
Cardboard waste							
5	6	30–193.5	52.2	193.5–349	27.2	349–545	14.5
10	6	30–193.3	53	193.3–366.5	24	366.5–539.6	17
15	7	30–209.5	50.3	209.5–353.1	27	353.1–532.9	15.7

Table 3 Activation energy for cardboard

Activation energy, E, kJ/mol			
Constituent	1st stage	2nd stage	3rd stage
Cardboard	128.15	133.65	83.74

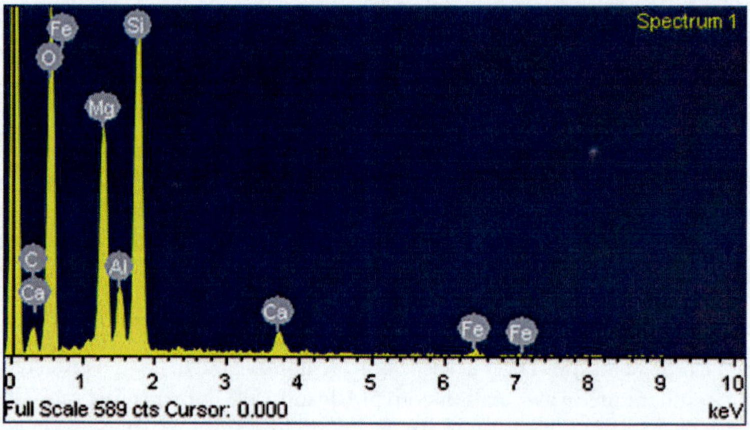

Fig. 3 Cardboard char (10 °C/min)

The elemental analysis of residues of lignocellulosic waste material, i.e., cardboard is performed with the help of SEM using Energy Dispersive Spectroscopy (EDS) technique. EDS is a scientific approach and it employs x-rays that are transmitted from the sample when blasted by the electron beam for identifying the elemental composition of the sample. The elemental analysis of residue with the help of scanning electron microscope at 10 °C/min of heating rate has been carried out. The importance of this study is to explore products and elements presence in the residues of cardboard [13]. Figures 3 and 4 indicate elemental analysis of cardboard ash obtained from TGA at 10 °C/min of heating rate.

5 Conclusion

For cardboard material, decomposition occurs in three stages. Stage I, II, and III represents decomposition of hemicellulose, cellulose, and lignin, respectively. The estimation of decomposition rate is possible through the assurance of kinetic parameters such as activation energy.

The activation energy for cardboard waste at the heating rate of 5 °C/min to 15 °C.

/min is obtained as 128.15 kJ/mol in first stage, 133.65 kJ/mol in second stage and 83.74 kJ/mol in third stage, respectively.

The significance of elemental analysis is to know the possible elements existing in the residue and how they will affect the environment from pollution point of view in later stage.

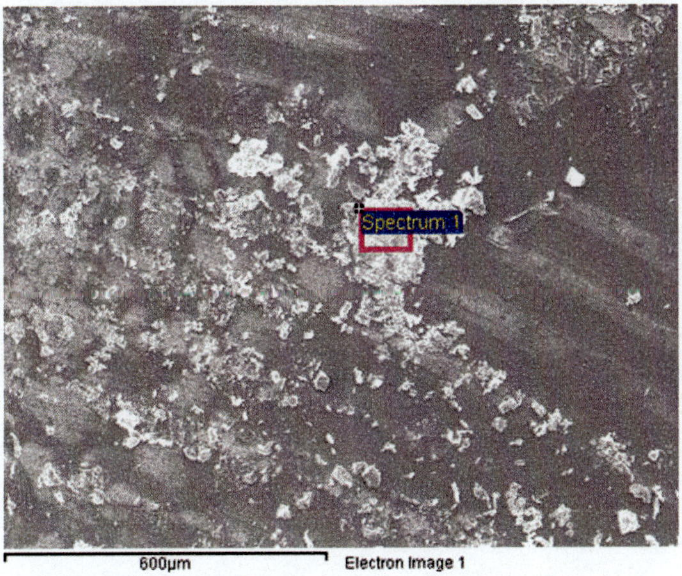

Fig. 4 Elemental spectrum

6 Future Scope of Work

The present research work can be extended for development of such reactor at macro-level of cardboard or any MSW constituents for various heating rate makes possible. It also yields products obtained by such reactor which establishes MSW as an alternative better fuel.

References

1. Garcia AN, Marcilla A, Font R (1995) Thermogravimetric kinetic study of the pyrolysis of mu- nicipal solid waste. Thermochim Acta 254:277–304
2. Font R, Marcilla A, Garcia AN, Caballero JA, Conesa JA (1995) Comparison between the pyrolysis products obtained from different organic wastes at high temperatures. J Anal Appl Pyrol 32:41–49
3. Wu C-H, Chang C-Y, Lin J-P, Hwang J- Y (1997) Pyrolysis kinetics of paper mixture in municipal solid waste. Fuel 76:1151–1157
4. Park IW, Sea Cheon O, Lee HP, Taik Kim H, Yoo KO (2000) A kinetic analysis of thermal degradation of polymers using a dynamic method. Polym Degrad Stabil 67:535–540
5. Sorum L, Gronli MG, Hustad JE (2001) Pyrolysis characteristics and kinetics of municipal solid waste. Fuel 80:1217–1227
6. David C, Salvador S, Dirion JLM (2007) Determination of a reaction scheme for cardboard thermal degradation using thermal gravimetric analysis. J Anal Appl Pyrolysis 67:307–323

7. Molto J,Font R, Conesa A, Martı´n-Gullo´n I (2005) Thermogravimetric analysis during the decomposition of cotton fabrics in an inert and air environment. J Anal Appl Pyrolysis (Article in Press)
8. Miranda R, Sosa_Blanco C,Bustos-Martı ´nez D, Vasile C (2007) Pyrolysis of textile wastes kinetics and yields. J Anal Appl Pyrolysis 80:489–495
9. Buah WK, Cunliffe AM, Williams PT (2007) Characterization of products from the py- rolysis of MSW. Process Saf Environ Prot 85:450–457
10. Yourlmaz S (2006) Investigation of emissions and combustion kinetics of waste wood samples with thermal and spectral methods. PG Dissertation, Middle East Technical University, Beirut
11. Liu Z (1998) Handbook of thermal analysis. John willy & Sons, pp 47–48
12. Flynn JH, Wall LA (1966) Polym Sci Part B, Polymer Letters 4:323
13. Miskolczi N, Angyal A, Bartha L, Valkai I (2009) Fuels by pyrolysis of waste plastics from agricultural and packaging sectors in a pilot scale reactor. Fuel Process Technol 90:1032–1040

Chapter 5
Wavelength Optimization in Gigabit Passive Optical Network by Proposed Quad Play Architecture

Md. Hayder Ali and Mohammad Hanif Ali

1 Introduction

Unruly modification in the enterprise resident area network is the prime cause to fulfill the requirements of swift progress of high bandwidth requirements and also changing the nature of enterprise solutions. To maintain a significant solution and to reduce the capEX and opEX, fulfilling the high bandwidth demand, the utmost solution is gigabit passive optical network (GPON) technology [1–2]. ITU-T G.984.x approvals offer GPON system prototypical and does not need any electrical power at the transitional nodes between the aggregation and last-mile device [3]. A big number of researches had been worked on network structures, propagation policies, power consumption budget, bandwidth distribution, and stability of GPON technology [4–6]. The efficiency of service for GPON triple play in IP network and also passive optical network and bandwidth distribution policies are conversed in [7–8]. Ricciardi et al. have exposed an investigation and contrast between EP2P (Ethernet point to point) and GPON solution structure [9].

SDH (E1) traffic with the present triple play explanation is the demand of today's technological solution. It is much more important to focus on some parameters like output signal range, propagation wavelength, power budget, and sensitivity threshold for convergence solution. In this paper, wavelength is optimized with a proposed quad play architecture by calculating different parameters like path loss, the last end receiving sensitivity, etc.

The rest of this paper is ordered as follows. The literature review is presented in Sect. 2. GPON triple play general structure is stated in Sect. 3. In Sect. 4, proposed

Md. H. Ali (✉) · M. H. Ali
Jahangirnagar University, Dhaka, Bangladesh
e-mail: hayder.iict@gmail.com

M. H. Ali
e-mail: hanif_ju03@juniv.edu

M. Shorif Uddin et al. (eds.), *Intelligent Energy Management Technologies*,
Algorithms for Intelligent Systems,
https://doi.org/10.1007/978-981-15-8820-4_5

quad play architecture is briefly defined. The simulation scenario is in Sect. 5 and performance analysis between GPON triple play and proposed quad play is stated in Sect. 6. Finally, Sect. 7 draws a conclusion to this paper.

2 Literature Review

Over the years especially in the last decade, a lot of research study have been passed out to find out the GPON network suitable structure and different type of FTTx model. Some of them are discussed below.

Koonen et al. [10] stated fiber-optic technologies and it enables the solution with minimum cost of broadband solution services to the last mile users in multiple-access network architectures [10]. They also designated that the dynamic network reconfiguration can improve system staging. Derek Nesset et al. [11] have investigated GPON structure with a 1300 nm semiconductor optical amplifier and it developed for long spread GPON services [11]. The high gain of −29 dB has empowered a profitable GPON system to function over 60 km and with 128-way split. Claudio Rodrigues et al. [12] have studied and associated the present GPON fiber to the home (FTTH) type service solution, from the point of view of enterprise operator, they also stated that future next-generation passive optical networks (PONs) (XGPONs) and wavelength division multiplexing PONs (WDM-PONs). They also consider some facts like considering into account standardization, wavelength planning, optical line terminal (OLT) as well as optical network terminal (ONT) equipment, and transmission convergence layer [12].

Marcelo Alves Guimaraes et al. [13] have projected a new system for E1 transmission and it allows the fair bandwidth policy, it is performed without circuit emulation techniques by fragmenting E1 signals. Andrej Chu et al. [14] presented an explanation by using ant cluster optimization to strategy GPON-FTTH networks with combining equipment [14]. Due to the reflection of many strategy issues such as the number, types, positions of network elements, the optical network development progression frequently parades numerous trials from the optimization point of view. Irfan imitation an estimation of 2.5 Gbps bi-directional GPON based FTTH link using innovative modulation setups [15]. In this research, the presentation of the future scheme was calculated when a single wavelength and two wavelengths were used for triple-play services with different modulation systems.

3 Present Structure of GPON Triple Play

In the present GPON triple play structure, the signals enter intoOLT in two separate ways. Firstly, EDFA receives a video signal and pass through. Secondly, data and voice signals enter via a layer-2 switch (ISP connection). These two signals modulated into OLT and passed through fiber, optical splitter up to ONT. Finally, end-users receive signals (data, video, and voice) from ONT's ethernet port (Fig. 1).

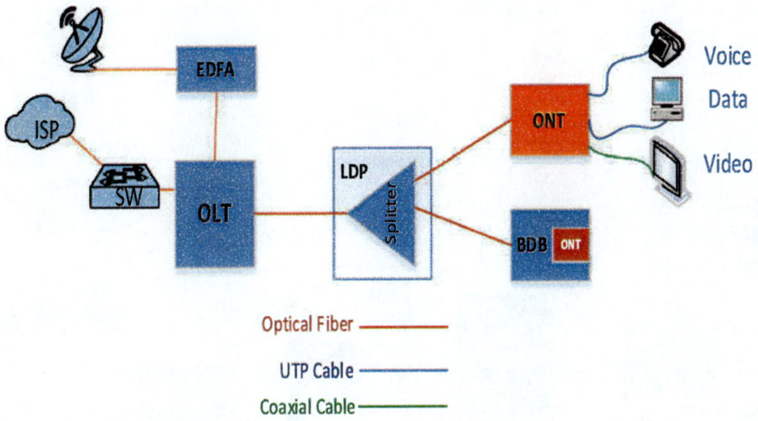

Fig. 1 Present triple play architecture

4 Projected Quad Play Architecture

The projected quad play design is not similar as present triple play design. All signals (video and ISP's data and voice) will enter into a MUX combinedly. These signals will be modulated into OLT. From the OLT, the modulated signal will propagate toward the ONT. The end-users will receive the signals from ONT's ethernet port as present triple play.

The main difference between the present triple play design and projected quad play design is, the modulation done into OLT in the present design and OLT becomes busier and OLT's CPU uses become high, On the other hand, modulation is done into MUX and that's why OLT does not disburse time for the modulation (Fig. 2).

5 GPON Quad Play Simulation Environment

For better results, the simulation experimented only for single-user connectivity. For this goal, the packet simulation software OptSIM has been used. OptSIM proposes a hierarchical structure of pattern for creating new simulation scenarios. Defined tree levels:

- Network model: It's the primary stage of design. It's the maximum intellectual and general level. The goals will be to describe network topology, to express network nodes, and to define the statement between each node.
- Node model: The second level, objectives are to express the functionalities for each node that is used in the network topology. For all nodes, we built a scheme for scheming inner purposes with the component offered by OptSIM.

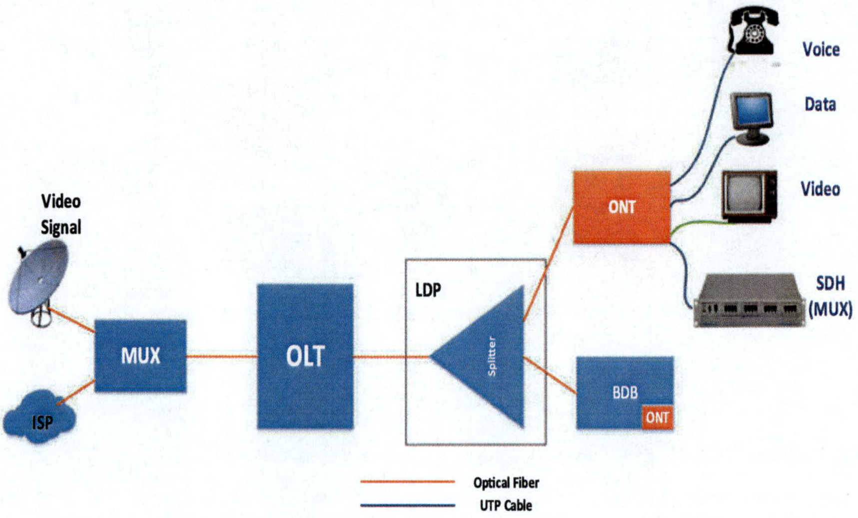

Fig. 2 Proposed quad play architecture

- Process model: The final level. In this level, a graph is defined for elements used in the second level. The graphs stipulate the jobs performed by the module with the info which will be processed.

5.1 Network Construction

To generate the simulation scenario, it is used the tree levels proposed by OptSim. We have agreed to use FTTH topology (no FTTC or others) because the optical fiber reaches straight at the user's home and it is the resolution that offers more bandwidth capacity.

It is considered only this type of control packets in our simulation and ignored others feature related to delays or protocols since the objectives of our analysis need only results about wavelength optimization due to intermediate entree control measures.

5.2 ONU/ONT Design

User's ONT contains splitter, and data earpiece, and video earpieces. Data earpiece is constructed with optical filter, PIN/TIA receiver, and BER Tester. The video signal receiver contains optical filter, PIN/TIA receiver, and electrical filters (Fig. 3).

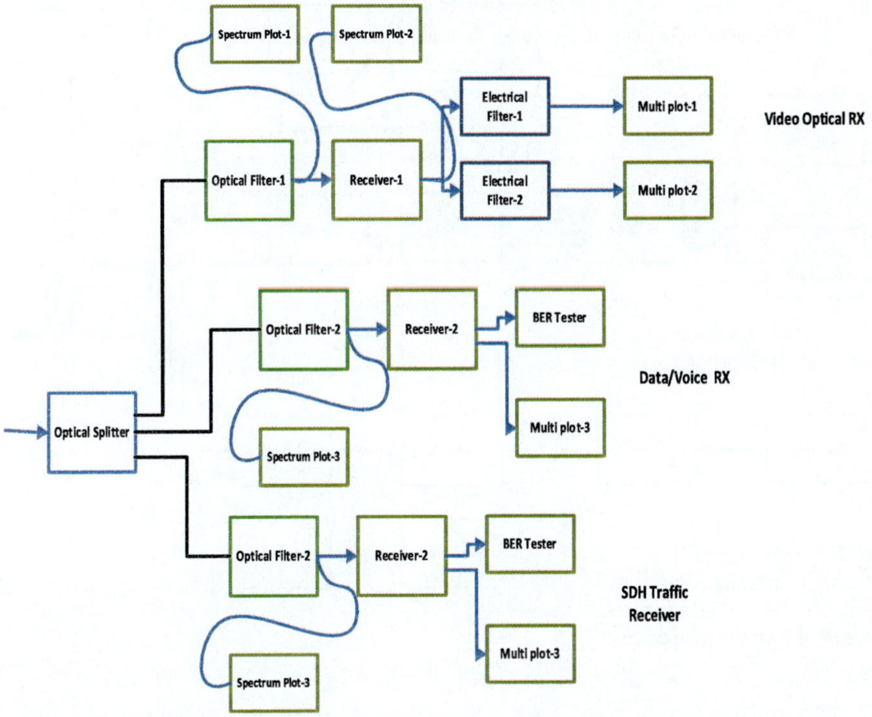

Fig. 3 Block diagram of ONU/ONT

5.3 OLT Design

First, we describe in detail the configurations for Central Office OLT and single end-user ONT and then, will generalize the treatment to all 16 uses. In this experiment, we consider the downstream configuration of GPON with bitrate 1.25 Gb/s and support for quad play. The quad play provision is appreciated as a blend of SDH traffic, data, voice, and video indications. The high-speed Internet element is denoted by a data link with 1.25 Gb/s downstream bandwidth. The voice element can be characterized as VOIP service (voice over IP, packet-switched protocol) and can be collected with data component in physical layer simulations. Finally, the video element can be denoted as an RF video signal (traditional CATV) or as an IPTV signal that also can be mutual with data. In this research, it is considered the former case with an RF video link. To improve the wavelength in PON, the propagation over the optical fiber pathway services the CWDM technique with data/voice component transmitted at wavelengths in the variety of 14,801,500 nm, and video within the 1550–1560 nm range [16] (Fig. 4).

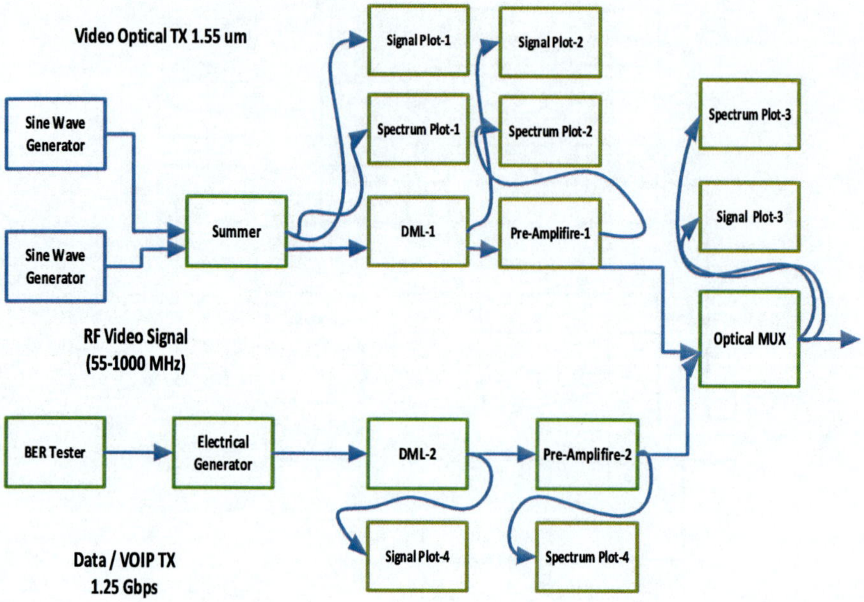

Fig. 4 Block diagram of OLT

The Central Office Optical Line Terminal block (Transmitter block) involves Data/VOIP and Video elements. The Data/VOIP spreader demonstrated with pseudo-random data originator (PRBS), NRZ modulator driver, direct-modulated laser, and admirer amplifier. The video section demonstrated as RF SCM (subcarrier multiplexed) link with only two tones (channels) for simplicity. The two channels we used are from standard NTSC analog CATV frequency plan—channel 2 and channel 78 at frequencies 55.25 MHz and 547.25 MHz, respectively. RF video source contains two electrical signal producers, summer, direct-modulated laser, and pre-amplifier. Following, data/voice and video signals are multiplexed at multiplexer and propagated into 20 km fiber span.

6 Result Analysis

Applying the loss calculation equation (in Sect. 5), the below table is founded. It shows the last end (end-user) devices (ONT/ONU) signal receiving sensitivity. Minimum receiving sensitivity is the desired one (Table 1).

Using 1:32 Splitter with 20 km distance, it is observed that the minimum sensitivity is −34.35 dB at 1550 nm wavelength and the highest sensitivity is −36.66 dB at 1310 nm wavelength (Fig. 5).

Table 1 1:32 Splitter with 20KM distance

Wave length	Attenuation (dB/Km)	Attenuation for 10m (dB)	Mux loss (dB)	Add Loss	Drop Loss	Demux Loss	Filter Loss	ODF (IN+OUT) Loss	LDP (IN+OUT) Loss	Splitter Loss	BDB (IN+OUT) Loss	ONT Loss	Totale	Remarks
1310	0.36	7.2	0.5	1	1	0.5	8.5	0.3+0.3=0.6	0.3+0.3=0.6	17.3	0.3+0.3=0.6	0.3	36.66	
1330	0.335	6.7	0.5	1	1	0.5	8.5	0.3+0.3=0.6	0.3+0.3=0.6	17.3	0.3+0.3=0.6	0.3	36.135	
1350	0.322	6.44	0.5	1	1	0.5	8.5	0.3+0.3=0.6	0.3+0.3=0.6	17.3	0.3+0.3=0.6	0.3	35.862	
1370	0.311	6.22	0.5	1	1	0.5	8.5	0.3+0.3=0.6	0.3+0.3=0.6	17.3	0.3+0.3=0.6	0.3	35.631	
1390	0.333	6.66	0.5	1	1	0.5	8.5	0.3+0.3=0.6	0.3+0.3=0.6	17.3	0.3+0.3=0.6	0.3	36.093	
1410	0.291	5.82	0.5	1	1	0.5	8.5	0.3+0.3=0.6	0.3+0.3=0.6	17.3	0.3+0.3=0.6	0.3	35.211	
1430	0.281	5.62	0.5	1	1	0.5	8.5	0.3+0.3=0.6	0.3+0.3=0.6	17.3	0.3+0.3=0.6	0.3	35.001	
1450	0.272	5.44	0.5	1	1	0.5	8.5	0.3+0.3=0.6	0.3+0.3=0.6	17.3	0.3+0.3=0.6	0.3	34.812	
1470	0.266	5.32	0.5	1	1	0.5	8.5	0.3+0.3=0.6	0.3+0.3=0.6	17.3	0.3+0.3=0.6	0.3	34.686	
1490	0.26	5.2	0.5	1	1	0.5	8.5	0.3+0.3=0.6	0.3+0.3=0.6	17.3	0.3+0.3=0.6	0.3	34.56	
1510	0.254	5.08	0.5	1	1	0.5	8.5	0.3+0.3=0.6	0.3+0.3=0.6	17.3	0.3+0.3=0.6	0.3	34.434	
1530	0.252	5.04	0.5	1	1	0.5	8.5	0.3+0.3=0.6	0.3+0.3=0.6	17.3	0.3+0.3=0.6	0.3	34.392	
1550	0.25	5	0.5	1	1	0.5	8.5	0.3+0.3=0.6	0.3+0.3=0.6	17.3	0.3+0.3=0.6	0.3	*34.35*	***Best***
1570	0.25	5	0.5	1	1	0.5	8.5	0.3+0.3=0.6	0.3+0.3=0.6	17.3	0.3+0.3=0.6	0.3	34.351	
1590	0.257	5.14	0.5	1	1	0.5	8.5	0.3+0.3=0.6	0.3+0.3=0.6	17.3	0.3+0.3=0.6	0.3	34.497	
1610	0.266	5.32	0.5	1	1	0.5	8.5	0.3+0.3=0.6	0.3+0.3=0.6	17.3	0.3+0.3=0.6	0.3	34.686	

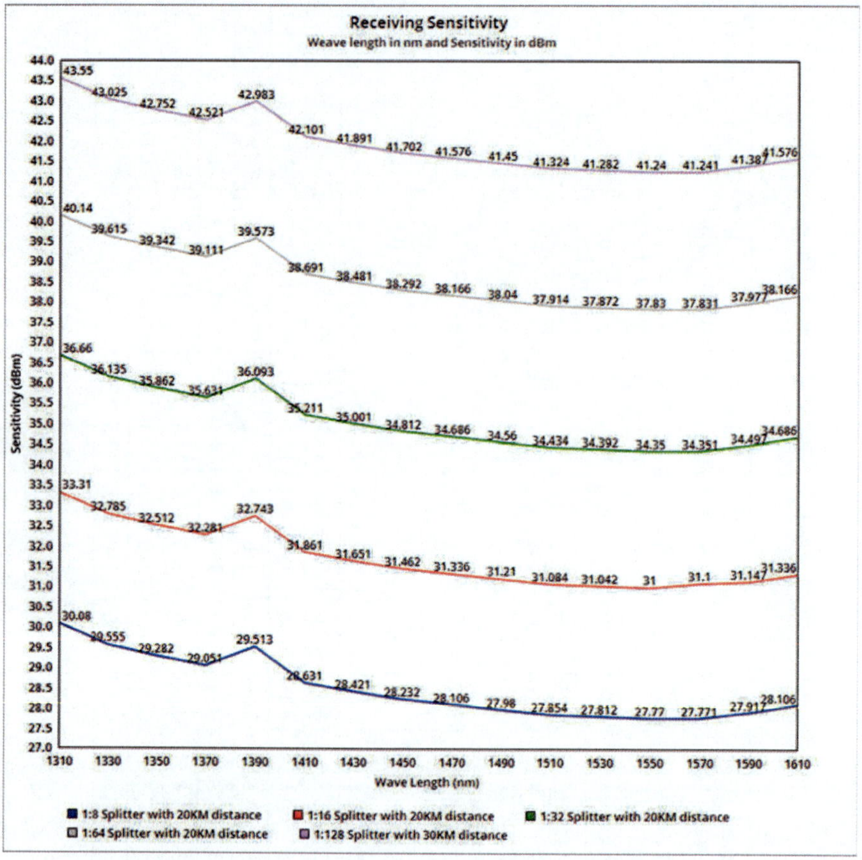

Fig. 5 Receiving sensitivity at different splitting ratio with various distance

Analyzing the sensitivity calculation table and graph, it is instituted that 1550 nm wavelength has sensitivity with minimum dB loss. That is, end-user will get better performance. Could use 1550 nm wavelength for GPON quad play instead of using 1310–1610 nm.

In general triple play services, the eye diagram could not measure, but in proposed quad play, can easily measure the eye diagram from ONT/ONU's ethernet port. Figure 6 shows that almost noise-free traffic could be captured from quad play architecture.

In Fig. 7, for data connectivity, the blue stripe displays the general triple play sensitivity and the green line displays the projected quad play sensitivity. General triple play has approximately −30 dBm power loss while projected quad play has approximately −25 dBm.

In Fig. 8, in the state of video connectivity, the blue line demonstrates the general triple play sensitivity and the green line displays the projected quad play sensitivity.

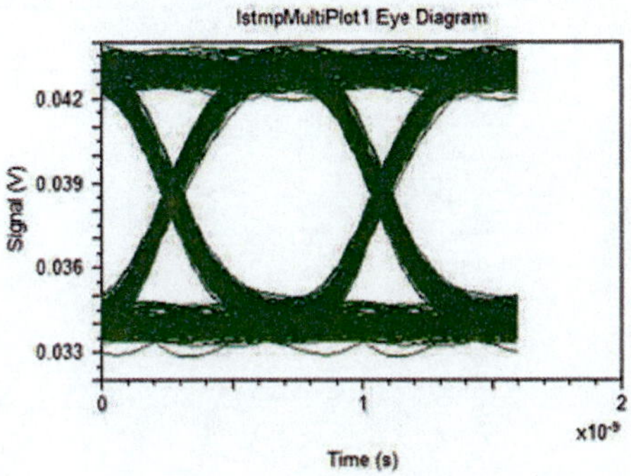

Fig. 6 Eye diagram of SDH traffic (QP)

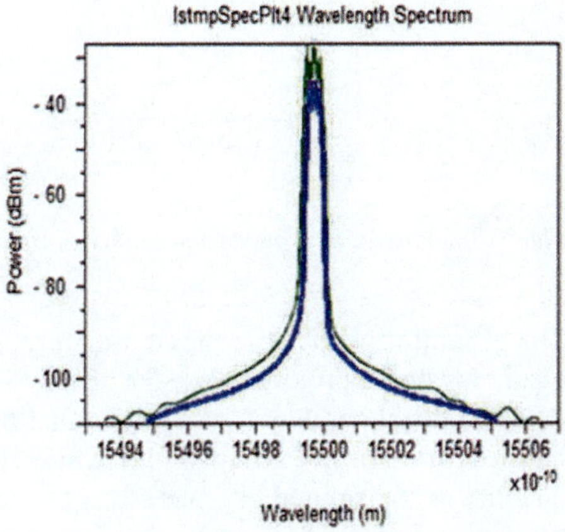

Fig. 7 Data output signal for quad play versus triple play for 1550 nm

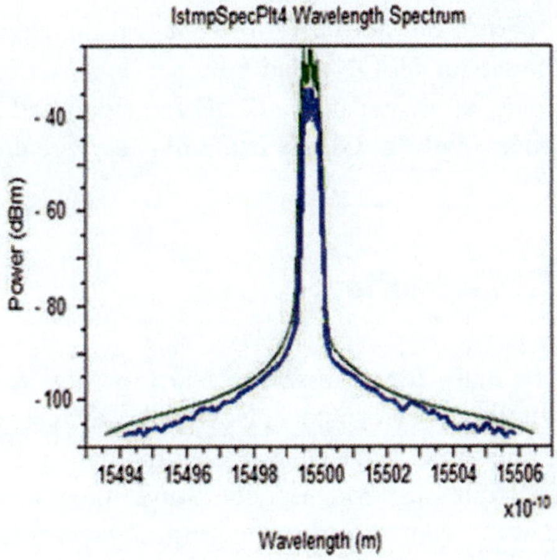

Fig. 8 Video output signal for quad play versus triple play for 1550 nm

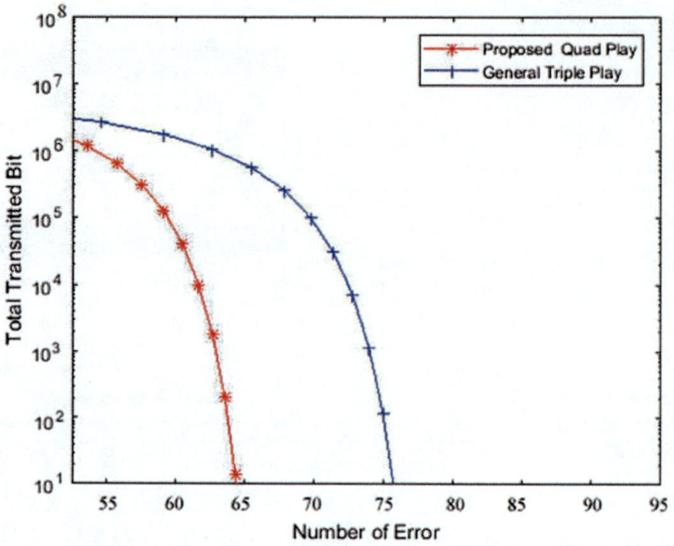

Fig. 9 Total transmitted bit versus number of error

The general triple play has approximately −35 dBm power loss while the proposed quad play has approximately −32 dBm.

From the above Fig. 9, it is found that for every 1 megabit transmission there is 64-bit error in projected quad play while 76-bit error in a general triple play. If the quantity of transmitted bit increased, then the number of errors will also increase in triple play, but in the projected quad play the error bit will not increase in the same ratio compare to triple play.

Form the above Table 2, it can be summarized that if we use 1550 nm wavelength for GPON quad play services, we gain approximately −5 dB for data and voice, approximately −3 dB for video and these are the best gain compare to other wavelengths at GPON triple play service area.

7 Conclusion

The main attention was to learn the best wavelength for GPON quad play solution with high performance. Nowadays, the usual triple play facility solution is operating by a varied range of wavelengths. Here, 1310 nm to 1550 nm wavelength is used for triple play connectivity. Sometimes, it makes some intrusion within consumers. It was keen to find out a single wavelength as an alternative of numerous wavelengths. Analyzing all possible combinations, it is initiated that for quad play architecture, 1550 nm wavelength gives a better eye diagram for SDH traffic and better performance in data, voice, and video service.

Table 2 Comparison of different wavelength with dB gain in Quad Play

SN	Wavelength (nm)			Input (dBm)	Output (dBm)		Gain (dBm)	Remarks
	Voice	Data	Video	V = Video	Triple Play (TP)	Quad Play (QP)	(TP-QP)	
1	1310	1310	1310	0, −10 (V)	−30, −26 (V)	−27, −26 (V)	3, 0 (V)	
2	1490	1490	1490	0, −10 (V)	−30, −32 (V)	−30, −30 (V)	0, 2 (V)	
3	1550	1550	1550	0, −10 (V)	−30, −32 (V)	−25, −29 (V)	5, 3 (V)	*Best*
4	1490	1490	1310	0, −10 (V)	−34, −29 (V)	−32, −35 (V)	2, 6 (V)	
5	1550	1550	1310	0, −10 (V)	−35, −29 (V)	−30, −27 (V)	5, 2 (V)	
6	1490	1490	1550	0, −10 (V)	−35, −34 (V)	−38, −27 (V)	3, 7 (V)	

References

1. Cale I, Salihovic A, Ivekovic M (2007) Gigabit passive optical network—GPON. In: Proceedings of the 29th international conference on information technology interfaces, pp 25–28
2. Leo M, Trotta M (2011) Performance evaluation of WDM-PON RSOA based solutions in NGAN scenario. In: Proceedings of the 50th FITCE congress (The Forum for European ICT and Media Professionals), pp 1–4
3. ITU-T Recommendation G.984.1 (2003), 2 (2003), 3 (2004), 5 (2006) and 6 (2007).
4. Frnda J, Voznak M, Fazio P, Rozhon J (2015) Network performance QoS estimation. In: Proceedings of the 38th international conference on telecommunications and signal processing (TSP), pp 1–6
5. Rokkas T (2015) Techno economic analysis of PON architectures for FTTH deployments. In: Proceedings of the conference of telecommunication, media and internet techno-economics (CTTE), pp 1–5
6. Lee SS, Li W, Wu MS (2016) Design and implementation of a GPON-based virtual open flow-enabled SDN switch. J Lightwave Technol 34(10):1–8
7. Lee J, Hwang I, Nikoukar A, Liem T (2013) Comprehensive performance assessment of bipartition upstream bandwidth assignment schemes in GPON. J Optic Commun Netw 5(11):1285–1295
8. Milanovic S (2014) Case study for a GPON deployment in the enterprise environment. J Netw 9(1):42–47
9. Ricciardi S, Santos-Boada G, Careglio D, Domingo-Pascual J (2012) GPON and EP2P: a techno-economic study. In: Proceedings of the 17th European conference on networks and optical communications (NOC), pp 1–6
10. Koonen AMJ (2006) Technologies and applications of FTTx. Published at 19th annual meeting of the IEEE, lasers and electro-optics society, LEOS 2006
11. Nesset D, Appathurai S, Davey R (2008) Extended reach GPON using high gain semiconductor optical amplifiers. In: Proceedings of the conference on optical fiber communication/national fiber optic engineers (OFC/NFOEC), pp 8–16

12. Rodrigues C, Gamelas A, Carvalho F, Cartaxo A (2011) Evolution of FTTH networks based on radio-over-fibre. In: Proceedings of the 13th international conference on transparent optical networks, pp 7–12
13. Guimarães MA, Rocha LD (2011) Fractional E1 transport in gigabit passive optical network. In: Proceedings of the SBMO/IEEE MTT-S international microwave and optoelectronics conference (IMOC 2011), pp 44–50
14. Chu A, Poon KF, Ouali A (2013) Using ant colony optimization to design GPON-FTTH networks with aggregating equipment. In: Proceedings of the IEEE symposium on computational intelligence for communication systems and networks (CIComms), pp 1–6
15. Róka R (2015) Analysis of possible exploitation for long reach passive optical networks. In: Proceedings of the 4th international conference on simulation and modeling methodologies, technologies and applications (SIMULTECH), pp 123–130
16. OptSIM user manual, Available online: https://www.lmd.polytechnique.fr

Chapter 6
A Priority-Based Deficit Weighted Round Robin Queuing for Dynamic Bandwidth Allocation Algorithm in Gigabit Passive Optical Network

Md. Hayder Ali and Mohammad Hanif Ali

1 Introduction

Accessing Internet is going to be a fundamental right like other basic human rights. It is not a matter that how often a user uses or surf Internet, waiting for a web page to load is certainly an irritation. To overwhelm this condition, Internet surfing speed has enlarged meaningfully in the earlier era to retain step with the petition of end-users, innovative amenities, and bandwidth-hungry claims. These anxieties embrace as hypermedia content-based e-commerce, video on claim, high classification TV, IPTV, online gaming, social media, etc. The communication protocols for hypermedia circulation have conventional a great transaction of attention in the earlier few ages. Since multimedia traffic must provision numerous types of traffic simultaneously, it is crucial to process data according to its characteristics. Thus, protocol originators have to grasp the features of traffic and select a processing method suitable for the performance requirements. For instance, actual acoustic traffic in a voice service requires rapid transmission, but the loss of a small amount of audio information is tolerable. On the opposite site, the transfer of a text file should guarantee 100% reliable transfer; real-time delivery is not of primary importance in this case. Real-time video service, such as video on demand (VOD), requires not only rapid transfer but also high reliability. When a piece of video information is lost, its quality of service (QoS) is degraded. Therefore, multimedia communication protocols should be premeditated to afford the performance requirements of a wide range of multimedia services [1–7].

Md. H. Ali (✉) · M. H. Ali
Jahangirnagar University, Dhaka, Bangladesh
e-mail: hayder.iict@gmail.com

M. H. Ali
e-mail: hanif_ju03@juniv.edu

M. Shorif Uddin et al. (eds.), *Intelligent Energy Management Technologies*, Algorithms for Intelligent Systems,
https://doi.org/10.1007/978-981-15-8820-4_6

The other parts of this paper are organized as follows. Proposed DBA poling mechanism for PMDWRR is introduced in Sect. 2. In Sect. 3, Pseudo code for PMDWRR is briefly described. Simulation scenario is in Sect. 4 and performance analysis between WRR, DWRR, and PMDWRR and stated in Sect. 5. Finally, Sect. 6 draws a conclusion to this paper.

2 Proposed DBA Poling Mechanism Design for Modified Deficit Weighted Round Robin

1. **Start**
2. ONUs/ONTs calculate the amount of received data
3. ONUs/ONTs predict the size of data in waiting time
4. ONUs/ONTs calculate the traffic for T-CONTs (T-CONT1 to T-CONT4)
5. ONUs/ONTs enhance the prediction of accuracy by calculating arriving in waiting time ($P^{T}{}_{i,n}$) for n-th Cycle.
6. OLT calculates the arithmetic mean for the data received in waiting time.
7. Repeat step 2 to 6 for n-th cycle.
8. **Then**
9. Calculate the difference for the size of data received in waiting time at ONT_i/ONU_i in the n-the cycle.
10. Calculate the size of traffic prediction arriving in waiting time at ONT_i/ONU_i in the n-th cycle by adding a weighting factor.
11. Increase the average order to ensure the prediction accuracy.
12. OLT grant frame size or allocate the frame size for priority traffic.
13. ONTs/ONUs send data according to OLT's priority grant frame.
14. **End**

To execute the above algorithm for GPON networks, it is necessary to maintain a queuing system. DWRR algorithm is suitable for this. As it provides protection among different flows, it overcomes the limitations of strict PQ (Fig. 1).

3 Pseudo Code for Priority-Based Deficit Weighted Round Robin

The pseudo code in this section does not describe the procedure of any exact DWRR execution. Although each execution will vary from this model, studying the samples and outlining the pseudo code will make it calmer to recognize the explicit enterprise results that are compulsory to make for executions.

The array flexible Deficit_Counter is reset to zero. In this sample, the queues are numbered 1 to n, where n is the extreme number of lines on the output port:

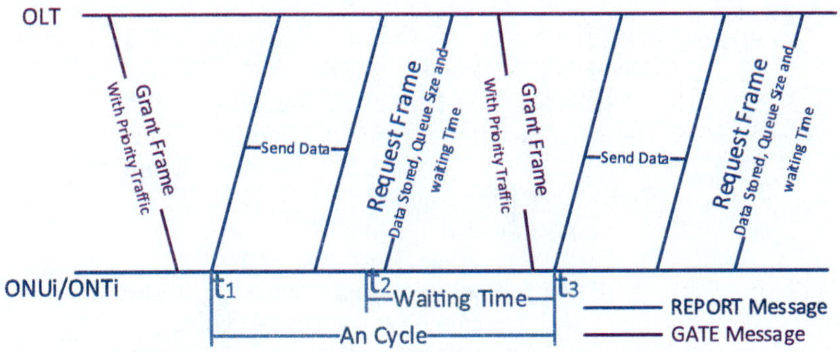

Fig. 1 DBA polling mechanism diagram for PMDWRR

```
FOR i = 1 to n                          /* Visit each queue index */
 Deficit_Counter[i] = 0    /* Initialize Deficit_Counter[i] to 0 */
ENDFOR

      i = the index of the queue that will grip the new package
        IF (ExistsInActiveList(i) = FALSE) THEN /*IF i not in
           ActiveList */ InsertActiveList(i)          /* Add i to the end
           of ActiveList */ Deficit_Counter[i] = 0       /*Reset
           queue Deficit_Counter[i] to 0*/
        ENDFOR
        Enqueue packet to Queue[i] /* Place packet at finish of queue i */
  END Enqueue
// Removes the element with the highest priority form the list
void pop (Node** head)
{
      Node* temp = *head;
      (*head) = (*head)->next;
      free(temp);
}

// Function to push according to priority
void push (Node** head, int d, int p)
{
      Node* start = (*head);
```

Whenever an index is at the head of the ActiveList, the function Dequeue () transmits up to Deficit_Counter[i] + Quantum[i] worth of bytes from queue.

```
Dequeue ()
      While (TRUE) DO
            IF (ActiveList is NotEmpty) THEN
                  i = the index at the head of the ActiveList
                  Deficit_Counter[i] = Deficit_Counter[i] +
                                                              Quantum[i]
                  WHILE (Deficit_Counter[i] > 0 AND NOT
                      Empty (Queue[i])) DO
                      PacketSize = Size (Head (Queue[i]))
                      IF (PacketSize <= Deficit_Counter[i]) THEN
                              Transmit packet at head of Queue[i]
                              Deficit_Counter[i] = Deficit_Counter[i] –
                              PacketSize
/* To find out the largest traffic volume */
                             /*Declare largest as integer*/
                        Set largest to 0
                          FOR EACH value in A DO
                                  IF A[n] is greater than largest THEN
                                      largest ← A[n]
                                  ENDIF
                          END FOR
                              Dequeue the largest first
                        ELSE
                                                      Break /*exit this while loop*/
                        ENDIF
             ENDWHILE
             IF (Empty (Queue[i])) THEN Deficit_Counter[i] = 0
                   RemoveFromActiveList(i)
             ELSE
                        InsertActiveList(i)
             ENDIF
         ENDIF
     ENDWHILE
END Dequeue
```

4 Simulation Design

The simulation is done by using OptSIM (RSoft System Suit, version-2016.06) Simulation software. The simulation scenario is like bellow Fig. 2. It is composed of one OLT (Optical Line Terminal), one 1:32 Splitter and then thirty-two ONUs/ONTs. Each ONU/ONT has four signal sources producing T-CONT1 to T-CONT4 services respectively. Before each ONT, there is Bandwidth Analyzer to measure the consumed bandwidth of individual ONU/ONT. The OLT's up and down link rates are 1.25 and 2.48 Gbps. Each ONU/ONT is 20 km far from the OLT.

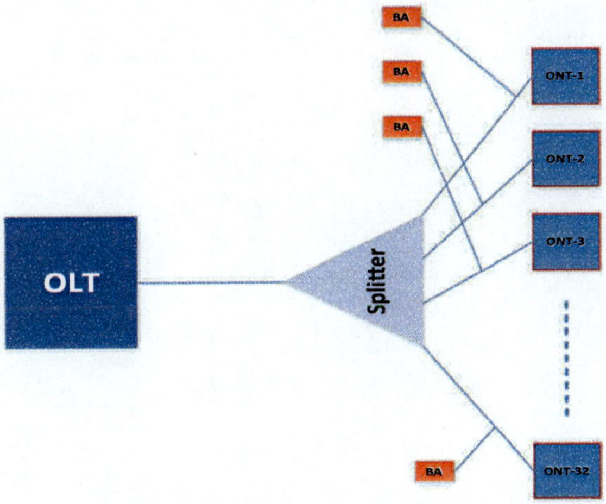

Fig. 2 Simulation scenario in OptSIM simulator

5 Result Analysis

The data has captured from OptSIM simulator and analyzed by MATLAB. Delay measurement and usage bandwidth are calculated by MATLAB coding.

Figure 3 stated that number of end-users is more in PMDWRR. Considering 2GBPS bandwidth allocation, DWRR is allowing 17 users, MDWRR is allowing 20 users, while PMDWRR is allowing 23 users. Bandwidth is shared with maximum users which satisfy the properties of GPON system.

Figure 4 shown that delay per user is minimum in PMDWRR. Considering 25 users in the system, DWRR shows 3 ms delay, MDWRR shows 2.4 ms delay while

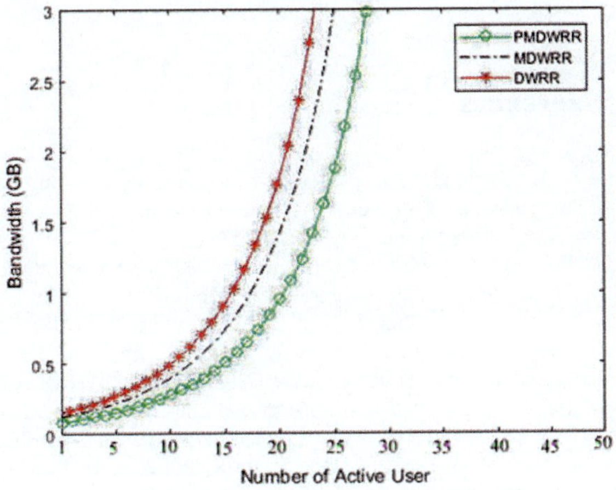

Fig. 3 Bandwidth allocation versus number of users

Fig. 4 Delay versus number of users

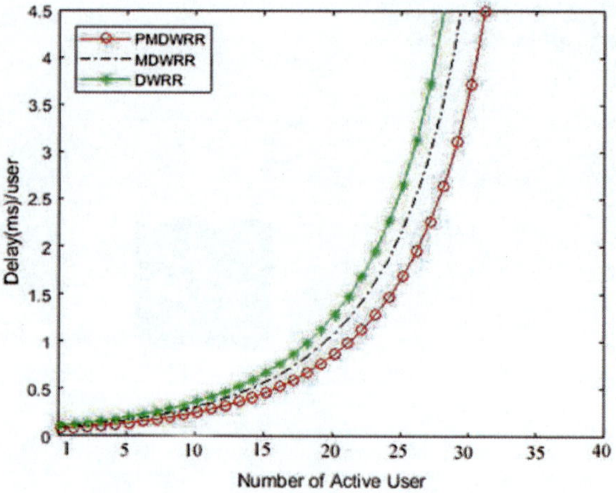

PMDWRR shows only 1.7 ms/user. The per user delay is less in PMDWRR and it increases the system performance.

6 Conclusion

In this paper, have partially modified to set priority into the existing MDWRR. In PMDWRR, ONU/ONT predict the data size and send allocation request to OLT, OLT increases the prediction accuracy. It could calculate the size of the queue and store the data size. By calculating the queue size, it can predict the priority traffic. It transmits the priority of the traffic and this way it could minimize the transmission delay and it satisfies the full service of QoS requirements of GPON system.

References

1. Tae Il J, Jae Ho J, Sung Jo K (1997) An efficient scheduling mechanism using multiple thresholds for multimedia traffic in ATM switching nodes. In: Proceedings of the 22nd IEEE conference on local computer networks (LCN'97)
2. Michael P Mc, Martin M, Martin R (2004) Ethernet PONs: a survey of dynamic bandwidth allocation (DBA) algorithms. IEEE Commun Mag 42(8), 1–15
3. Taeck-Geun K, Sook-Hyang L, June-Kyung R (1998) Scheduling algorithm for real-time burst traffic using dynamic weighted round robin. In: Proceedings of the IEEE international symposium on circuits and systems (Cat. No.98CH36187), pp 506–509
4. Patel Z, Dalal U (2014) Design and implementation of low latency weighted round robin (LLWRR) scheduling for high speed networks. Int J Wirel Mob Netw (IJWMN) 6(4):59–71
5. Ji-Young K, Ji-Seung N, Doo-Hyun K (2002) A modified dynamic weighted round robin cell scheduling algorithm. ETRI J 24(5):360–372

6. Ouni R, Bhar J, Torki K (2013) A new scheduling protocol design based on deficit weighted round robin for QoS support in ip networks. J Circ Syst Comput 22(3):1–21
7. Lenzini L, Mingozzi E, Stea G (2006) Bandwidth and latency analysis of modified deficit round robin scheduling algorithms. In: Proceedings of the 1st international conference on performance evaluation methodologies and tools (VALUETOOLS 2006)

Chapter 7
Protection of Six-Phase Transmission Line Using Demeyer Wavelet Transform

Gaurav Kapoor

1 Introduction

An increase in the inevitability of electrical power has been perceived by the people of the modern generation. The electrical power transfer potentiality of the currently operating power transmission systems ought to be augmented in order to assist the significant increase in the necessity of electrical energy.

Thus, accurate recognition of the faults in the SPTL turns out to be very decisive for mitigating the loss of gain and providing fast renovates.

Numerous newly reported research works addressed the issue of FD and FC in TL's. The discussion of few researches is shown in brief here in this section. In [1], FSM-based technique has been used for TL protection. In [2, 4, 7], WT and MM have been applied for the protection of SPTL. VMD has been employed for disturbance recognition in power system [3]. POVMD and WPNRVFLN have been used for FD and FPR in SCDCTL [5]. DM and MM have been used for high impedance fault recognition in TL [6]. Alienation-based technique has been proposed for TTTL protection [8]. In [9], the theory of TW has been reported and the combination of VMD and TEO has been applied for HVDC TL protection. VMD and HT have been used for micro-grid protection [10].

In this work, the demeyer wavelet transform (DMWT) has been used for the protection of six-phase transmission line (SPTL). No such type of work has been reported yet to the best of the knowledge of author. The results exemplify that the DMWT efficiently recognizes and categorizes the faults and the consistency of the DMWT is not affected by variation in the fault factors of SPTL.

G. Kapoor (✉)
Modi Institute of Technology, Kota, India
e-mail: gaurav.kapoor019@gmail.com

M. Shorif Uddin et al. (eds.), *Intelligent Energy Management Technologies*, Algorithms for Intelligent Systems,
https://doi.org/10.1007/978-981-15-8820-4_7

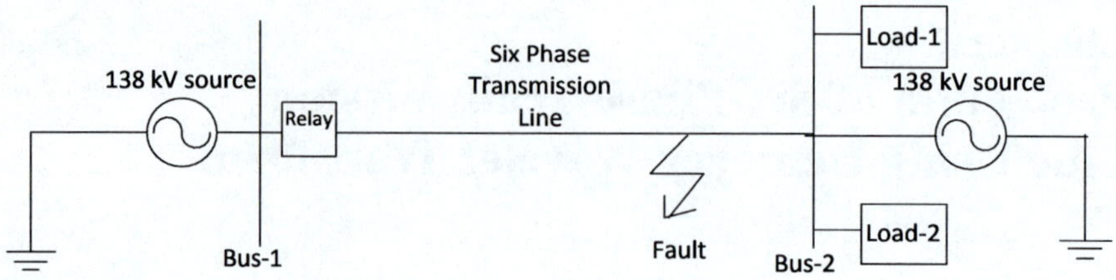

Fig. 1 The schematic of SPTL

This paper is structured as follows: The specifications of SPTL are presented in Sect. 2. The flow-diagram for DMWT is presented in Sect. 3. Section 4 is dedicated to the discussion of results. Section 5 concludes the paper.

2 The Specifications of SPTL

Figure 1 shows the schematic of SPTL. The schematic consists of 138 kV, 60 Hz SPTL of 68 km, connected to a 138 kV voltage source at one end and loads at the other end. The SPTL is divided into two parts of length 34 km each. The CT's and relay are connected at bus-1 to protect the full length of SPTL.

3 The Flow Chart of DMWT

Figure 2 illustrates the process for the DMWT. The steps are shown beneath.

Step 1 Simulate the SPTL for creating faults and produce the post-fault currents.
Step 2 Use DMWT to examine the post-fault currents for characteristics retrieval and determine the range of DMWT output.
Step 3 The phase will be proclaimed as the faulted phase if its DMWT output has a larger amplitude as compared to the healthy phase in a faulty situation.

4 Performance Assessment

The simulations have been performed for the near-in relay faults, far-end relay faults, converting faults, inter-circuit faults, and multi-position faults with the objective of verifying the feasibility of the DMWT. In the separate subcategories, the outcomes of the work are investigated.

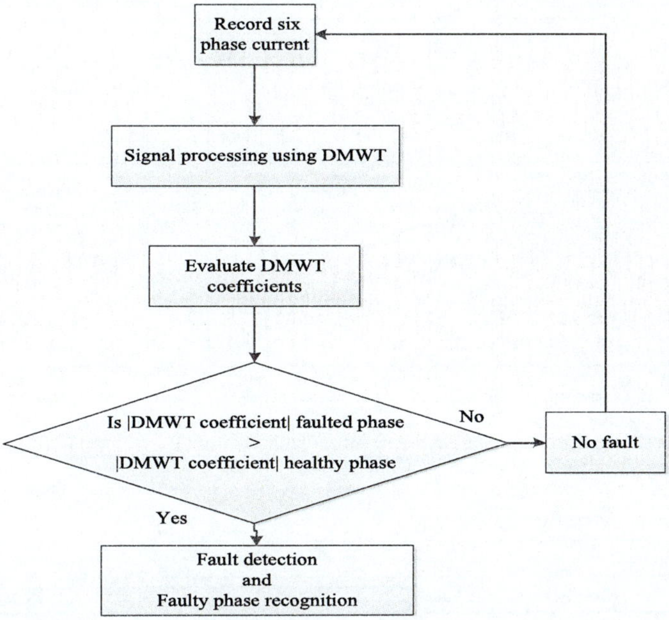

Fig. 2 Flow diagram showing DMWT process

4.1 The Efficacy of DMWT for Healthy Condition

Figure 3 shows the SPTL currents and voltages for no-fault. The DMWT outputs for no-fault are shown in Fig. 4. Table 1 reports the results of DMWT for healthy situation.

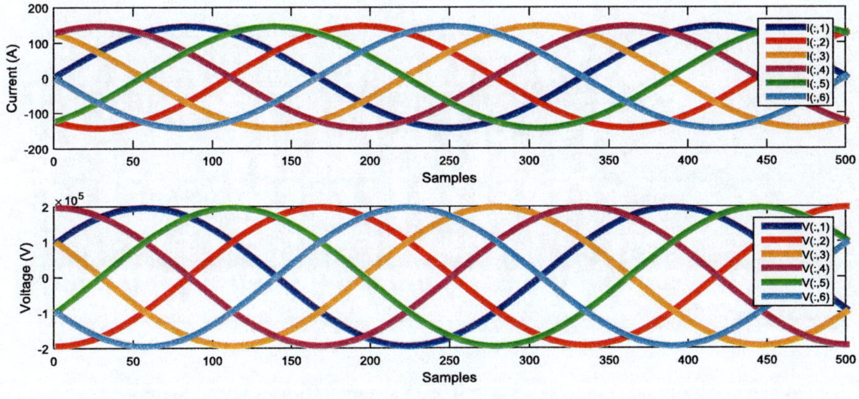

Fig. 3 Six-phase current and voltage waveforms for no-fault

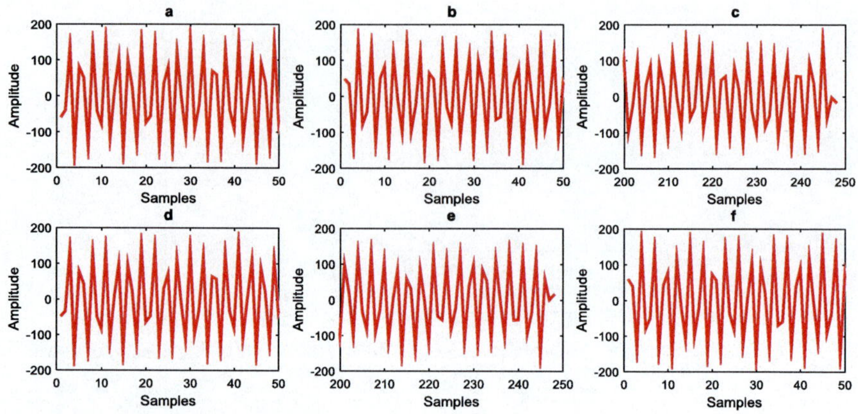

Fig. 4 DMWT outputs of six-phase currents for no-fault

Table 1 Results of DMWT for no-fault

DMWT output					
Phase-A	Phase-B	Phase-C	Phase-D	Phase-E	Phase-F
175.9988	152.7358	155.3587	154.5936	133.9446	188.1193

4.2 *The Efficacy of DMWT for Near-in Relay Faults*

The efficiency of the DMWT is investigated for the near-in relay faults on the SPTL. Figure 5 exemplifies the SPTL currents for ABDFG near-in relay fault at 5 km at 0.0535 s among $R_F = 2\ \Omega$ and $R_G = 4\ \Omega$. Figure 6 illustrates the DMWT output for ABDFG fault. The fault factors for the fault cases are: T = 0.0535 s, $R_F = 2\ \Omega$, and

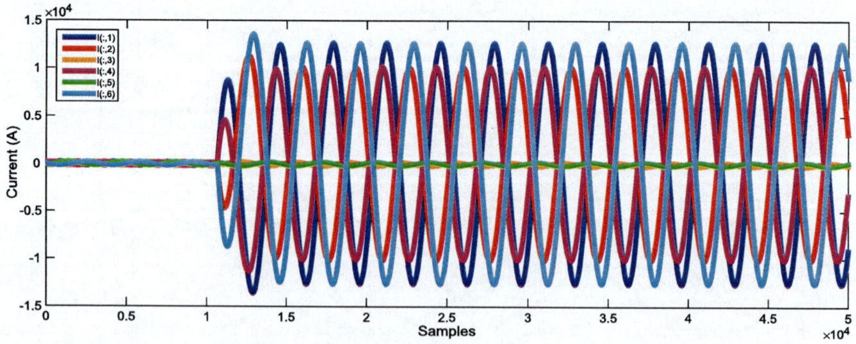

Fig. 5 ABDFG near-in relay fault at 5 km at 0.0535 s among $R_F = 2\ \Omega$ and $R_G = 4\ \Omega$

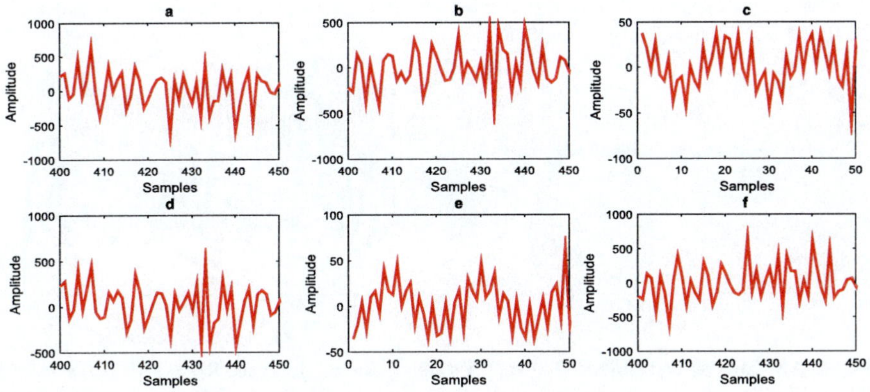

Fig. 6 DMWT output for ABDFG near-in relay fault at 5 km at 0.0535 s

Table 2 Results of DMWT for near-in relay faults

Fault type	DMWT output					
	Phase-A	Phase-B	Phase-C	Phase-D	Phase-E	Phase-F
ABDFG (5 km)	975.5428	1.5645×10^3	82.1565	905.7343	62.6620	1.4003×10^3
ABCG (6 km)	1.1302×10^3	1.0896×10^3	1.2082×10^3	86.5100	44.6221	50.1690
BCEFG (7 km)	208.6318	1.1057×10^3	1.0163×10^3	151.0041	1.3309×10^3	1.0599×10^3
ABCDEG (8 km)	1.4257×10^3	1.1870×10^3	1.1547×10^3	1.1865×10^3	993.3149	208.7100
DEFG (9 km)	53.2673	46.4882	60.4460	1.9163×10^3	1.8024×10^3	1.1828×10^3

$R_G = 4\ \Omega$. Table 2 details the results of the DMWT for near-in relay faults. It is seen from Table 2 that the DMWT precisely detects the near-in relay faults.

4.3 The Efficacy of DMWT for Far-End Relay Faults

The DMWT has been explored for the far-end relay faults. Figure 7 illustrates the SPTL currents for ABCEFG far-end relay fault at 63 km at 0.1 s among $R_F = 1.15\ \Omega$ and $R_G = 2.15\ \Omega$. Figure 8 shows the DMWT output for ABCEFG fault. Table 3 reports the results for far-end relay faults. It is inspected from Table 3 that the effectiveness of the DMWT remains unaffected for the far-end relay faults.

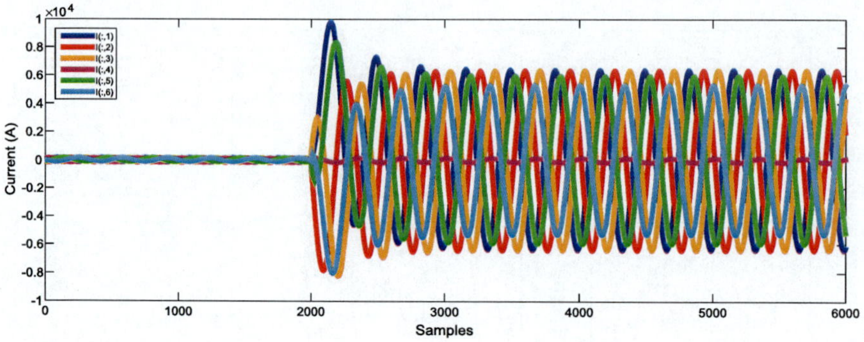

Fig. 7 ABCEFG far-end fault at 63 km at 0.1 s among $R_F = 1.15\ \Omega$ and $R_G = 2.15\ \Omega$

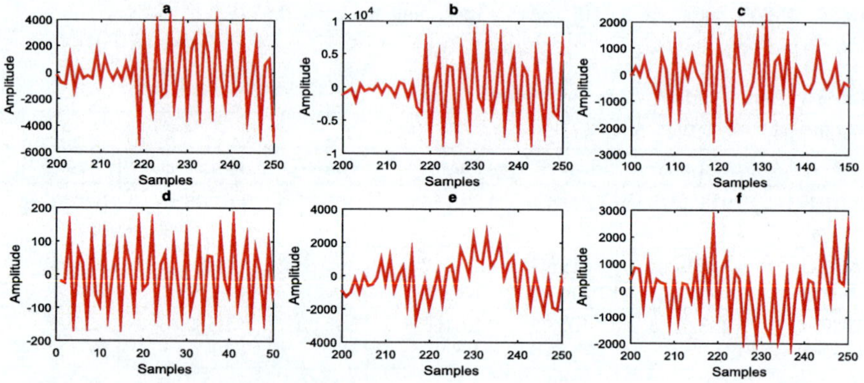

Fig. 8 DMWT output for ABCEFG far-end relay fault at 63 km at 0.1 s

Table 3 Results of DMWT for far-end relay faults

Fault type	DMWT output					
	Phase-A	Phase-B	Phase-C	Phase-D	Phase-E	Phase-F
ABCEFG (63 km)	3.8060×10^3	7.9079×10^3	4.0745×10^3	158.1244	3.4767×10^3	2.5117×10^3
ABCG (64 km)	4.8194×10^3	8.5085×10^3	4.3048×10^3	165.9761	157.4620	197.3903
DEFG (65 km)	142.1903	137.1558	124.3455	5.1810×10^3	2.7625×10^3	4.8779×10^3
ABDEG (66 km)	5.8668×10^3	6.2243×10^3	166.0415	5.8480×10^3	4.5843×10^3	153.9397
ABCDG (67 km)	3.0659×10^3	6.4599×10^3	3.7578×10^3	3.1028×10^3	197.5127	211.9889

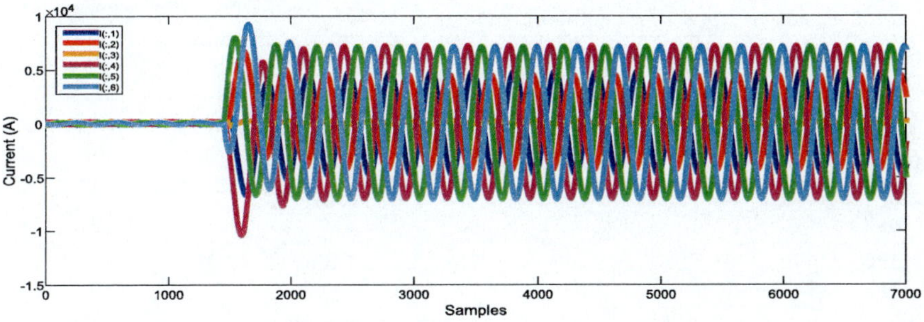

Fig. 9 SPTL currents for multi-position ABG fault at 42 km and DEFG fault at 26 km

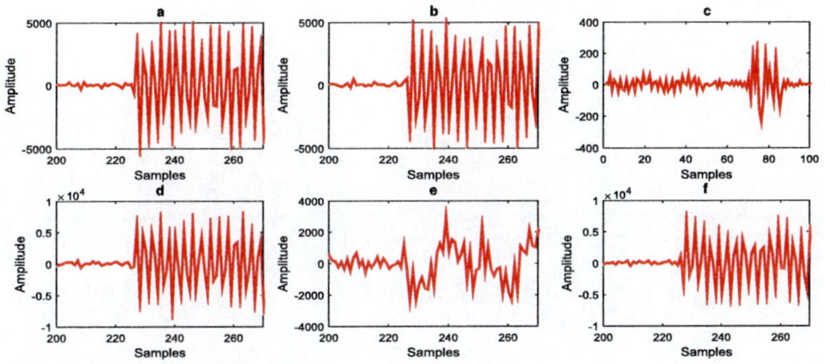

Fig. 10 DMWT output for multi-position ABG fault and DEFG fault

4.4 The Efficacy of DMWT for Multi-position Faults

The DMWT is tested for different cases of multi-position faults. Figure 9 depicts the currents when the SPTL is simulated for the multi-position ABG fault at 42 km and DEFG fault at 26 km at 0.0725 s among $R_F = 1.5\ \Omega$ and $R_G = 1.25\ \Omega$. Figure 10 shows the DMWT output for ABG and DEFG multi-position fault. Table 4 presents the results for different position-position faults.

4.5 The Efficacy of DMWT for Inter-circuit Faults

The DMWT is tested for different cases of inter-circuit faults. Figure 11 depicts the currents when the SPTL is simulated for the inter-circuit ABCG and FG fault at 30 km at 0.1325 s among $R_F = 2.75\ \Omega$ and $R_G = 1.85\ \Omega$. Figure 12 depicts the DMWT output for ABCG and FG inter-circuit fault. Table 5 reports the results for

Table 4 Results of DMWT for multi-position faults

FT-1 (km)	FT-2 (km)	DMWT output					
		Phase-A	Phase-B	Phase-C	Phase-D	Phase-E	Phase-F
ABG (42)	DFFG (26)	4.1866×10^3	4.4410×10^3	203.0695	6.5372×10^3	2.9300×10^3	6.4875×10^3
CG (20)	DFG (48)	248.1627	247.6852	1.4210×10^3	2.4488×10^3	179.0262	2.7339×10^3
DEFG (35)	BG (33)	93.2050	2.7384×10^3	114.9325	4.1065×10^3	2.7835×10^3	4.5857×10^3
ACG (44)	EFG (24)	2.4514×10^3	111.0308	2.0142×10^3	132.6308	2.8471×10^3	4.2729×10^3
DG (38)	ABCG (30)	4.8020×10^3	6.5308×10^3	2.4552×10^3	2.6415×10^3	97.7596	163.331

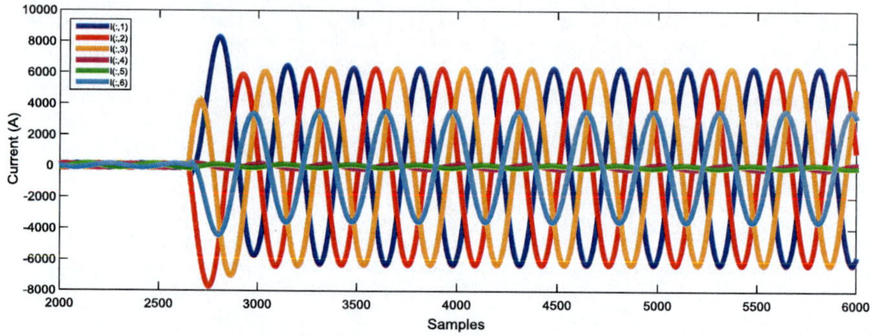

Fig. 11 SPTL currents for ABCG and FG inter-circuit fault at 30 km at 0.1325 s

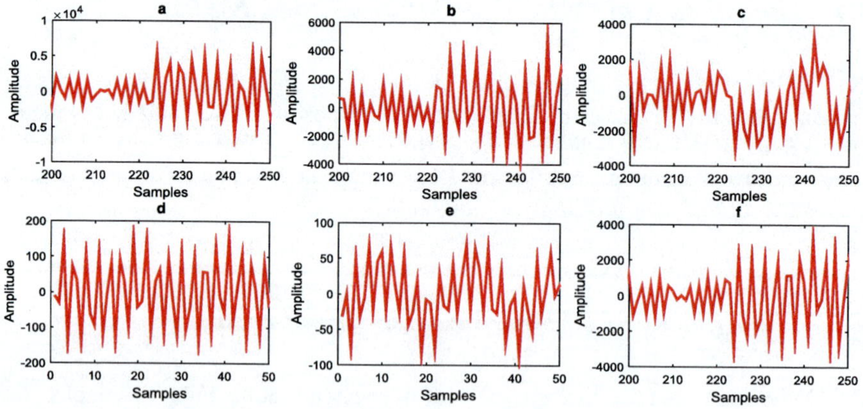

Fig. 12 DMWT output for ABCG and FG inter-circuit fault at 30 km at 0.1325 s

Table 5 Results of DMWT for inter-circuit faults

Fault-1	Fault-2	DMWT output					
		Phase-A	Phase-B	Phase-C	Phase-D	Phase-E	Phase-F
ABCG	FG	5.4287×10^3	5.0592×10^3	3.2656×10^3	159.5320	95.4246	3.3819×10^3
BG	DG	264.7818	4.0431×10^3	260.5588	3.2120×10^3	244.3075	245.0540
ACG	DEG	3.7037×10^3	176.3243	2.6696×10^3	4.8352×10^3	3.5412×10^3	131.1642
ABG	EFG	5.7039×10^3	5.8627×10^3	195.1929	201.4383	2.6028×10^3	4.5607×10^3
CG	DEFG	271.4223	202.1157	3.9056×10^3	1.0169×10^3	8.9809×10^3	4.5267×10^3

the inter-circuit faults. It is examined from Table 5 that the DMWT performs well for the recognition of inter-circuit faults.

4.6 *The Efficacy of DMWT for Converting Faults*

The DMWT has been investigated for the converting faults. Figure 13 exemplifies the currents of the SPTL when initially the ABG fault at 35 km at 0.05 s is converted into the DEFG fault at 35 km at 0.15 s among $R_F = 2.50\ \Omega$ and $R_G = 1.50\ \Omega$. Figure 14 exemplifies the DMWT output for the same fault case. The fault factors preferred for the additional fault cases are: $F_L = 35$ km, $R_F = 2.50\ \Omega$ and $R_G = 1.50\ \Omega$. Table 6 reports the results for different converting faults.

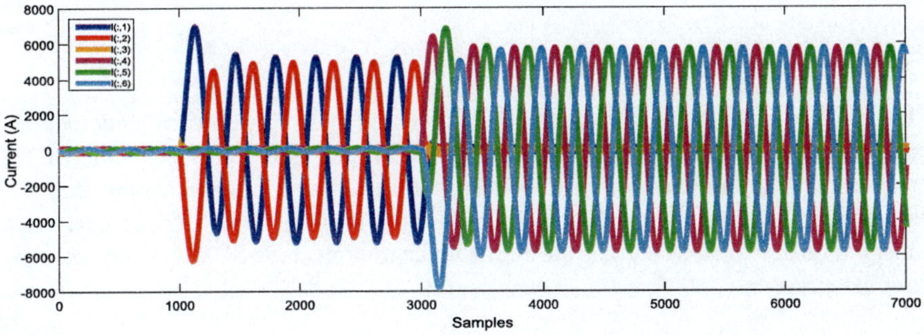

Fig. 13 SPTL current when ABG fault at 0.05 s is converted into DEFG fault at 0.15 s at 35 km among $R_F = 2.50\ \Omega$ and $R_G = 1.50\ \Omega$

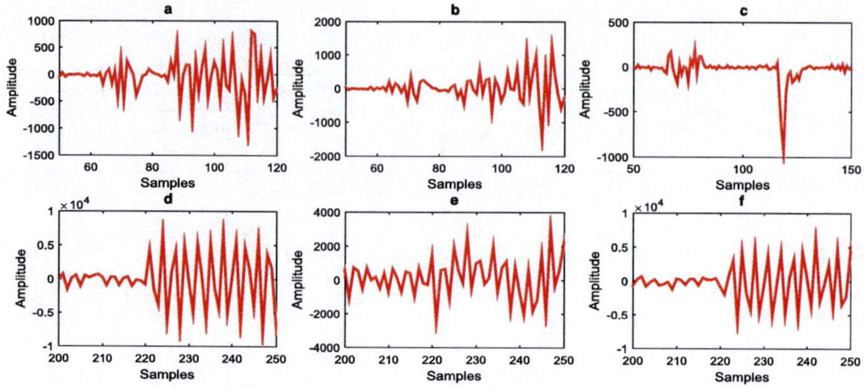

Fig. 14 DMWT output when ABG fault is converted into DEFG fault

Table 6 Results of DMWT for converting faults

Fault type	Converted fault	DMWT output					
		Phase-A	Phase-B	Phase-C	Phase-D	Phase-E	Phase-F
ABG (0.05)	DEFG (0.15)	816.6582	1.2847×10^3	196.2113	7.2091×10^3	3.1614×10^3	6.2976×10^3
DG (0.1)	DEFG (0.2)	159.9643	136.8911	198.3681	5.6853×10^3	1.7459×10^3	4.2094×10^3
ABCG (0.07)	EG (0.16)	1.5407×10^3	2.1744×10^3	1.6289×10^3	197.2483	1.3585×10^3	173.1067
ACG (0.05)	DEG (0.2)	2.0178×10^3	171.0564	1.1652×10^3	3.2852×10^3	2.0547×10^3	110.2217
EFG (0.1)	ABG (0.15)	5.8620×10^3	6.6895×10^3	208.7799	645.9092	1.4708×10^3	912.3268

5 Conclusion

The demeyer wavelet transform (DMWT) is seemed to be very efficient under varied fault categories for the SPTL. The DMWT output of fault currents of the SPTL is assessed. The fault factors of the SPTL are varied and it is discovered that the variation in fault factors do not influence the fidelity of the DMWT. The outcomes substantiate that the DMWT has the competence to protect the SPTL beside different fault categories.

References

1. Yadav A, Swetapadma A (2016) A finite-state machine based approach for fault detection and classification in transmission lines. Elect Power Component Syst (Taylor and Francis) 44(1):43–59
2. Kapoor G (2018) Six phase transmission line boundary protection using wavelet transform. In: Proceedings of the 8th IEEE India international conference on power electronics (IICPE), Jaipur, India (2018)
3. Jena MK, Samantaray SR, Panigrahi BK (2017) Variational mode decomposition-based power system disturbance assessment to enhance WA situational awareness and post-mortem Analysis. IET Gener Transm Distrib 11(13):3287–3298
4. Kapoor G (2018) Fault detection of phase to phase fault in series capacitor compensated six phase transmission line using wavelet transform. Jordan J Elect Eng 4(3):151–164
5. Sahani M, Dash PK (2019) Fault location estimation for series-compensated double-circuit transmission line using parameter optimized variational mode decomposition and weighted P-norm random vector functional link network. Appl Soft Comput J (Elsevier), 1–18 (2019)
6. Sekar K, Mohanty MK (2018) Data mining-based high impedance fault detection using mathematical morphology. Comput Elect Eng (Elsevier) 69:129–141
7. Kapoor G (2018) Six phase transmission line boundary protection using mathematical morphology. In: Proceedings of the IEEE international conference on computing, power and communication technologies (GUCON), pp 857–861, Greater Noida, India (2018)
8. Rathore B, Shaik AG (2018) Fault analysis using alienation technique for three-terminal transmission line. In: Proceedings of the IEEE 2nd international conference on power, energy and environment: towards smart technology (ICEPE), pp 1–6, Shillong, India (2018)
9. Wang L, Liu H, Dai LV, Liu Y (2018) Novel method for identifying fault location of mixed lines. Energies 11(1529):1–19
10. Chaitanya BK, Yadav A, Pazoki M (2019) An improved differential protection scheme for micro-grid using time frequency transform. Elect Power Energy Syste (Elsevier) 111:132–143

Chapter 8
Analysis of Voltage Sag and Swell Problems Using Fuzzy Logic for Power Quality Progress in Reliable Power System

Ankit Tandon and Amit Singhal

1 Introduction

The electricity generation from conventional and non-conventional sources, its transmission from DC or AC system, its distribution to domestic and commercial consumers, and its utilization from human being and for industrial requirements makes a power system network. In a smart power system, the difficulties during this complete process are solved in a smart manner. Such type of disturbances includes mainly VAR controlling and power quality issues. Managerial controlling from load dispatch centers makes a power system better. Additionally, the fuzzy-based expert systems can be implemented so as to make the existing system smarter with higher accuracy.

The problems due to power quality and problems arising are penalty of more utilization of solid-state switching devices, nonlinear load, electronic load, and switching load. The arrival of high rating semiconductor switches in distribution and transfer leaves current to be non-sinusoidal [1]. Electronic load causes voltage distortion and harmonic distortion. Power quality problems create onset of systems, sensitive equipment, data loss in the commuter, and MAL functioning in the memory, like computers, programmable logic controller, the protective and relay apparatus [1].

A voltages sag or voltage dip is a short duration decrease in rms voltage. It is caused by short circuit phenomena, overload, and starting of electric motors. If the value of RMS voltage reduces from 10 to 90 percent of the nominal voltage for a cycle to a minute duration, then phenomena of voltage sag is produced. Voltage sag is termed as sustained sag if this low voltage remains in the system for a longer duration up to a minute. Voltage swell is the contradictory of voltage sag. It is the phenomena of an increase in the level of voltage which occurs due to turn off of a

A. Tandon (✉) · A. Singhal
Department of Electrical Engineering, JIET, Jodhpur, India
e-mail: ankit.tandon@jietjodhpur.ac.in

M. Shorif Uddin et al. (eds.), *Intelligent Energy Management Technologies*, Algorithms for Intelligent Systems,
https://doi.org/10.1007/978-981-15-8820-4_8

heavy load in an electrical power system. Similar to sag, swell happens for a very short duration, and if it is of longer duration, then termed as sustained swell.

Voltage sag mainly increases the problems of power quality. It affects the electricity distribution system and various industries which causes very high loss of power. Short duration voltage dips can cause the whole industry out from their normal operating condition. Generally, it is considered that the voltage dip as a source 10–90%. The main reasons include short circuit condition and breakdown voltage dips, lightning phenomena, and power surges. In overhead transmission and distribution systems, lightning phenomena is the main reason for producing voltage sag, with an approximate occurrence of 50% of total cases [2]. The following pie chart shows the main reasons for poor power quality and their approximate percentage.

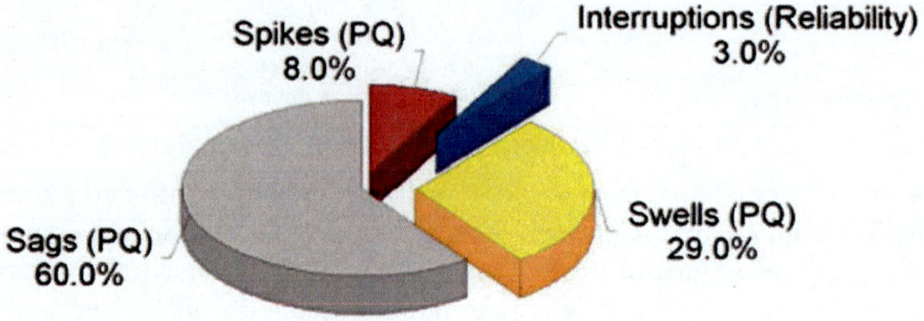

To minimize conflict of power quality in electrical devices, some well-organized detection and categorization techniques are necessary in the electrical power systems. Types of power quality disturbances which are based on visual check of waveforms by human being operators is a time consuming and laborious task. Also, it's not always necessary that we can remove some important information from simple visual inspection [2]. The categorization of Power quality disturbances in an electrical power system has now become a very important job for proper development and design of preventive and corrective actions.

In the mid of 1950s, artificial intelligence has emerged as a new stream of computer science engineering. There are many powerful tools produced by artificial intelligence, which are used practically in different engineering streams so as to resolve those complex engineering problems which require human intelligence. In this paper, fuzzy logic tool is used for the analysis of disturbances like voltage sag and swell [2, 3].

Fuzzy Controller

Fuzzy controllers are very simple and it is a new addition to control theory. They consist of an input stage, a processing stage, and an output stage. Its design philosophy differs from all previous methods by containing dexterous knowledge in controller design. These FLCs are attractive choices where specific mathematical modeling formulas are not possible. It has good control robustness as compared to traditional control scheme. It modifies a dexterous knowledge-based control strategy into automatic control strategy in essence [3].

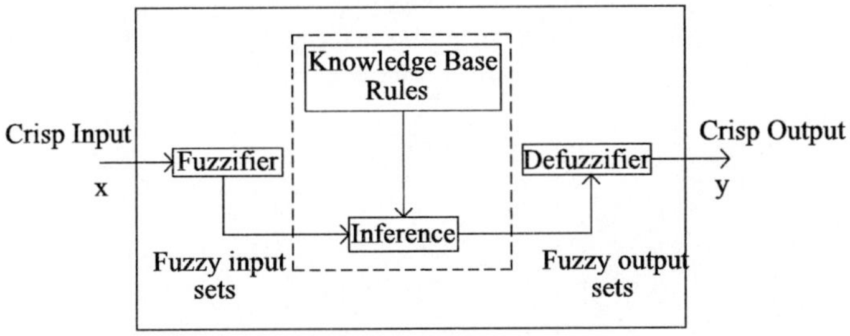

Fig. 1 Architecture of a Fuzzy Controller

A Mamdani-based fuzzy controller is used with maximum-minimum interference method. It performs the function in the following 4 stages which is shown in the following Fig. 1.

(1) Fuzzification process
(2) Rule base
(3) Inference
(4) Defuzzification process.

Fuzzification is the procedure of changing real scalar value into a fuzzy value which is achieved with the different types of fuzzifiers. (Membership functions).

Rule base scheme is used as a way to store and manipulate knowledge to understand the information in a useful manner. They are used in AI.

FIS is a system that uses the theory of fuzzy set for mapping of inputs. The input is features while the output is classes of fuzzy classification.

Defuzzification is the process of getting a scientific result in the Crisp logic, available fuzzy sets, and related membership degrees. So, the process of mapping of fuzzy to crisp set is called as Defuzzification.

The planned fuzzy-based expert system is developed to classify short voltage disturbances which can be defined as instantaneous and temporary sag, swell, and interruption, which are shown in Fig. 2 [3]. In the study, the disturbance data are obtained from PQ monitoring in which the monitoring software by default has three different sampling frequencies of 0.4 kHz (128 cycle), 1.6 kHz (32 cycle), and 6.4 kHz (8 cycles) and each frame has 1024 samples.

To process the raw disturbance data so as to remove features of the various disturbances, preprocessing of the disturbance signals is required. Initially, fast Fourier transform analysis is used to separate the 8, 32, and 128 cycle waveforms. Then root mean square (rms) method is applied by first approximating the fundamental frequency profile of actual voltage waveform and determining the maximum and minimum voltage magnitudes. An advantage of this method is its simplicity, fast calculation, and less requirement of memory because rms voltage can be stored periodically instead of per sample [4].

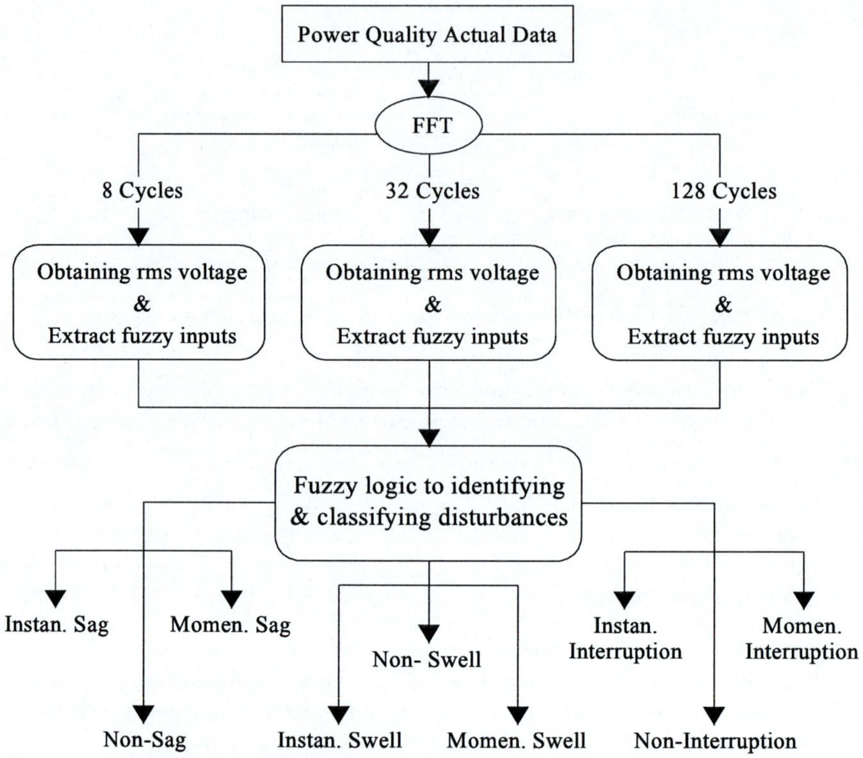

Fig. 2 Design of the proposed fuzzy-expert system

2 Fuzzy Logic Inputs and Outputs

The Mamdani-type fuzzy system with five inputs and three outputs has been considered in the proposed fuzzy-expert system. The five inputs include maximum voltage in p.u (Max-V), sag duration in second (Sag Durat), swell duration in second (Swell Durat), transient duration in second (Tran Durat), and minimum voltage in p.u (Min-V) [5]. The maximum voltage has been chosen as a fuzzy input variable so as to classify instantaneous and momentary swell whereas the minimum voltage is for differentiating between interruption and voltage sag.

The sag duration is used as an input variable for classifying instantaneous and momentary sag whereas the swell duration is used to classify swell disturbance to instantaneous and momentary swell. The transient duration is chosen as a fuzzy input for distinguishing between instantaneous swell and impulsive transient disturbance [5]. The three FL outputs are Output1, Output2, and Output3 in which Output1 is designated for classifying instantaneous sag, non-sag, and momentary sag, Output2 for classifying instantaneous swell, non-swell, and momentary swell, and Output3 for classifying instantaneous interruption, non-interruption, and momentary interruption [6].

3 Membership Functions

The input and output variables are represented by the most common shape of membership functions which are either in trapezoidal or triangular forms and bell curves are also used but the shape is less important than no. of curves. The range of input variables and thresholds are chosen in accordance with the respective disturbance definition as defined in the IEEE Std. 1159-1995. The following Tables 1 and 2 represent fuzzy sets for Input and Output Variables [7, 8].

The membership functions defined for the five input variables are as shown in Figs. 3, 4, 5, 6, and 7. The output variables are defined by three membership functions as shown in Figs. 8, 9 and 10.

If-Then rules (30) have been generated for classifying sag, swell, and interruption disturbances. These rules are represented in the following form: IF premise THEN consequent.

Table 1 Fuzzy sets defined for the input variables

Membership Function	Input 1: Maximum voltage (p.u.)	Input 2: Sag duration (s)	Input 3: Swell Duration (s)	Input 4: Transient Duration (s)	Input 5: Absolute Minimum voltage (p.u.)
1.	L Low	ESH Extremely short	ESH Extremely short	ESH Extremely short	VL Very Low
2.	M Medium	VSH Very Short	VSH Very Short	SH Short	L Low
3.	H High	SH Short	SH Short	SH Short	M Medium
4.	VH Very High	M Medium	M Medium	–	H High
5.	EH Extremely High	–	–	–	–

Table 2 Fuzzy sets defined for the output variables

Membership function	Output 1	Output 2	Output 3
1	I sag Instantaneous sag	I swell Instantaneous swell	I interrupt Instantaneous interruption
2	N sag Non sag	N swell Non swell	N interrupt Non interruption
3	M sag Momentary sag	M swell Momentary swell	M interrupt Momentary interruption

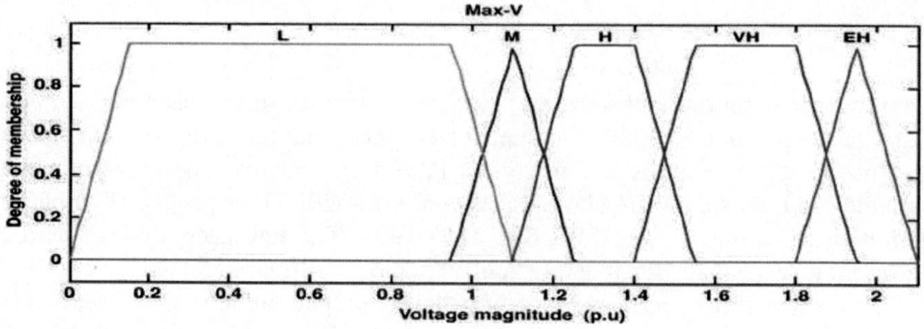

Fig. 3 Maximum voltage input membership function

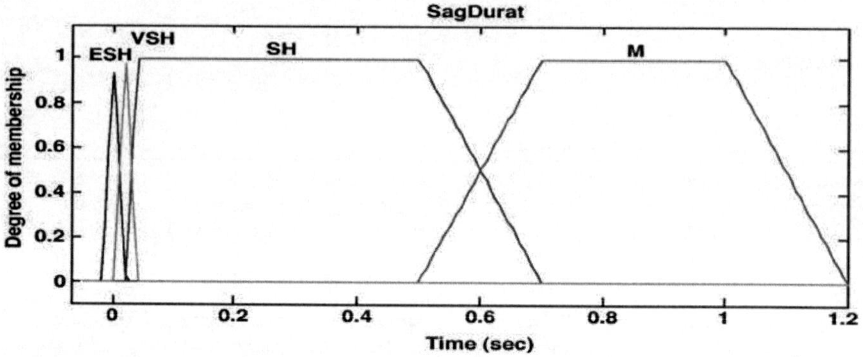

Fig. 4 Sag duration input membership function

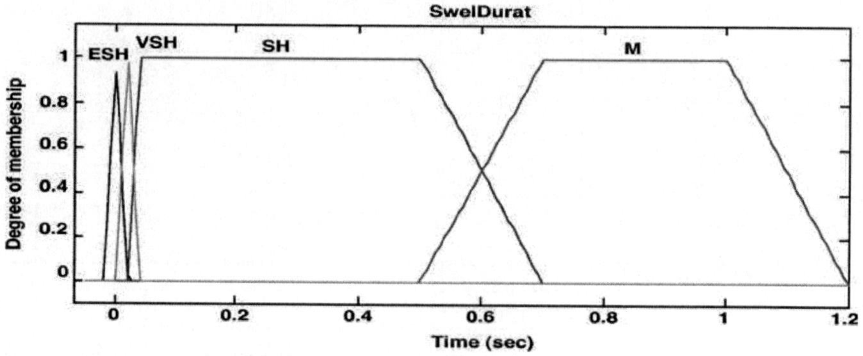

Fig. 5 Swell duration input membership functions

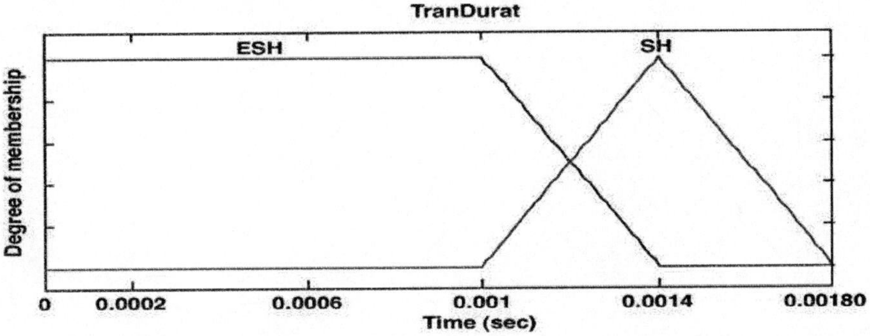

Fig. 6 Transient duration input membership functions

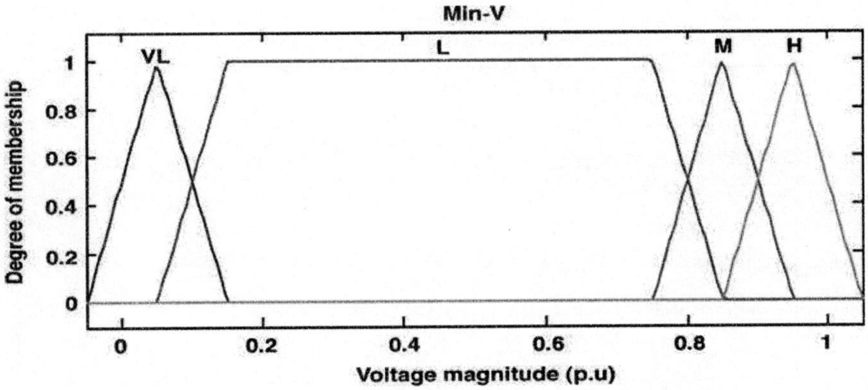

Fig. 7 Absolute minimum voltage input membership function

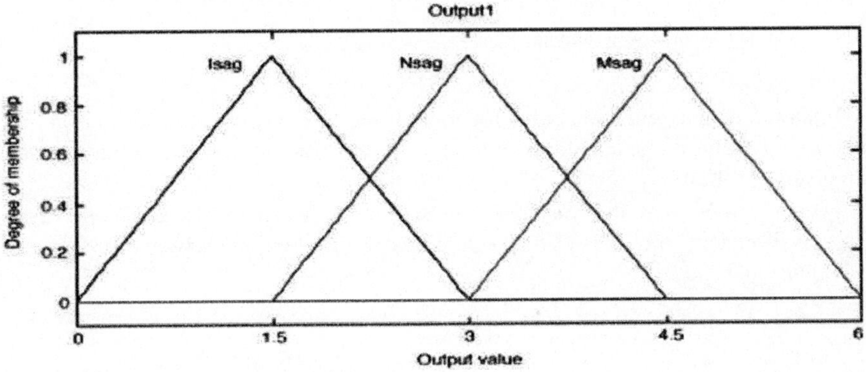

Fig. 8 Output1 membership function

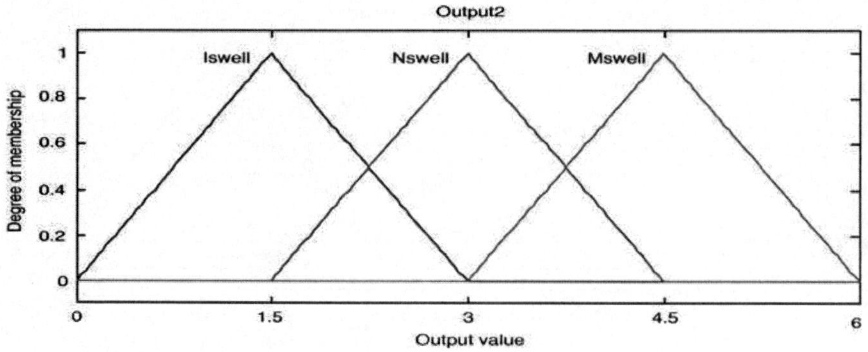

Fig. 9 Output2 membership function

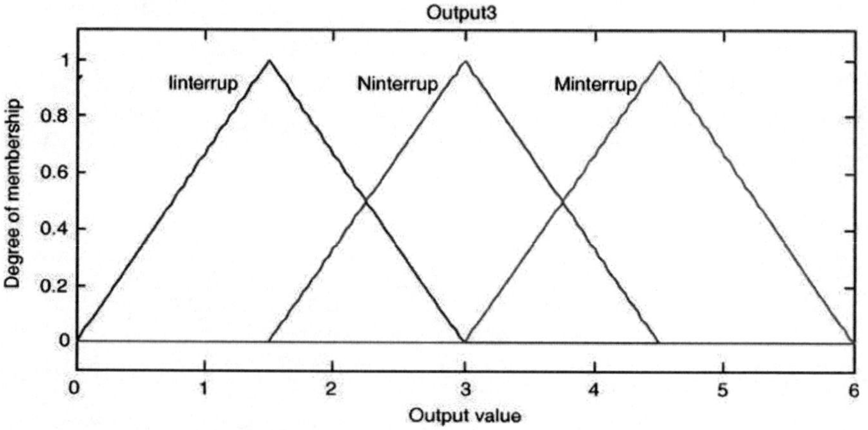

Fig. 10 Output3 membership function

Example of the generated rules for identifying sag, swell, and interruption and classifying them to instantaneous, momentary, and non-sag, swell and interruption are given as follows:

(1) If (Max-V is L) and (Sag Durat is SH) and (Swell Durat is ESH) and (Tran Durat is ESH) and (Min-V is L) then (Output1 is I sag) (Output2 is N swell) (Output3 is N interrupt). etc.

4 Conclusion

During the contingency analysis of an electrical power system, the variation in the level of voltage and current are the two important issues that affects the reliability a lot. In this paper, a fuzzy logic-based system is developed so as to analyze such types of disturbances like sag, swell, transient signal, etc. If such types of disturbances are detected and corrected, the reliability of existing power system network will increase automatically. Five fuzzy inputs and three outputs are used with If-Then rules in the fuzzy inference system (FIS). In this, a technique based on FFT with RMS average method is used to describe the features of many kinds of disturbances. The disturbances in the voltage are applied to the proposed fuzzy system to test the accuracy, sag, swell, and other disturbances. Hence, better results are obtained with the proposed fuzzy-based expert system that confirms the power quality progress and makes the power system reliable.

References

1. “Power quality issues standards &guide lines”, IEEE, Vol 32, May 96
2. Ebron S, Lubkeman D, White M (1990) A neural network power distribution feeders. IEEE Trans Power Deliver 5(2):905–914
3. Sultan AF, Swift GW, Fedirchu DJ (1992) Detection of high impedance arcing faults using a multi-layer perceptron. IEEE Trans Power Delivery 7(4):1871–1877
4. Amouzad Mahdiraji, G, Mohamed A (2006) A fuzzy-expert system for classification of Short duration voltage disturbances, Jurnal Teknologi, 45(D) Dis. 2006: 41–57, © Universiti Teknologi Malaysia
5. Liao Y, Lee JB (2004) A Fuzzy-expert system for classifying power quality disturbances. J Elect Power Energy Syst 26:199–205
6. Classification of Power Disturbances using Fuzzy LogicBoris Bizjak, Peter Planins University of Maribor - FERI, Smetanova 17, Maribor, SI - 2000, Slovenia Giang Z.L., 2004
7. Gaouda AM, Kanoun SH, Salama MMA, Chikhani AY (2002) Pattern recognition application for power system disturbance classification. IEEE Trans Power Deliver 17:677–683
8. Dash PK, Mishra S, Salama MMA, Liew AC (2000) Classification of power system disturbances using a fuzzy expert system and a Fourier linear combiner. IEEE Trans Power Deliver 15:472–477

Chapter 9
Vehicle Detection and Its Speed Measurement

Morium Akter, Jannatul Ferdous, Mahmuda Najnin Eva, Sumaita Binte Shorif, Sk. Fahmida Islam, and Mohammad Shorif Uddin

1 Introduction

In Bangladesh, road accidents are unfortunately increasing in an alarming rate day by day. For example, in the year 2018 around 4317 dangerous accidents occurred in Bangladesh which caused 10828 people are injured and 4580 people died according to the statistics of road accidents in Bangladesh [1]. The most common reasons for these road accidents are reckless driving, unlicensed and incompetent drivers, distracted driving, drunk driving, driver's unconsciousness, overspeed, and many more [2]. However, most of these road accidents are caused due to overspeed of the vehicles. Automated overspeeded vehicle detection and controlling may solve this problem.

Some overspeed detection systems have been developed [3–9]. Ushakiranmai and Thirumal [3] developed a system for vehicle speed measurement in traffic junctions using magnetic sensors. The system consists of wireless anisotropic magnetic devices

M. Akter (✉) · J. Ferdous · M. N. Eva · S. B. Shorif · Sk. Fahmida Islam · M. S. Uddin
Department of Computer Science and Engineering, Jahangirnagar University, Savar, Dhaka, Bangladesh
e-mail: ecs_morium@yahoo.com

J. Ferdous
e-mail: polyjannat040@gmail.com

M. N. Eva
e-mail: mneva2015@gmail.com

S. B. Shorif
e-mail: sumaita.bs@gmail.com

Sk. Fahmida Islam
e-mail: jharnaju@yahoo.com

M. S. Uddin
e-mail: shorifuddin@gmail.com

M. Shorif Uddin et al. (eds.), *Intelligent Energy Management Technologies*, Algorithms for Intelligent Systems,
https://doi.org/10.1007/978-981-15-8820-4_9

which are placed next to the roadway and measure traffic in the immediate adjacent lane. The accuracy of the system is good, but error is somehow higher. Shedbalkar et al. [4] described a technique for vehicle speed estimation using permanent magnet synchronous motors based on extended Kalman filter. However, the detection rate as well as vehicle speed is not very good.

Benjamin et al. [5] developed a deep neural network-based method for car speed estimation which gives 91% accuracy. But their developed system is slow. Abbas [6] developed a system which estimates the vehicle speed using image processing. The system performance suffers from poor lighting or bad weather. The system also requires high performance signal processing hardware, large amount of memory, and high bandwidth data links. Arash Gholami et al. [7] developed a system for the detection of vehicle speed using CVS method, which works accurately in measuring speed, but the system is sensitive to the thresholds. Budi Setiyono et al. [8] proposed a Gaussian mixture model to detect vehicle speed using a sequence of images. The speed is estimated by computing the movement of centroid in the sequence of frames. Haque et al. [9] developed a primitive velocity estimation method by three-frame differencing method.

The above methods are somehow expensive and their performances not up-to the mark. Therefore, in this paper we are trying to develop a simple and cost-effective complete system. Our system uses highly enhanced open source TensorFlow [10] machine learning library object detection API (Application Program Interface) to detect a vehicle and estimate its speed as well as overspeeded situation. The TensorFlow models have different pre-trained models for various machine learning tasks and their one important resource is the object detection API, which uses neural network-based SSD (Single-Shot Detector) algorithm [11].

The rest of the paper is organized as follows. Section 2 deals with designing of the system. Section 3 presents the experimental results and finally Sect. 4 concludes the paper.

2 System Design

Figure 1 describes the flow diagram of our system. At first, the acquired input video is fragmented into sequential frames by OpenCV. We marked a fixed line, known as ROI (Region of Interest) line from which we perform our system computation. Then the ROI line approaching vehicle is detected by using TensorFlow API. There are many models which can be used to train our detection system, but we have selected "ssd_mobilenet_v1_coco_2017_11_17", as it consumes less memory. The API model creates a bounding box including ROI (region of interest) line after detecting the location of the vehicle. After that, it measures the speed of each vehicle, and, also determines whether the vehicle is running at overspeed or not. The speed is calculated in our system using the usual velocity formula, as there is a pre-mapping between pixel displacement in image domain and real displacement in vehicle domain.

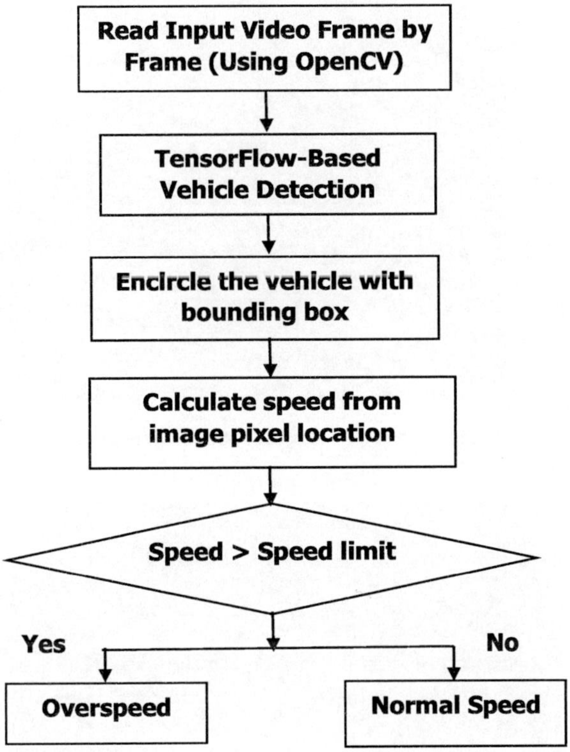

Fig. 1 Flow diagram of the system

$$\text{Speed} = \text{distance}/\text{time} \quad (1)$$

We then check this estimated speed crossed the speed limit or not along with annotation. The annotation is performed by using "LabelImg" tool from GitHub [12] and the annotated data are saved as XML/CSV file.

3 Experimental Results and Analysis

We have tested the system with several sample videos taken by a mobile phone (OPPO F1) camera from real scenes. If a vehicle's speed is more than 50 km/hr, then we have taken it as "overspeed". Figure 2 shows an experimental scene of a vehicle where we estimated its speed. The estimated speed is found normal. Figure 3 shows another experimental scene of a vehicle where we estimated its speed crossed the speed limit. Our method failed to estimate speed if the scene contains multiple vehicles. Figure 4 shows such a scene, where our system could not estimate vehicle velocity. We have compared the estimated speed with the correct speed of a vehicle as shown in Table 1. Here, we have taken the measured speed by "Speed Gun App"

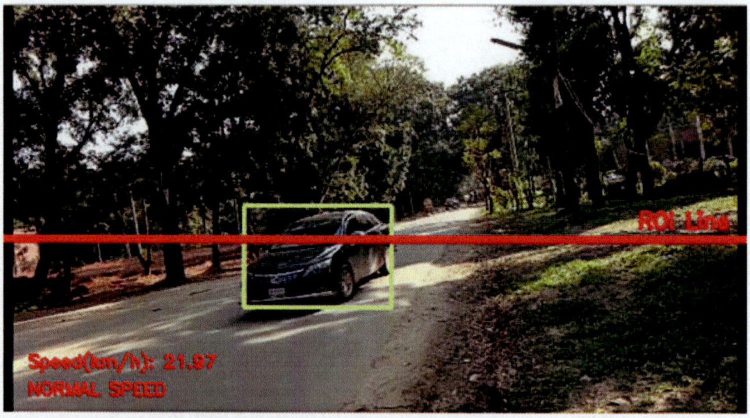

Fig. 2 Vehicle detection and speed estimation (normal speed)

Fig. 3 Vehicle detection and speed estimation (overspeed)

as correct speed. Our system's speed measurement is almost matched with the speed of the "Speed Gun App". However, we have seen that error is higher for higher speed.

4 Conclusions and Future Works

This paper presents a TensorFlow-based system that detects the vehicle in a scene and estimates its speed for the detection of overspeeded vehicle. The obtained results

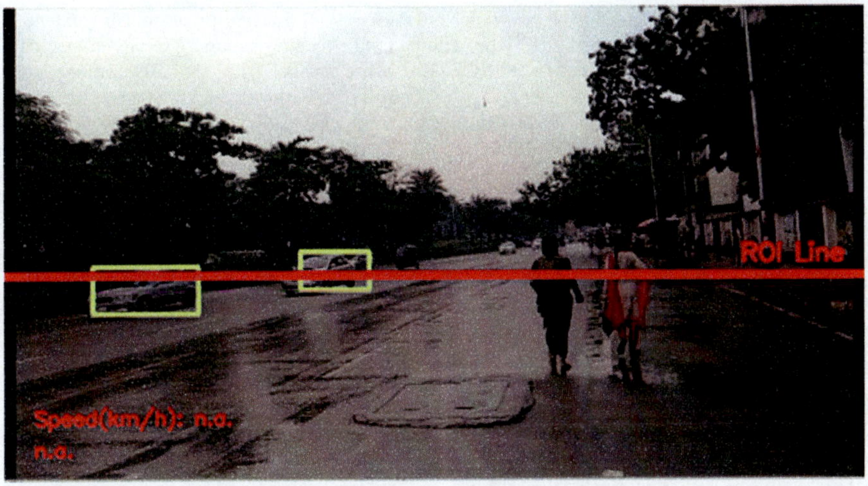

Fig. 4 The system could not estimate the vehicle speed due to multiple vehicle

Table 1 Comparison of real speed with obtained speed

Real speed (km/h)	Obtained speed (km/h)	Accuracy (%)
23.6	21.97	93.00
58.01	53.57	92.30
49.80	45.88	92.14
62.80	68.77	91.30
106.50	89.92	84.40

are quite satisfactory. This system will help in reducing road accidents by detecting overspeeded vehicles. The main limitation of our system is that it cannot estimate the speed if the scene contains mutiple vehicles. So, our system should be placed in a fixed location to estimate the vehicle speed, so that the scene should contain only one vehicle. As our future work, we expect to develop a system which can estimate the speed of multiple vehicles in a single shot.

References

1. Tribune D (2018) saw surge in road accidents. https://www.dhakatribune.com/bangladesh/nation/2019/01/16/2018-saw-surge-in-road-accidents
2. Sivaraman S, Manubhai Trivedi M (2013) Looking at vehicles on the road: a survey of vision-based vehicle detection, tracking, and behavior analysis. IEEE transactions on intelligent transportation systems, vol 14, no 4, December 2013, pp 1773–1795
3. Ushakiranmai M, Sai Thirumal G (2015) Design Portable sensors system for vehicle counting, classification, and speed measurement in traffic junctions. Int J Prof Eng Stud V(5):174–179

4. Shedbalkar K, Dhamangaonkar AP, Walchand ABP (2012) Speed estimation using extended Kalman filter for PMSM. In: 2012 international conference on emerging trends in electrical engineering and energy management (ICETEEEM), Chennai, India, 13–15 December 2012, pp 433–435, IEEE Explore
5. Penchas B, Bell T, Monteiro M (2019) A deep learning approach to vehicle speed estimation. http://cs229.stanford.edu/proj2017/final-reports/5244226.pdf. Accessed 30 July 2019
6. Awni Abbass A (2010) Estimating vehicle speed using image processing. AL-Mansour J 14:127–141
7. Gholami Rad A, Dehghani A, Rehan Karim M (2010) Vehicle speed detection in video image sequences using CVS method. Int J Physical Sci 5(17)–2555–2563
8. Setiyono B, Ratna Sulistyaningrum D, Soetrisno, Fajriyah F, Wahyu Wicaksono D. Vehicle speed detection based on Gaussian mixture model using sequential of images. J Phys: Conf Ser 890(1):012144. IOP Publishing. https://iopscience.iop.org/article/10.1088/1742-6596/890/1/012144/pdf
9. Reduanul Haque M, Golam Moazzam Md, Islam S, Das R, Shorif Uddin M (2016) Vehicle speed determination from video streams using image processing. In: International workshop on computational intelligence (IWCI), 12–13 December 2016, Dhaka, Bangladesh, pp 252–255, (IEEE Explore)
10. Francis J (2019) Object detection with tensor flow. https://www.oreilly.com/ideas/object-detection-with-tensorflow. Accessed 30 July 2019
11. Liu W, Anguelov D, Erhan D, Szegedy C, Reed S, Fu C-Y, Berg AC (2016) SSD: single shot multi box detector. ECCV 2016, Amsterdam, Netherlands, October 11–14, 2016, Lecture Notes in Computer Science, vol 9905, Springer, pp 21–37, 2016
12. A Graphical Image Annotation Tool—"LabelImg,". https://github.com/tzutalin/labelImg

Chapter 10
Fuel Cells as Naval Prime Movers: Feasibility, Advances and Implications

Akshat Mathur and Sushma Dave

1 Introduction

Traditionally, the guidelines for the selection of naval prime movers focused on the demands for greater speed, better shock absorption, low noise generation and low infrared signature; but modern warships require a more versatile package which should be flexible enough for broadly varying operational requirements. Modern propulsion systems are expected to operate in a greater range of speed with longer durations of deployment and cleaner emissions [1] while aiming to operate with a numerically inferior manpower compliment than ever before.

Fuel cells inherently offer reduced maintenance costs, emissions and infrared signatures, acoustic signatures, radar cross-section and increased ship survivability due to distributed power reduction and greater flexibility of design and operation due to their modular nature [2]. Fuel cells produce such low levels of noise that the fuel cell cabin on board a ship is expected to have noise levels similar to that of any regular office [3]. Thus, they promise to be a more promising approach towards a clean and efficient power solution as compared to other power sources or prime movers in both stationary and mobile applications. However, their foray into naval propulsion systems is relatively new and is still in its infancy in both maturity and deployment.

The first fuel cells to enter production for naval propulsion as prime movers are Siemens' PEM fuel cells employed by German-Italian Type 212A submarines along with its export derivative: the Type 214 as a part of their Air Independent Propulsion mechanism. These boats are in service with various navies across the globe including the German, Italian, Hellenic (Greek), Portuguese and South Korean navies.

A. Mathur (✉) · S. Dave
Jodhpur Institute of Engineering and Technology, Jodhpur, RJ, India
e-mail: mathur.akshat@outlook.com

M. Shorif Uddin et al. (eds.), *Intelligent Energy Management Technologies*,
Algorithms for Intelligent Systems,
https://doi.org/10.1007/978-981-15-8820-4_10

Furthermore, on the economic front, the U.S. Navy estimated that if a ship service fuel cell plant, having an efficiency of 37–52%, replaces the existing gas turbines on board a DDG 51 Arleigh Burke class destroyer, then, considering a 3,000-h time window, it would consume only a paltry 33% of the fuel relative to what would have otherwise been consumed by the warship. This, the U.S. Navy estimated, could save them more than $1 million per ship per year in the form of ship service fuel costs [2].

Despite the proliferation of fuel cell-powered Air Independent Propulsion equipped submarines around the world in the last few years, the technology has not gained much turf in the realm of surface vessels: especially naval. In [4], Jing Sun, et al. note that there is a lack of published studies focusing on larger vessels with greater power requirements. They propose a notional Solid Oxide Fuel Cell module and analyse the fuel saving and machinery arrangements by using a hybrid SOFC-gas turbine module for a U.S. Navy Sealift vessel. They concluded the same benefits of employing fuel cell prime movers but also noted that large technology gaps exist and such power density levels, as they have considered in their study, have not been yet achieved by modern state-of-the-art fuel cell technologies.

However, there have been many attempts to construct viable fuel cell modules; the developments and deployments of which are discussed further in this paper.

2 Naval Fuel Cell Technologies

2.1 *The Notional 5 MW SOFC Unit*

Fuel cell technology has been making rapid strides with the worldwide effort to develop cleaner and more efficient power sources. A study conducted by Jing Sung, et al. identified the baseline desirable parameters of a marine fuel cell prime mover. Four such notional fuel cell units were coupled with a gas turbine to obtain a total output power of 24 *MWe* [4]. Principal parameters of this hybrid SOFC-GT system are listed in Tables 1 and 2.

Using this system, a mission endurance analysis was performed on a U.S. Sealift Command's Roll on–Roll off (Ro-Ro) cargo vessel, and the mission endurance was determined by the amount of distance travelled by the ship with 90% of the full fuel capacity at the peak speed of 24 *knots* [4].

Table 1 Principal SOFC parameters of the hybrid SOFC-GT system [4]

SOFC Parameter	Magnitude
Output Power (MW_e)	5
Efficiency (%)	50
Fuel Utilisation (%)	85
Air Utilisation (%)	25
Air/fuel inlet temperature (°C)	600

Table 2 Principal GT parameters of the hybrid SOFC-GT system [4]

GT Parameter	Magnitude
Compressor Efficiency (%)	5
Turbine Efficiency (%)	50
Pressure Ratio	85
Generator Power (MW_e)	25

In this analysis, the fuel cell-powered ship was found to have an endurance advantage of 25% over diesel systems and 57% over gas turbine systems while consuming 20% and 50% less fuel than the said prime movers, respectively. Furthermore, amongst all the conditions evaluated, the power output of the combined system never fluctuated by more than 2% which indicated a reliable design and arrangement [4].

However, this analysis was performed with an auxiliary vessel of the U.S. Sealift Command and, therefore, does not adequately reflect the demands of a frontline warship. Also, it was acknowledged by the authors that present technology does not allow us to field such systems, however, they predicted that they can be developed in the near future, i.e. within a span of 15–20 years [4].

2.2 *Siemens SINAVY PEM Fuel Cells*

Siemens offers two kinds of SINAVY Proton Exchange Membrane fuel cells: the FCM 34 and the FCM 120. The FCM 34, developed in 1984 at the request of the German Ministry of Defence, is employed by Type 212A submarines which are in service with the German and the Italian navies. The FCM 120, developed at a later date, is deployed in Type 214 submarines which are Type 212A's export variant currently in service with various navies across the globe including the Hellenic (Greek), Portuguese and South Korean navies [5]. Table 3 lists the technical specifications of FCM 34 and FCM 120 SINAVY fuel cells.

The voltage of SINAVY PEM Fuel Cells with respect to the operating time is quite stable. The degradation rate of the system is less than 2 $\mu V/h$ *per cell* for a FCM 34 module and significantly lower than the same for a FCM 120 module [5]. This property is particularly desirable as naval vessels have stringent operational requirements wherein all the deployed equipment is expected to adhere to their specifications for the duration of a reasonably long service life along with extended operating time durations.

These fuel cells are employed by the aforementioned submarines as a part of their Air Independent Propulsion system which allows them to say submerged for relatively longer durations than what their diesel engines would have allowed for. However, under the fuel cells' power, the boats can only cruise at a speed of approximately 5 *knots* (submerged).

In comparison to this, the cruising speed of naval surface escort ships, which tend to be of significantly greater displacements and thus requiring greater power

Table 3 Specifications of SINAVY FCM 34 and FCM 120 SINAVY fuel cell modules [5]

Technical data	FCM 34	FCM 120
Rated power (kW_e)	Approximately 34	Approximately 120
Voltage range (V_{DC})	Approximately 54	Approximately 215
Efficiency at rated load (%)	>58	>53
Efficiency at 20% load (%)	Approximately 71	Approximately 68
Operating temperature (°C)	70–75	Approximately 70
H_2 pressure (bar)	2.3 (absolute)	2.4 (absolute)
O_2 pressure (bar)	2.6 (absolute)	2.7 (absolute)
Dimensions (cm)	H = 47 W = 47 L = 143	H = 50 W = 53 L = 176
Weight (kg) (without module electronics)	630	930
Maximum applied current in continuous operation (A)	650	560
Technical data	FCM 34	FCM 120

densities than their sub-surface counterparts, is about 10–15 *knots* while peak power allows them to achieve 25–30 knots conventionally in the open seas. These requirements pose a challenge for fuel cells as significant technology gaps exist which need bridging if such systems are to be successfully incorporated in surface vessels as prime movers.

However, due to inherent flexibilities offered by fuel cells, it is possible to refit operational submarines with SINAVY PEM fuel cell modules and, thus, extend AIP capabilities to those boats. Furthermore, as shipboard reformer technology advances suitably, fuel cell technology's field of application can be expanded and it may become the preferred prime mover for sub-surface and surface vessels [5].

2.3 Fuel Cell Energy Inc. Direct Carbonate Fuel Cell

In 1997, the Office of Naval Research of the United States sponsored a three-phase project to demonstrate the viability of commercial fuel cell technology in ship service applications on board its warships whose conceptual design criteria are listed in Table 4. By 2003, FuelCell Inc. completed the conceptual design and critical component testing of a 2.5 MW fuel cell module for future surface combatants. To achieve this, FuelCell Inc. used its Direct carbonate Fuel Cell (DFC) technology which provided thermal efficiencies in excess of 50% [6].

The device is fuelled by reforming methane and steam to produce carbon dioxide and hydrogen which is then fed to the anode of the cell. The power-producing anode reaction converts hydrogen to steam which allows reforming reaction to tend towards the product side at a relatively low temperature of 650 °C by continuously shifting the

Table 4 Fuel cell energy Inc. direct carbonate fuel cell specifications [5]

Parameter	Magnitude
Power (MW_e)	2.5
Voltage range (V_{AC})	450
Efficiency at 50% load (%)	40
Spatial allowance (m^3/kW_e)	0.057
Weight allowance (kg/kW_e)	18
Production cost ($\$/kW_e$)	1,500

reaction equilibrium to obtain the desired products. Such a commercial fuel cell stack typically comprises 350 individual cells arranged in a cuboidal fashion measuring 1.2 m × 0.6 m [6].

PEM fuel cells, in general, require storage of highly flammable and dangerous hydrogen gas which can be a potential safety concern on board any ship. An error, accident or battle damage (in case of warships) may result in uncontrolled discharge of the stored hydrogen gas which can be a significant health hazard on board any vessel. Therefore, internal reforming in the direct carbonate fuel cell at a relatively moderate temperature is a desirable capability on board naval vessels as it only requires the storage of relatively safer hydrocarbon fuels.

The efficiency, size and weight constraints, as indicated in Table 4, were determined by the U.S. Office of Naval Research so as to be comparable with marine diesel engines and offshore power generators. The production cost of the system was determined as such to be "somewhat" lesser than that of large diesel engines [7]. These parameters, therefore, indicate the absolute minimum respective magnitudes that ship bound fuel cell technology should attain to be operationally and economically feasible as compared to conventional diesel prime movers.

3 Integration

Conventionally, naval propulsion systems comprise internal combustion engines, viz. gas turbines and diesel engines. These systems are employed in a combination in the propulsion system with different prime movers often being employed for different purposes such as propulsion, services and auxiliary demands. The integration of notional fuel cells in such an arrangement is outlined in [4] and is illustrated in Fig. 1.

In modern propulsion systems, however, the concept of Integrated Full Electric Propulsion (IFEP) is gaining turf. IFEP is defined as the use of a common power system for both propulsion and ship's services. In such an arrangement, efficient operation is obtained by the use of the minimum number of prime movers necessary to meet the load which all run at their optimum efficiency [8].

It is lucrative for applications in naval propulsion because of the inherent benefits that it carries. A lesser number of prime movers would imply lesser sources of noise

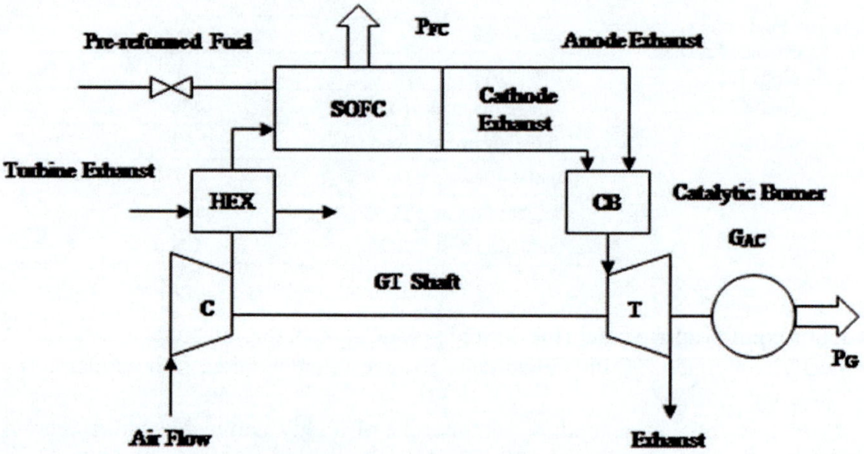

Fig. 1 Combined Fuel Cell and Gas Turbine arrangement for marine applications [4]

generation which is a very desirable outcome for any warship. Furthermore, such a system will be less maintenance intensive and would, thus, require a lesser manpower compliment which would result in reductions in life cycle costs [9]. The first warships to wield an Integrated Full Electric Propulsion system are the Royal Navy's Type 45 air defence destroyers. Their IEP system is illustrated in Fig. 2 [9]. This concept is also known as that of an All Electric Ship (AES).

In [10], the authors consider that in an AES, power can be provided by the means of diesel generator motors through transformers or by battery banks connected to the motors and evaluate three "options" of integrating hydrogen-oxygen fuel cells on board ships. The outcome of the analysis of these 3 arrangements, as determined by the authors, is presented in Table 5.

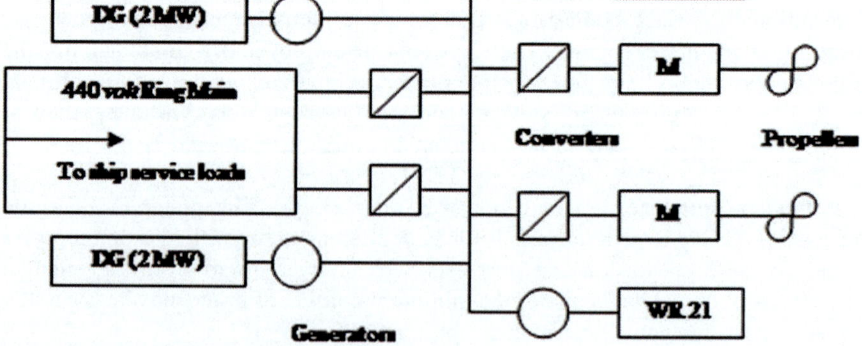

Fig. 2 Electric Propulsion System of Type 45 destroyers [9]

Table 5 Comparison of options for fuel cell integration on board ships [10]

Parameter	Option 1	Option 2	Option 3
Advantage(s)	Less dependence on fuel cells	Without batteries or battery charge system	Power system totally operated with renewable energies
Disadvantage(s)	Fuel cells in parallel with batteries; the diesel generator is still in the system	The diesel generator is still in the system	Requires 8 fuel cell modules and a 12 V DC-DC converter
Complexity of the control system	High	Low	Low
Space	Not sufficient	Not sufficient	Adequate
Emissions	High	High	Low
Harmonic distortions	High	Low	Low

It is noted that diesel generators pose the problem of limited energy production and are sensitive to harmonic current consumption. Furthermore, the batteries have the disadvantage of relatively longer charging time as compared to the subsequent duration of operation. Also, the impedance of the diesel generator in combination with the non-sinusoidal current feedback of the battery chargers produces voltage distortion that limits the functionality of the power system [10].

Option 1 comprised of an arrangement in which fuel cells are arranged in parallel with batteries. This arrangement adversely affected the system's response time due to the different sources. Currently, fuel cells are not directly controlled causing disarrangement between the fuel cell and battery impedances, which makes it difficult to implement power systems that have fuel cells in parallel with the batteries [10].

Furthermore, an elaborate control system would be required to match the fuel cell's voltage, which varies from 440 to 800 V, to that of the batteries. However, the space that would be gained by eliminating a series of batteries to house the fuel cell modules and power conditioning system would be insufficient for the said task [10].

In option 2, the batteries and the continuous current charge mechanism were completely eliminated from the system. This option benefits by the removal of the limited use time and the presence of harmonic distortions due to these eliminations. Also, an increase in the overall power density of the system is thus achieved. However, the space constraints still do not allow for the accommodation of the fuel cells and their fuel and cooling systems in the space created by the elimination of battery stacks [10].

In option 3, the entire power system of the vessel is replaced by fuel cells and a DC-DC converter. Such an arrangement will be able to run completely on renewable energy and would operate below 80 °C making it "highly reliable". Furthermore, such an arrangement would be highly efficient with low emissions and would also be able to avoid the problem of harmonic distortions [10].

For such an arrangement to succeed, fuel cell modules would need to be connected in parallel with each other to satisfy the system's needs. However, upon connecting the fuel cells to a bus in parallel, a high possibility exists that the current flows back into the fuel cell modules which can reduce their useful life and service efficiency [10]. All of these analyses are summarised in Table 5.

4 Fuel

Like the prime movers, it has the potential to replace, fuel cells also have the ability to utilise hydrocarbon fuels, amongst other kinds of fuels. However, since the production of consumable energy, in the form of heat and electricity, only takes place through electrochemical reactions, and not through combustion, as opposed to the internal combustion engines in widespread use, emissions from fuel cells are mostly negligible in nature. Table 6 summarises the various compounds which may be used by a particular kind of cell with a potential for naval application as a fuel, and those which may act as a poison for them beyond certain concentrations.

Alkaline fuel cells as in contrast to the listed kinds of cells, they necessarily require pure hydrogen as a fuel for their operation and cannot operate on a reformed fuel as noted in [12], which was a study conducted by Arctic Energies Ltd. for the feasibility analysis of replacing the diesel-electric propulsion of a U.S. Coast Guard ship which later resulted into the development of the (Molten Carbonate) Direct Fuel Cell technology (2.3).

On board reforming of fuel is desirable for fuel cell operations as it avoids the shipboard storage of hydrogen which is dangerously flammable. The same influenced the Russian Navy's AIP choices and, in the light of the records of frequent fires onboard its ships, it chose not to develop fuel cells similar to the ones employed by German submarines (2.2) for its new Lada class diesel-electric boats, but instead went forward with developing technology that would "convert diesel fuel into hydrogen for power" [13], i.e. reform it.

Table 6 Various gaseous compounds as fuel and poison for various kinds of fuel cells for potential naval applications [11]

Gas	Phosphoric Acid Fuel Cells	Molten Carbonate Fuel Cells	Solid Oxide Fuel Cells	Proton Exchange Membrane Fuel Cells
H_2	Fuel	Fuel	Fuel	Poison (>10 ppm)
CO	Poison (>0.5%)	Fuel	Fuel	Diluent
CH_4	Diluent	Diluent	Fuel	Diluent
CO_2 & H_2O	Diluent	Diluent	Diluent	Diluent
S as H_2S & COS	Poison (>50 ppm)	Poison (>0.5 ppm)	Poison (>1.0 ppm)	No studies

In [3], the authors conduct a thermodynamic analysis of a 120 kW_e diesel-fuelled SOFC system and conceive it as an auxiliary engine while comparing it with a diesel-electric generator set for surface ships. They report a reformer efficiency of 85%, fuel cell efficiency of 68.47% and, thus, a net efficiency of 55.28% while using the NATO F-76 marine diesel fuel.

Also, the higher operating temperatures of MCFCs and SOFCs, in conjunction with their ability to use reformed fuels, can be particularly beneficial for warships which have limited power supply as the waste heat from the said cell stacks can be utilised for fuel reforming which also requires high temperatures to occur.

5 Conclusion

We discussed various aspects of deploying fuel cells as naval prime movers covering the assessment of their feasibility, review of the recent advances and understanding of their implications. Furthermore, their fuelling and integration within the existing marine propulsion systems were discussed.

We conclude that fuel cells offer a promising option for the future of naval propulsion with numerous environmental, economic and combat benefits. However, it is notable that no fuel cell systems are presently deployed as naval prime movers with the notable exception of the German-Italian Type 212 submarines with their Air Independent Propulsion systems. However, a significant number of studies and attempts have been consistently made around the world which were discussed in the review.

It was observed that the principal hurdle in the path of realising fuel cells as naval prime movers is their poor spatial efficiency and weight-to-power ratios (specific power). Furthermore, certain types of fuel cells, such as the SOFC and MCFC, which are gaining the turf for naval applications, operate at very high temperatures.

All the fuel cells discussed in this study prefer hydrogen as fuel which requires its on-board storage. This poses significant risks in a marine environment due to the dangerously flammable nature of hydrogen. It is, therefore, safer and cheaper to employ shipboard fuel reforming technologies to produce the necessary fuel for the fuel cells on board in real time from the regular marine diesel fuel which would save considerable monetary resources in the terms of on board and off board storage, handling, transportation and processing. With the notable exception of alkaline fuel cells, most other fuel cells can comfortably operate on reformed fuels with both SOFCs and MCFCs benefitting particularly from this arrangement.

Further innovations in the area, therefore, need to be directed towards improving the spatial efficiency, the specific power and the reforming techniques for fuel cells so that their numerous benefits, especially environmental, can be suitably exploited for the better.

References

1. Ohmayer HF (2012) Propuslion system choices for modern naval vessels
2. O'Rourke R (2006) Navy ship propulsion technologies: options for reducing oil use - Background for Congress
3. SINAVY PEM Fuel Cells: For Submarines. https://assets.new.siemens.com/siemens/assets/public.1535009488.28615cde70250d0e81b68ba466bd77d7f5c68c73.sinavy-pem-fuel-cells.pdf. Accessed 08 April 2019
4. Abens S, Ghezel-Ayagh H, Steinfeld G, Sanderson R (2000) Development of a Ship Service Fuel Cell. AES 2000/All Electric Ships
5. Application of fuel cells in surface ships. https://www.osti.gov/etdeweb/servlets/purl/20249928. Accessed 13 April 2019
6. Fuentes DI, De Lavalle PÃ©rez H (2015) Fuel cell integration on board surface ships. Ship Sci Technol 8(6):19–28
7. Fuels for fuel cells, http://www.nfcrc.uci.edu/3/FUEL_CELL_INFORMATION/FCexplained/Fuels.aspx. Accessed 27 April 2019
8. Feasibility Study of Repowering USCGC VINDICATOR (WMEC-3) With Modular Diesel Fueled Direct Fuel Cells. https://apps.dtic.mil/dtic/tr/fulltext/u2/a337901.pdf. Accessed 27 April 2019
9. Why Russia's New 'Stealth' Submarines Have a Big Problem. https://nationalinterest.org/blog/why-russias-new-stealth-submarines-have-big-problem-22941?nopaging=1. Accessed 27 April 2019
10. Fuentes DI, De Lavalle PÃcrez H (2015) Ship science and technology
11. Fuels for fuel cells (National Fuel Cell Research Centre, University of California, Irvine), http://www.nfcrc.uci.edu/3/FUEL_CELL_INFORMATION/FCexplained/Fuels.aspx. Accessed 27 Apr 2019
12. Kumm WH, Lisle Jr HL (1997) Feasibility study of repowering USCGC VINDICATOR (WMEC-3) with modular diesel fueled direct fuel cells. Arctic Energies LTD. https://apps.dtic.mil/dtic/tr/fulltext/u2/a337901.pdf. Accessed 27 Apr 2019
13. Roblin S (2017) Why russia's new 'Stealth' submarines have a big problem. The National Interest. https://nationalinterest.org/blog/why-russias-new-stealth-submarines-have-big-problem-22941?nopaging=1. Accessed 27 Apr 2019

Chapter 11
Overview of Eco-Friendly Construction Materials

Nidhi Sharma and Aashish Kumar Jha

1 Introduction

Central Europe reflects a prominent example of eco-materials in the form of the Mostar Bridge being constructed. Suleman is a magnificent sixteenth century authorized new bridge as a substitution for a wooden bridge. [1]. In Spain, according to the Spanish power system advanced report in 2016, the author of the business group "Red Electra de Espana" said that electricity generated from renewable sources was 41.1% in 2016, compared to 58.9% from non-renewable sources. There has been a decrease compared to the year 2015, in which 63.1% of electricity was generated through non-renewable sources. The common and basic sources of non-renewable generation of electricity is the nuclear power, the thermoelectric power, coal plants, and the plans of combined cycles, manifested the excessive production, with the created 55,546 GWH, 37,038 GWH and 29,787 GWH, respectively, in 2016 [2]. The paramount challenges today are the management of wastes, which is generated on a large scale and greenhouse gas emission generated by energy generation, industrial production, and consumption models, green technologies or ecofriendly production models should be the basic pillars to develop a circular economy. All efforts must focus on flourishing the technology of ecology that helps reduce environmental adverse effects and also improve society's expansion and create enhancement for its betterment.

Road has been the most useful infrastructure that resulted from transport activities and has helped in stimulating human civilization and development. There has been a long history of road construction in the twentieth century BCE, the Arab Republic of Egypt constructed roads to transport large amounts of rocks from quarries to

N. Sharma (✉) · A. K. Jha
Jodhpur Institute of Engineering and Technology, Jodhpur, Rajasthan, India
e-mail: nidhisharma9862@gmail.com

A. K. Jha
e-mail: aashishjha003@gmail.com

M. Shorif Uddin et al. (eds.), *Intelligent Energy Management Technologies*, Algorithms for Intelligent Systems,
https://doi.org/10.1007/978-981-15-8820-4_11

sites where the rocks were indulged in pyramid building and the Great Sphinx [3, 4]. In the reflecting history of Rome, people constructed an advanced road network being centered in Rome had played a noteworthy role in the prosperity of the ancient Roman. The Empire and the proverb had it: "all roads lead to Rome" [5]. In addition, the "Silk Road", which had been in operation since the second century BCE from the thirteenth and fourteenth centuries BCE, is characterized by economic, cultural, and technological exchanges between East and West of the Asian continent, had been promoted a lot. It was great contribution to the economic development and social progress of the world [6]. Then came the first study of the use and reuse of waste glass as a building material in 1963 by Schmidt and Sia. They converted the discarded glass into useable glass chips for making wall panels. Subsequently, extensive studies came into being to reuse waste glass in construction materials as glass being the most versatile materials across the globe owing to the excellent characteristics that are being optically transparent, chemically inert, high intrinsic in strength and its permeability is quite low. Glass powder is a shallow and promising eco-complementary cement material. These are beneficial in geofilmers, forming eco-friendly cement, artificial lightweight aggregates, and composite phase change materials. It has also helped in sustainable development and betterment of industries by producing durable building materials (Fig. 1).

In the twenty-first century being engulfed with global warming and climate change era, the earth's surface temperature is expected to rise approximately 1.1–6.4 °C; until

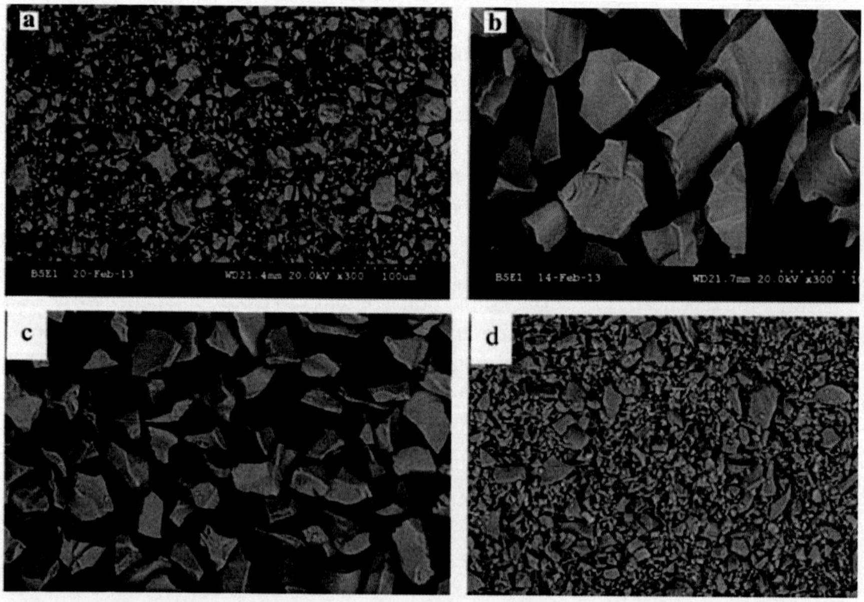

Fig. 1 SEM images representing **a** cement grain, **b** green glass (63–75 μm), **c** green glass (25–38 μm), and **d** green glass (0–25 μm) [7]

2100 [8]. And a 2 °C rise in the global temperature is noticeable because of the rise in total emissions caused as a result of the combustion of fossil fuels to produce and supply electricity to buildings [9].

Steered by the high energy demand owing to make use of electricity for both heating and cooling purposes, the sector of the building has an extreme dependence on fossil fuels by being its consumer and emits greenhouse gases extremely (GHGs) [10]. The magnification and the development in space cooling in the building have been expected to occur in developing economies due to the rise in the standard of living [11, 12].

Many other pre-eminent eco-friendly materials can be manufactured and their usage can be indulged more prominently in the constructional sector, such as AAC Blocks, Low VOC paints, Low E glass windows, and also PPC and Bamboos being discussed separately.

2 Objectives and Fact-Finding Views of the Contemporary Study

The contemporaneous study presents the gist of the feasibility of eco-friendly materials. This study flourishes the remarkable extent of the theories and work that has been handled and tackled in this field in different forms. The study was aimed at manifesting the current knowledge as a base and has been attributed by featuring its scope for further research in this field.

3 Coal Fly Ash

Coal fly ash is used as a raw material for the manufacture of two building materials, which are clay bricks that include silica-calcareous non-foamed bricks. Clay bricks are prepared with the help of commercial clay and varying waste ratios, taken as 0–50%. At 10 mpa and is being molded and fired at 1000 °C. For the preparation of Non-Fired clay bricks; basically, two materials are used in its formation, CFA and Geosilex where Geosilex is a lime residue that is obtained from the acetylene industry waste. Its mold is also prepared at 10 mpa and is cured at room temperature for 28 days. These results indicate the concretization of CFA (20 wt%) that produces fired clay bricks with the physical and mechanical properties similar to the control bricks without any wastage. Because of the increase in open porosity; the mechanical properties are reduced by 25–50%. Now, secondly in silica calcareous fired bricks, on a decrease in fly ash content; a decrease in bulk density and water absorption is also observed. These bricks contain 40–60 wt% of CFA had the highest compressive strength up to 46–43 mpa. Therefore, both kinds of bricks helped in optimal technological properties that attain quality standards.

4 Glass Powder

Then came the first study of utilizing and reusing waste glass as construction materials which were carried out by Schmidt and Saia in 1963. In construction industries, recycling of waste glass helped in reducing the burden on landfills and contributed to resource preservation and

carbon footprint reduction. Considering the case like Hong Kong; where out of total waste glass bottles, 10% is recycled and 90% is landfilled. These discarded glasses were converted into usable glass chips for utilizing it to construct the wall panels. Due to the limitations originated from the researches, waste glass was quickly converted into powder. This GP had angular shape and sharp edges with the particle size below 600 μm. These include 70% SiO_2, 13% Na_2O, 10% CaO, and smooth texture as compared to ordinary Portland cement grains. Therefore, being amorphous in structure, it served as a pozzolan in cement concrete.

5 Eco-Friendly Road Construction Material

The study adopts the art of research on the applications, demonstrations, and challenges of six environmentally friendly functional road materials, which are noise-reducing pavement materials, permeable asphalt concrete, low heat-absorbing pavement materials, de-icing pavement materials, energy harvesting pavement materials, and exhaust gas dissolved pavement materials. In relation to this study, we focus on providing a combination of the latest relevant works of literature on the development and design of environmentally friendly functional pavements and to promote innovation in materials science and pavement design principles. Porous noise reduction asphalt concrete is used to reduce pavement noise which is based on the noise reduction principle by pores. Hence, by considering and utilizing these adequate pavement materials prevention of noise from horizontal propagation can be obtained by the application of soundproof structures as well [13].

6 The Adoption of Eco-Block as a Green Construction Material

In the Mauritius construction industry, cellular blocks were complete concrete masonry units. These blocks are used in the construction of buildings due to its durability and structural strength. With the worldwide energy crisis and the constraint of carbon footprints, stakeholders are moving toward ecological materials to reduce energy consumption. Being a Paramount, it is a concrete block embedded with a polystyrene core which means that the Eco-block has been introduced since 2015 with its thermal and acoustic insulation.

7 Portland Pozzolanic Cement [PPC]

It is particularly a contrast to ordinary Portland cement (OPC) which consists of a mixture of pozzolanic materials known for their increased strength of concrete and reduced cement content. Pozzolanic materials include fly ash, rice bran ash, volcanic tuffs. After burning coal in thermal power plants, fly ash is obtained as a waste product.

8 AAC Blocks

Autoclave Aerated Concrete Blocks are an alternative to red bricks. These are manufactured from cement, sand, lime, water, and additives. The use of fly ash as a substitute instead of sand

being a fine aggregate is inculcated to produce AAC blocks. The characteristics defining these blocks are: they are easy to handle, less fuel in transportation, lightweight, decrease cost of construction by saving structural steel requirements for high RCC framed structures.

9 Bamboo

Bamboos are the best eco-friendly material on the planet. They possess a very high self-generation rate and grows up to 3 ft. in 24 h. These are perennial grasses that are replicated after harvesting and therefore have a higher strength to weight ratio, higher compressive strength, and exceptional durability than solid bricks therefore used for flooring and cabinet purpose. Since ancient times it was used as a building medium and has also gained popularity due to the lack of natural wood. They have five times the strength of concrete and are lighter than steel. Therefore, it has been proved to be good moisture resistance and scratch-resistant. They are available in various shades and is cheaper than hardwood. Also, it enrolls a drawback that requires treatment of insect resistance.

10 Low E Glass Windows

Low emissivity glass known as low E glass has an excellent insulation property. There has been a special coating being applied to glass by which only the visible light can pierce into the room and accords protection from UV and infrared rays.

11 Controlled Low-Strength Material

Modern hardening properties of controlled low-strength material (CLSM) which is a clay-based controlled low strength material have been investigated according to current studies. CLSM contains cementitious material like the stainless steel and this form of CLSM fundamentally is being used as a trench-fill material for construction purposes. Four levels of cement replacement (i.e., 0%, 10%, 20%, and 30%) are characterized by four levels and three quantities of binder content (80-, 100-, and 130 kg/m^3), Generated for investigation work. The modern properties of hardening of this recommended CLSM have been inspected in the laboratory. Therefore, the proposed CLSM can be employed for environmentally friendly trench fill based on the result being shown. In the present day, CLSM have been commonly used as a subgrade, granular compacting soil, and trench fills. As it successfully consumes a gigantic quantity of waste materials that is combustion fly ash, rubber, foundry sand so it can be used as an EFM. CLSM typically contains small amounts of Portland cement, mixing water, and large amounts of supplementary fine aggregates, CLSM has 8.3 MPa or less compressive strength and is higher than compacted soils [14]. Subsequently, recent studies have also described that the CLSM's maximum strength of up to about 1.4 MPa is suitable for almost all backfilling applications, with future re-excavation by simple equipment expected and desired. [14, 15].

12 Eco-Friendly Polymer Composites for Green Packaging

Green packaging is based on biodegradable composite materials that have a unique property and has gained attention in the current scenario in comparison with the classical photochemical-based plastics [16]. In the wide range of applications, its uses can be featured and it can be used in intelligent nano-food packaging [17], bio-membranes for wastewater, and composting purposes. Thus, the prominent function for package material is particularly to achieve the benefits of improved food quality and also ensure safety with prolonging and increased shelf life. Figure 2 represents the classification of green reinforcements.

13 Conclusion

These paper and examples explain that the conceptualization of "eco-materials" provide an interesting technological solution to enormous manufacturing products that prove to help in the establishment of sustainability. The expansion of the eco-friendly sustainable materials will help to address pollution due to the issues raised by the environment and participate in the different outcomes and solutions of social appropriateness. The progress in these eco-friendly systems will help to lessen the

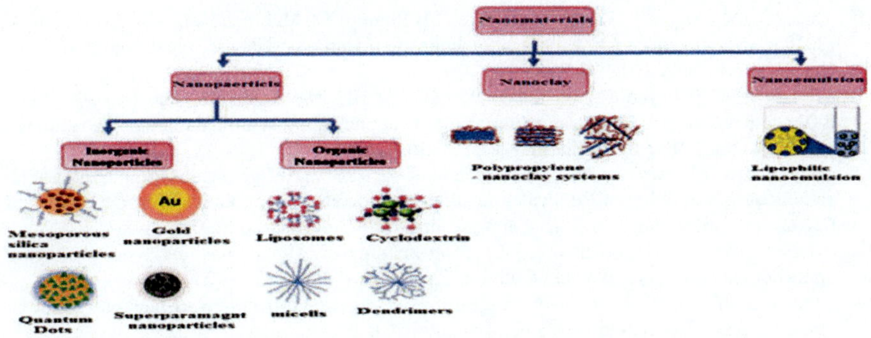

Fig. 2 Classification of nanomaterials based on their categories [16]

impacts of pollution efficiently. Also, the motive is to work on recyclable products that help in inducing sustainable requirements and needs. Hence, the above materials provide the prominent use of eco-friendly materials and thereby fabricate an advanced and empirical country.

References

1. Mirzahosseini M, Riding KA (2015) Influence of different particle sizes on reactivity of finely ground glass as supplementary cementitions material (SCM). Cement Concr Compos 56:95–105
2. IPCC, Pachauri RK (Ed), Climate Change 2007: Contribution of ... To the Fourth Assessment Report of the Intergovernmental Panel on Climate Change, IPCC, Geneva, 2008, p 4 Synthesis report: [a report of the Intergovernmental Panel on Climate Change]
3. Ong HC, Mahlia TMI, Masjuki HH (2011) A review on energy scenario and sustainable energy in Malaysia. Renew Sustain Energy Rev 15:639–647
4. Aditya L, Mahlia TMI, Rismanchi B, Ng HM, Hasan MH, Metselaar HSC, Muraza O, Aditiya HB (2017) A review on insulation materials for energy conservation in buildings. Renew Sustain Energy Rev 73:1352–1365
5. Huiothers SC (2000) Building energy efficiency standards in Hong Kong and mainland China. In: Proceedings of the 2000 ACEEE summer study on energy efficiency in buildings, 2000, pp 20–25
6. Choi K, Kim JH, Shin K (2004) Economic feasibility analysis of roadway capacity expansion with accounting traffic noise barrier cost. KSCE J Civil Eng 8:117–127
7. Mirzahosseini M, Riding KA (2015) Influence of different particle sizes on reactivity of finely ground glass as supplementary cementitious material (SCM). Cement Concr Compos 56:95–105
8. Lachemi M, Şahmaran M, Hossain KMA, Lotfy A, Shehata M (2010) Properties of controlled low-strength materials incorporating cement kiln dust and slag. Cement Concr Compos 32(8):623–629
9. Sheen YN, Zhang LH, Le DH (2013) Engineering properties of soil-based controlled low-strength materials as slag partially substitutes to Portland cement. Constr Build Mater 48:822–829

10. Youssef AM, Assem F, Essam M, Elaaser M, Ibrahim O, Mahmoud M, Abd El- Salam M (2019) Development of a novel bionanocomposite material and its use in packaging of Rams cheese. Food Chem 270:467–75
11. Youssef AM, El-Sayed SM, El-Sayed HS, Salama HH, Mohamed HF, Abd El- Salam MA (2018) Novel bionanocomposite materials used for packaging skimmed milk acid coagulated cheese (Karish). Int J Biol Micromoles 115(2018):1002–11
12. Youssef AM, El-Nahrawy AM, Abou Hammad AB (2017) Sol-gel synthesis and characterizations of hybrid chitosan-PEG/calcium silicate nanocomposite modified with ZnO-NPs and (E102) for optical and antibacterial applications. Int J Biol Macromol 97:561–567
13. Youssef AM, El- Sayed SM (2018) Bionanocomposites materials for food packaging applications: concepts and future Outlook. Carbohydr Polym 193:19–27
14. BumbudsanpharokeN Ko S (2015) Nano-food packaging: an overview of market, migration research, and safety regulations. J Food Sci 80:R910
15. Youssef AM (2013) Polymer nanocomposites as a new trend for packaging applications. Polym Plast Technol Eng 52:635–660
16. Anbazhagan M, Ramachandran S, Subramanian P, Nachimuthu R, Gothandam KM, Ranjan S, Dasgupta N, Lichtfouse E (2016) Nanoscience in food and agriculture "nanomaterials: classification, biological synthesis and characterization", 3rd ed. Springer, pp 31–71
17. Youssef AM, El- Sayed SM (2018) Bionanocomposites materials for food packaging applications: concepts and future Outlook. Carbohydr Polym 193:19–27

Chapter 12
Protection of Nine-Phase Transmission Line Using Biorthogonal-2.2 Wavelet Transform

Gaurav Kapoor

1 Introduction

An increase in the inevitability of electrical power has been perceived by the people of the modern generation. The electrical power transfer potentiality of the currently operating power transmission systems ought to be augmented in order to assist the significant increase in the necessity of electrical energy. In the literature, NPTL's have been suggested as an imminent replacement of the prevalent configuration of the electrical power transmission system which has the prospective for transferring the large extent of electrical energy.

The feasibility of fault occurrence on the NPTL is more as compared with the DCTL. Thus, accurate recognition of the faults in the NPTL turns out to be very decisive for mitigating the loss of gain and providing fast renovates.

Several newly reported research works addressed the issue related to fault recognition and categorization in TL's. Some important research attempts are presented in concise here in this section. Recently, fuzzy logic has been employed for micro-grid protection [1]. The TPTL has been protected using WT in [2]. In [3], ConvNet has been applied for fault recognition in micro-grid. In [4, 5], WT has been used for fault categorization in SPTL and SCCDCTL, respectively. Mathematical morphology has been applied for SPTL protection in [6]. ANN and phasor data have been used for islanding recognition in the smart grid [7]. HHT has been used for SPTL protection in [8]. In [9], POVMD and WPNRVFLN have been employed for fault recognition in SCDCTL. Mathematical morphology and data mining-based techniques have been used for high impedance recognition [10].

In this work, a novel tool, i.e., the biorthogonal-2.2 wavelet transform (BWT) is used for NPTL protection. No such type of work has been reported yet to the

G. Kapoor (✉)
Modi Institute of Technology, Kota, India
e-mail: gaurav.kapoor019@gmail.com

M. Shorif Uddin et al. (eds.), *Intelligent Energy Management Technologies*, Algorithms for Intelligent Systems,
https://doi.org/10.1007/978-981-15-8820-4_12

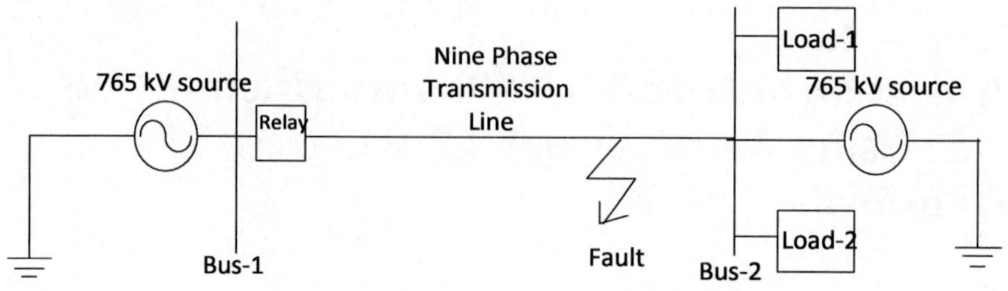

Fig. 1 The graphic of 765 kV nine-phase power system

best of the knowledge of the author. The results show that the BWT powerfully recognizes and categorizes the faults and the consistency of BWT is not perceptive to the variation in various fault factors.

This article is structured as: The specifications of NPTL are reported in Sect. 2. Section 3 shows the flow diagram of BWT. Section 4 is dedicated to the discussion of the response of BWT. Section 5 completes the paper.

2 The Specifications of NPTL

The system of a 765 kV NPTL is designed using MATLAB. Figure 1 shows the illustration of NPTL. The NPTL has a rating of 765 kV, 50 Hz, and has a total length of 200 km. The NPTL is separated into two zones of length 100 km both. Two loads of 300 MW and 150 MVAr each are connected at the receiving end of NPTL. The combination of relay and transducers is connected near bus-1 for the relaying of the total length of NPTL.

3 BWT-Based Protection Technique

Figure 2 shows the process of BWT with the following steps:

Step-1 Nine-phase currents are recorded through transducers installed at bus-1.
Step-2 BWT is employed to estimate the BWT outputs of phase currents.
Step-3 The phase will be declared as the faulty phase if its BWT output has a larger amplitude as compared to the output of the healthy phase under a faulty situation.

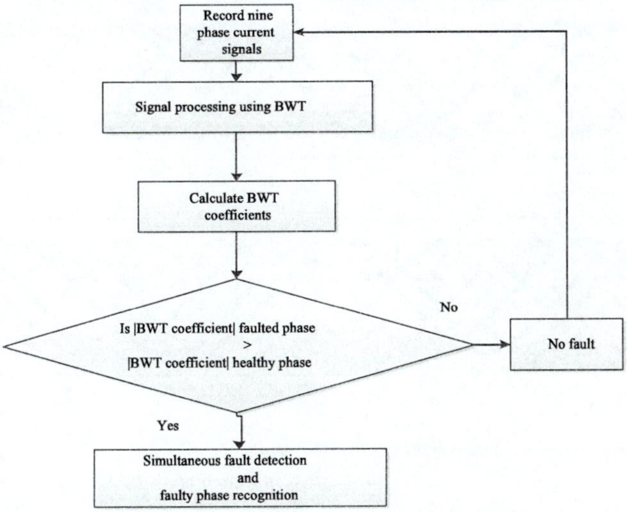

Fig. 2 Flow chart of BWT

4 Performance Assessment

To substantiate the ability of the BWT, the simulation effort has been carried out for numerous faults. The simulation outcomes of the study are examined in the successive subsections.

4.1 Response of BWT for Healthy Situation

Figure 3 shows the nine-phase currents and voltages for no-fault. Figure 4 exemplifies the output of BWT for no-fault. Table 1 reports the results of BWT for no-fault.

4.2 Response of BWT for Fault Switching Time Variation

The BWT is investigated for variation in fault switching time (FST). Figure 5 depicts the ABCEGHI-g fault at 100 km at 0.05 s among $R_F = 2.15\ \Omega$ and $R_G = 3.15\ \Omega$. Figure 6 shows the output of BWT for the ABCEGHI-g fault simulated at 0.05 s. The fault factors for all the fault cases are set as: T = 0.05 s, $F_L = 100$ km, $R_F = 2.15\ \Omega$, and $R_G = 3.15\ \Omega$. Tables 2, 3, 4, 5, and 6 tabularizes the results for variation in FST. It is inspected from Tables 2, 3, 4, 5, and 6 that the variation in FST does not manipulate the operation of BWT.

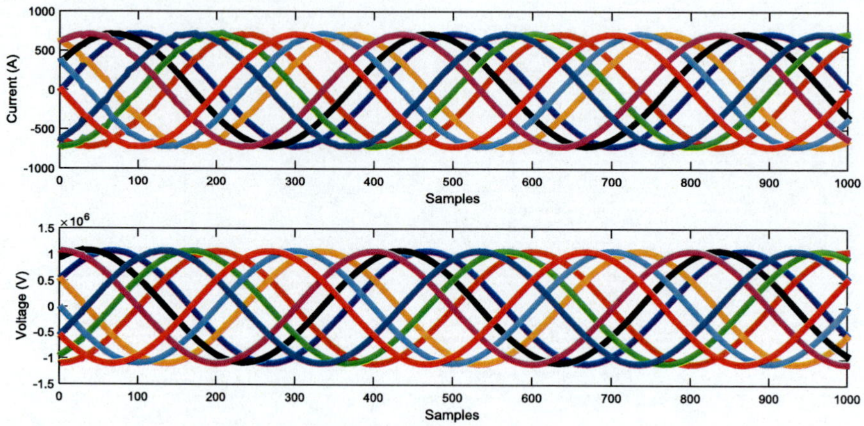

Fig. 3 Nine-phase currents and voltages for no-fault

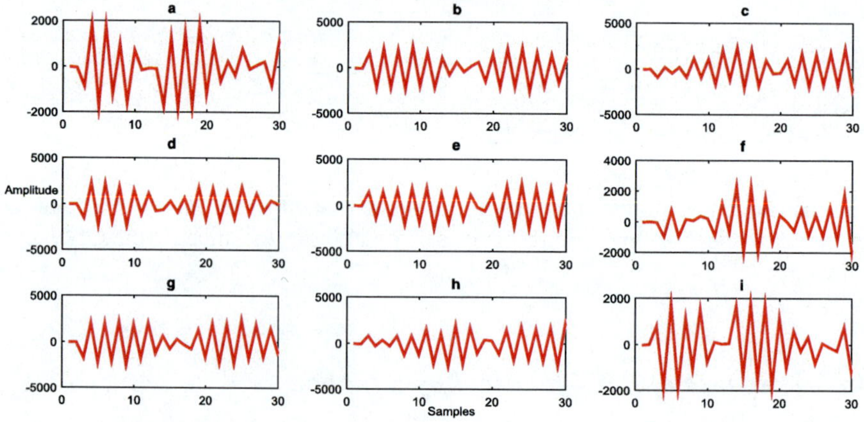

Fig. 4 BWT outputs for no-fault

Table 1 Response of BWT for no-fault

BWT outputs		
Phase-A	Phase-B	Phase-C
2.1264×10^3	2.4245×10^3	2.3967×10^3
Phase-D	Phase-E	Phase-F
2.3212×10^3	2.3690×10^3	2.3704×10^3
Phase-G	Phase-H	Phase-I
2.4514×10^3	2.1949×10^3	2.1032×10^3

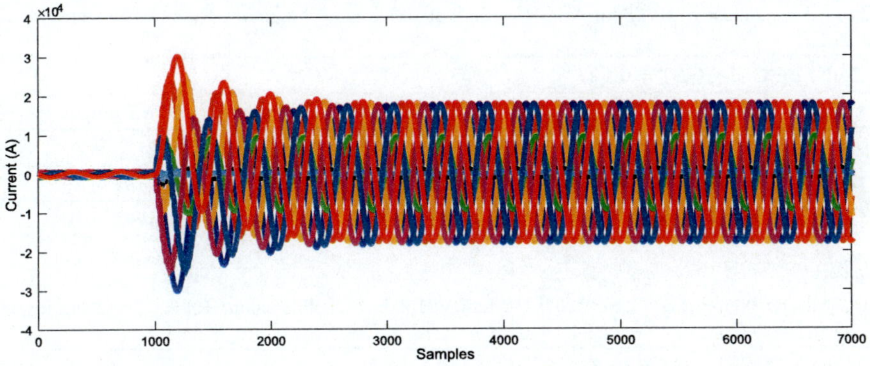

Fig. 5 ABCEGHI-g fault at 100 km at 0.05 s among $R_F = 2.15\ \Omega$ and $R_G = 3.15\ \Omega$

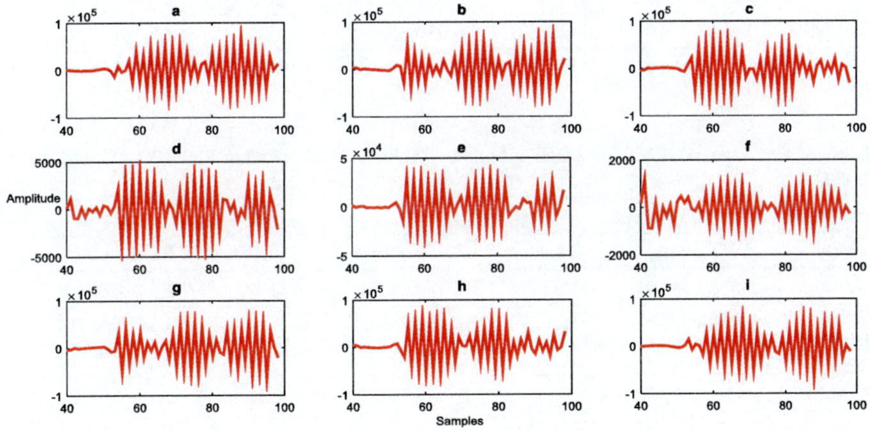

Fig. 6 BWT outputs for ABCEGHI-g fault at 100 km at 0.05 s among $R_F = 2.15\ \Omega$ and $R_G = 3.15\ \Omega$

Table 2 Response of BWT for ABCEGHI-g fault at 100 km at 0.05 s among $R_F = 2.15\ \Omega$ and $R_G = 3.15\ \Omega$

Fault—ABCEGHI-g (FST = 0.05 s)					
Phase	BWT output	Phase	BWT output	Phase	BWT output
A	6.7582×10^4	D	3.8703×10^3	G	5.4475×10^4
B	6.7227×10^4	E	3.0215×10^4	H	6.1760×10^4
C	5.7644×10^4	F	2.3550×10^3	I	5.8108×10^4

Table 3 Response of BWT for ABGHI-g fault at 100 km at 0.1 s among $R_F = 2.15\ \Omega$ and $R_G = 3.15\ \Omega$

Fault—ABGHI-g (FST = 0.1 s)					
Phase	BWT output	Phase	BWT output	Phase	BWT output
A	5.7034×10^4	D	2.1756×10^3	G	6.2186×10^4
B	5.4816×10^4	E	2.3311×10^3	H	5.4787×10^4
C	2.5569×10^3	F	2.4938×10^3	I	6.1941×10^4

Table 4 Response of BWT for BCDEF-g fault at 100 km at 0.17 s among $R_F = 2.15\ \Omega$ and $R_G = 3.15\ \Omega$

Fault—BCDEF-g (FST = 0.17 s)					
Phase	BWT output	Phase	BWT output	Phase	BWT output
A	2.8030×10^3	D	6.3673×10^4	G	2.3346×10^3
B	5.4067×10^4	E	6.0301×10^4	H	2.5046×10^3
C	5.2134×10^4	F	5.9313×10^4	I	2.3395×10^3

Table 5 Response of BWT for ABHI-g fault at 100 km at 0.08 s among $R_F = 2.15\ \Omega$ and $R_G = 3.15\ \Omega$

Fault—ABHI-g (FST = 0.08 s)					
Phase	BWT output	Phase	BWT output	Phase	BWT output
A	5.5787×10^4	D	2.3102×10^3	G	2.6451×10^3
B	4.9309×10^4	E	2.3134×10^3	H	5.3814×10^4
C	2.6365×10^3	F	2.4714×10^3	I	4.5725×10^4

Table 6 Response of BWT for DEF-g fault at 100 km at 0.2 s among $R_F = 2.15\ \Omega$ and $R_G = 3.15\ \Omega$

Fault—DEF-g (FST = 0.2 s)					
Phase	BWT output	Phase	BWT output	Phase	BWT output
A	2.3810×10^3	D	5.6832×10^4	G	2.3544×10^3
B	2.1445×10^3	E	5.9226×10^4	H	2.2560×10^3
C	2.1369×10^3	F	5.7041×10^4	I	2.4192×10^3

4.3 Response of BWT for Near-in Relay Faults

The efficiency of the BWT is tested for various near-in relay faults on the NPTL. Figure 7 depicts the DEFGHI-g near-in relay fault current at 5 km at 0.12 s among $R_F = 3.5\ \Omega$ and $R_G = 2.5\ \Omega$. Figure 8 shows the BWT coefficients for the DEFGHI-g fault simulated at 5 km. The fault factors for all the fault cases are: T = 0.12 s, $R_F = 3.5\ \Omega$ and $R_G = 2.5\ \Omega$. Tables 7, 8, 9, 10, and 11 details the results of the BWT for

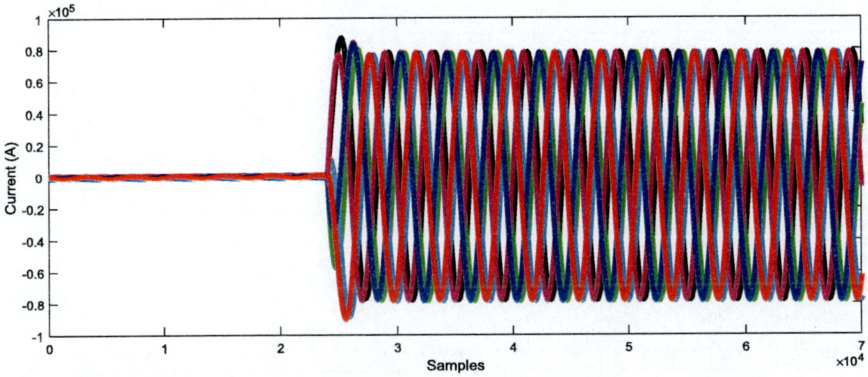

Fig. 7 DEFGHI-g near-in relay fault at 5 km at 0.12 s among $R_F = 3.5\ \Omega$ and $R_G = 2.5\ \Omega$

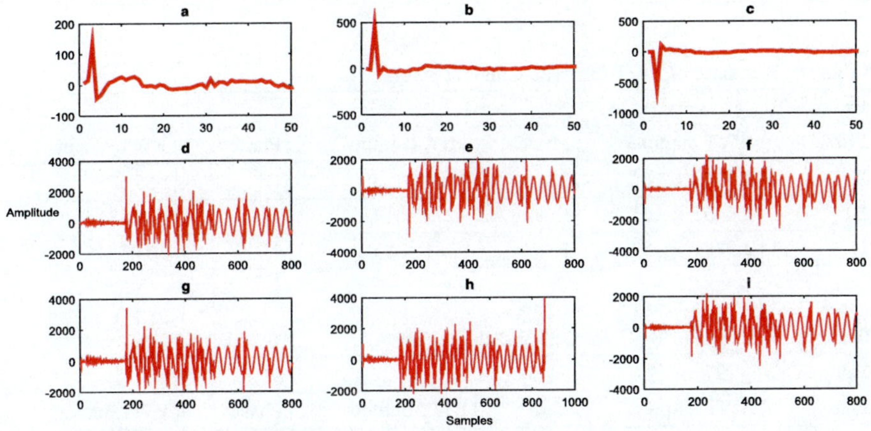

Fig. 8 BWT outputs for DEFGHI-g near-in fault at 5 km among $R_F = 3.5\ \Omega$ and $R_G = 2.5\ \Omega$

Table 7 Response of BWT for DEFGHI-g fault at 5 km

Fault—DEFGHI-g (5 km)					
Phase	BWT output	Phase	BWT output	Phase	BWT output
A	131.6663	D	2.6131×10^3	G	3.0297×10^3
B	456.0025	E	6.5031×10^3	H	3.9445×10^3
C	108.0168	F	1.8497×10^3	I	3.3792×10^3

five different near-in relay faults. It is confirmed from Tables 7, 8, 9, 10, and 11 that the BWT has the ability to detect the near-in relay faults precisely.

Table 8 Response of BWT for ABCEF-g fault at 6 km

Fault—ABCEF-g (6 km)					
Phase	BWT output	Phase	BWT output	Phase	BWT output
A	3.0852×10^3	D	818.7633	G	26.7045
B	9.0172×10^3	E	6.2284×10^3	H	194.3373
C	2.6214×10^3	F	2.5241×10^3	I	93.5313

Table 9 Response of BWT for GHI-g fault at 7 km

Fault—GHI-g (7 km)					
Phase	BWT output	Phase	BWT output	Phase	BWT output
A	86.5409	D	30.4280	G	5.1182×10^3
B	299.8279	E	394.6009	H	5.2485×10^3
C	89.1173	F	87.1063	I	3.5345×10^3

Table 10 Response of BWT for ABC-g fault at 8 km

Fault—ABC-g (8 km)					
Phase	BWT output	Phase	BWT output	Phase	BWT output
A	2.9549×10^3	D	61.3253	G	65.5750
B	1.0461×10^3	E	272.2014	H	249.4814
C	2.9781×10^3	F	81.4085	I	104.3835

Table 11 Response of BWT for DEF-g fault at 9 km

Fault—DEF-g (9 km)					
Phase	BWT output	Phase	BWT output	Phase	BWT output
A	106.9428	D	5.0679×10^3	G	94.3637
B	312.8468	E	8.8515×10^3	H	389.2334
C	173.8100	F	3.4420×10^3	I	103.6978

4.4 Response of BWT for Far-End Relay Faults

The BWT has been explored for different far-end relay faults. Figure 9 illustrates the ABCDEFG-g far-end relay fault at 195 km at 0.06 s among $R_F = 4.5\ \Omega$ and $R_G = 3.5\ \Omega$. Figure 10 shows the BWT coefficients for the ABCDEFG-g fault simulated at 195 km. The fault factors chosen for all the fault cases are: T = 0.06 s, $R_F = 4.5\ \Omega$ and $R_G = 3.5\ \Omega$. Tables 12, 13, 14, 15, and 16 report the results for various far-end relay faults. It is inspected from Tables 12, 13, 14, 15, and 16 that the effectiveness of BWT remains impassive for different far-end relay faults.

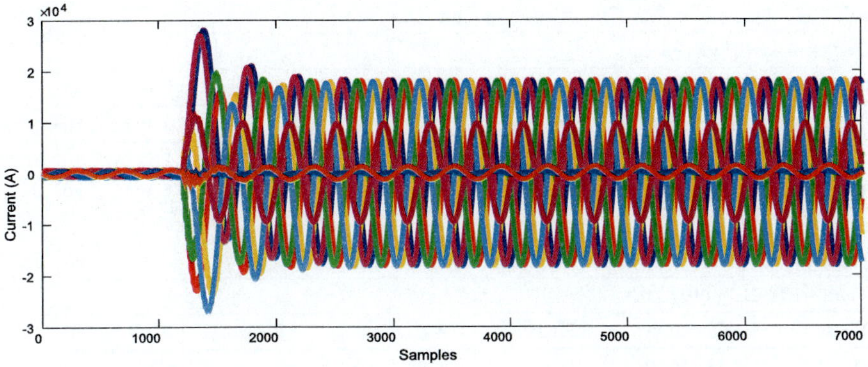

Fig. 9 ABCDEFG-g fault at 195 km at 0.06 s among $R_F = 4.5\ \Omega$ and $R_G = 3.5\ \Omega$

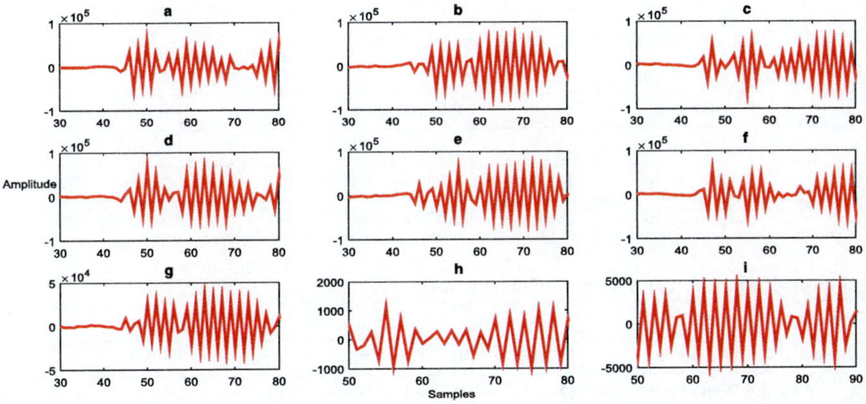

Fig. 10 BWT outputs for ABCDEFG-g fault at 195 km at 0.06 s among $R_F = 4.5\ \Omega$ and $R_G = 3.5\ \Omega$

Table 12 Response of BWT for ABCDEFG-g fault at 195 km

Fault—ABCDEFG-g (195 km)					
Phase	BWT output	Phase	BWT output	Phase	BWT output
A	6.1249×10^4	D	6.6442×10^4	G	3.4487×10^4
B	5.9615×10^4	E	6.1556×10^4	H	2.1853×10^3
C	5.3307×10^4	F	5.7471×10^4	I	4.2411×10^3

4.5 Response of BWT for Variation in Fault Resistance

The BWT is investigated for variation in fault resistances. Figure 11 depicts the ABCDEFHI-g fault at 100 km at 0.07 s among $R_F = 10\ \Omega$ and $R_G = 0.001\ \Omega$.

Table 13 Response of BWT for ABDEF-g fault at 196 km

Fault—ABDEF-g (196 km)					
Phase	BWT output	Phase	BWT output	Phase	BWT output
A	5.3989×10^4	D	5.7792×10^4	G	2.4971×10^3
B	4.7741×10^4	E	5.9487×10^4	H	2.3224×10^3
C	2.8607×10^3	F	5.9446×10^4	I	2.4280×10^3

Table 14 Response of BWT for EFGHI-g fault at 197 km

Fault—EFGHI-g (197 km)					
Phase	BWT output	Phase	BWT output	Phase	BWT output
A	2.3752×10^3	D	2.7792×10^3	G	6.1323×10^4
B	2.3674×10^3	E	5.0953×10^4	H	5.4694×10^4
C	2.1085×10^3	F	4.8050×10^4	I	5.5860×10^4

Table 15 Response of BWT for ABDE-g fault at 198 km

Fault—ABDE-g (198 km)					
Phase	BWT output	Phase	BWT output	Phase	BWT output
A	5.6799×10^4	D	5.5657×10^4	G	2.2046×10^3
B	5.1894×10^4	E	4.9065×10^4	H	2.4423×10^3
C	2.5343×10^3	F	2.5271×10^3	I	2.3698×10^3

Table 16 Response of BWT for ABC-g fault at 199 km

Fault—ABC-g (199 km)					
Phase	BWT output	Phase	BWT output	Phase	BWT output
A	5.9040×10^4	D	2.1843×10^3	G	2.1728×10^3
B	5.5764×10^4	E	2.3957×10^3	H	2.2799×10^3
C	5.1070×10^4	F	2.2733×10^3	I	2.1238×10^3

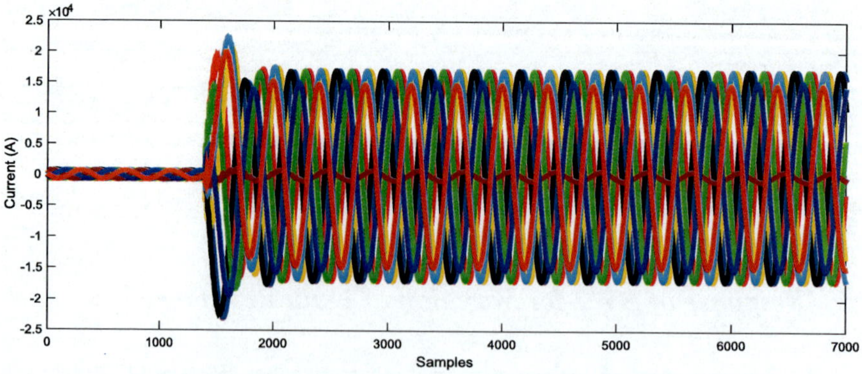

Fig. 11 ABCDEFHI-g fault at 0.07 s at 100 km among $R_F = 10\ \Omega$ and $R_G = 0.001\ \Omega$

Figure 12 shows the BWT coefficients for the ABCDEFHI-g fault simulated among $R_F = 10\ \Omega$. The fault factors for all the fault cases are set as T = 0.07 s, $F_L = 100$ km, and $R_G = 0.001\ \Omega$. Tables 17, 18, 19, 20, and 21 tabularizes the results for variation in fault resistances. It is inspected from Tables 17, 18, 19, 20, and 21 that variation in the fault resistances does not manipulate the working of the BWT.

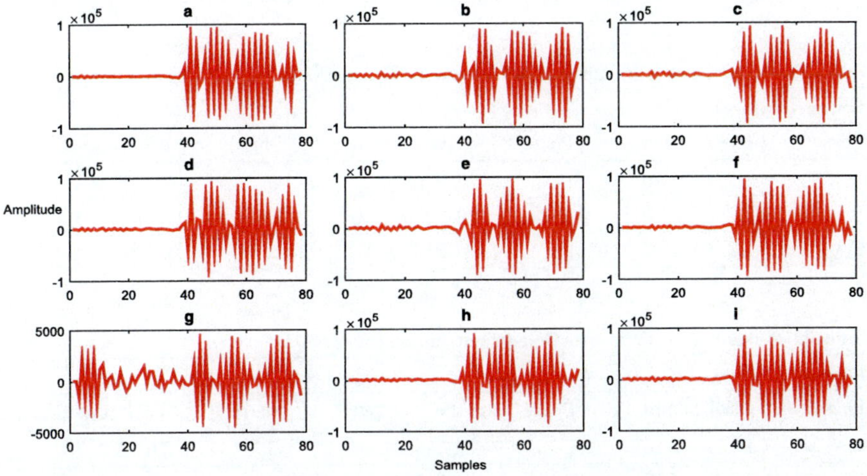

Fig. 12 BWT outputs for ABCDEFHI-g fault at 0.07 s at 100 km among $R_F = 10\ \Omega$ and $R_G = 0.001\ \Omega$

Table 17 Response of BWT for ABCDEFHI-g fault among $R_F = 10\ \Omega$

Fault—ABCDEFHI-g (10 Ω)					
Phase	BWT output	Phase	BWT output	Phase	BWT output
A	6.8829×10^4	D	6.5690×10^4	G	3.2895×10^3
B	6.5526×10^4	E	6.8908×10^4	H	6.4465×10^4
C	6.7075×10^4	F	6.7971×10^4	I	5.9149×10^4

Table 18 Response of BWT for BCDEF-g fault among $R_F = 40\ \Omega$

Fault—BCDEF-g (40 Ω)					
Phase	BWT output	Phase	BWT output	Phase	BWT output
A	2.7621×10^3	D	3.8084×10^4	G	2.2360×10^3
B	3.2730×10^4	E	3.6160×10^4	H	2.3975×10^3
C	3.3935×10^4	F	3.9879×10^4	I	2.3335×10^3

Table 19 Response of BWT for ACGH-g fault among $R_F = 70\ \Omega$

Fault—ACGH-g (70 Ω)					
Phase	BWT output	Phase	BWT output	Phase	BWT output
A	2.6259×10^4	D	2.4778×10^3	G	2.4334×10^4
B	3.0921×10^3	E	2.5364×10^3	H	2.8309×10^4
C	2.5700×10^4	F	2.2849×10^3	I	2.6976×10^3

Table 20 Response of BWT for DEF-g fault among $R_F = 100\ \Omega$

Fault—DEF-g (100 Ω)					
Phase	BWT output	Phase	BWT output	Phase	BWT output
A	2.2718×10^3	D	1.9765×10^4	G	2.3222×10^3
B	2.2663×10^3	E	1.8950×10^4	H	2.3157×10^3
C	2.3288×10^3	F	1.9448×10^4	I	2.1198×10^3

Table 21 Response of BWT for AGHI-g fault among $R_F = 150\ \Omega$

Fault—AGHI-g (150 Ω)					
Phase	BWT output	Phase	BWT output	Phase	BWT output
A	1.7154×10^4	D	2.4446×10^3	G	2.0733×10^4
B	2.2694×10^3	E	2.2100×10^3	H	1.9942×10^4
C	3.1200×10^3	F	2.4650×10^3	I	2.1785×10^4

5 Conclusion

This work showed an improved revelation using BWT to recognize and categorize the faults occurring on NPTL. The BWT coefficients of the nine-phase currents of the NPTL which are measured at one-end are employed by BWT for fault recognition, categorization, and faulty phase identification. The value of fault resistance is varied from 10 to 150 Ω, the position of fault for the near-in relay faults is varied from 5 to 9 km, and the position of fault for the far-end relay faults is varied from 195 to 199 km. The simulation studies support the consistency of BWT under extensive variations in fault type, location, resistance, and switching time. From the results, it is noticeable that there is an evident intolerance between the fault and no-fault situations and establishes the potential of the BWT-based fault recognition and faulty phase categorization technique by recognizing the faults correctly.

References

1. Chaitanya BK, Soni AK, Yadav A (2018) Communication assisted fuzzy based adaptive protective relaying scheme for microgrid. J Power Technol 98(1):57–69
2. Kapoor G (2018) Wavelet transform based detection and classification of multi-location three phase to ground faults in twelve phase transmission line. Majlesi J Mechatron Systems 7(4):47–60
3. Manohar M, Koley E, Ghosh S (2019) Enhancing resilience of PV-fed microgrid by improved relaying and differentiating between inverter faults and distribution line faults. Electric Power Energy Syst (Elsevier) 108:271–279
4. Kapoor G (2018) Six phase transmission line boundary protection using wavelet transform. In: Proceedings of the 8th IEEE India international conference on power electronics (IICPE). IEEE. Jaipur, India (2018)
5. Gautam N, Ali S, Kapoor G (2018) Detection of fault in series capacitor compensated double circuit transmission line using wavelet transform. In: Proceedings of the IEEE international conference on computing, power and communication technologies (GUCON), pp 769–773. IEEE. Greater Noida, India (2018)
6. Kapoor G (2018) Six phase transmission line boundary protection using mathematical morphology. In: Proceedings of the IEEE international conference on computing, power and communication technologies (GUCON), pp 857–861. IEEE. Greater Noida, India (2018)
7. Kumar D, Bhowmik PS (2018) Artificial neural network and phasor data-based islanding detection in smart grid. IET Gener Transm Distrib 12(21):5843–5850
8. Kapoor G (2019) Detection and classification of single line to ground boundary faults in a 138 kV six phase transmission line using Hilbert Huang transform. i-manager's J Electric Eng 12(3):28–41 (2019)
9. Sahani M, Dash P K (2019) Fault location estimation for series-compensated double-circuit transmission line using parameter optimized variational mode decomposition and weighted P-norm random vector functional link network. Appl Soft Comput J (Elsevier), 1–18 (2019)
10. Sekar K, Mohanty NK (2018) Data mining-based high impedance fault detection using mathematical morphology. Comput Electric Eng (Elsevier) 69:129–141

Chapter 13
An Intelligent Hybrid Model for Forecasting of Heart and Diabetes Diseases with SMO and ANN

Shalini, Pawan Kumar Saini, and Yatendra Mohan Sharma

1 Introduction

Nowadays, among several human life killer diseases, heart and diabetes are the two most widely recognized causes. Related literature of medical science illustrates that only forecasting such diseases at the initial period is only a single hope for saving a life. However, verification of diseases with its causes is an effort of the physician but a huge amount of common and related symptoms make the decision step harder. Additionally, incorrect or deferred forecast information may cause of threat or more harm to a patient. However, to boost disease recognition process and treat patients untimely huge medical experts consume different practices but due to an intricate system of the heart and huge common symptoms of diabetic disease active schemes have its fashionable restraint. Therefore, the field is still open for modernization. To improve disease recognition quality with elevated accuracy a naïve hybrid forecasting model based on neural network and SMO theorem is employed in this paper. The experimental fallouts have shown the importance of the proposed method in the classification of heart and diabetic diseases.

Shalini
Arya Institute of Engineering Technology & Management, Jaipur, Rajasthan, India

P. K. Saini
Modern Institute of Technology & Research Centre, Alwar, Rajasthan, India

Y. M. Sharma (✉)
Biff and Bright College of Engineering & Technology, Dudu, Jaipur, Rajasthan, India
e-mail: er.shalini.pathak7@gmail.com; alka.choubey2@gmail.com

M. Shorif Uddin et al. (eds.), *Intelligent Energy Management Technologies*, Algorithms for Intelligent Systems,
https://doi.org/10.1007/978-981-15-8820-4_13

2 Related Efforts

Among several obtainable techniques, a good amount of investigators has demonstrated the rewards of ANN and SMO theorem in the skillful classification of heart and diabetes diseases, some modern study has discussed in [1–7]. To improve the forecasting accuracy of heart and diabetes diseases a naïve hybrid model that consists of combine features of ANN and FNN in sole form has illustrated in [8]. For demonstrating the effectiveness of the intended approach, the proposed methods have evaluated with two benchmarked dataset, Pima Indians diabetes and Heart disease dataset. The model exploits k-fold cross-validation scheme to crack the selected dataset into a k part and each time one part is consumed for training and rest remains has utilized for testing purposes. The proportional fallouts demonstrate that the presented approach attain high accuracy level when match up to the records at UCI and related studies for heart disease. With the integration of NN and GA (genetic algorithm) into a single model, a naïve approach has made an endeavor to improve forecasting accuracy of heart diseases [9]. GA-based procedure utilized for determined an appropriate preliminary weight of neural network. To show the effectiveness of presented method, the investigators have evaluated an act of proposed scheme with five different benchmarked dataset with utilizing the functionality of 10-fold cross-validation scheme. Experimental upshots depicted the efficiency of presented scheme. With the comparative estimation of SVM, DT, NB, KNN, and ANN methods for forecasting heart diseases a naïve investigation demonstrates that NB and SVM are the most suitable methods for recognizing heart diseases in a speedy form [10]. A naïve model with MLPNN (Multilayer Perceptron Neural Network) and BP (Back Propagation) has utilized to improve the attained accuracy level of heart diseases [11]. To demonstrate the uses of presented approach it has evaluated with the 303 patients' records and results prove the efficiency of published approach.

Additionally, the above-discussed methods, a number of algorithms, have been presented by different investigators [12–16] for forecasting of heart and diabetes diseases but elaboration of each and every approach is not possible to comprise in this paper. On the other hand, much of published work has demonstrated that ANN-based approaches are significantly well performed and can be utilized for prediction and classification purposes but in front of ensemble classification-based approaches the simple ANN-based algorithm has shown stumpy performance, hence there is appalling need of optimization.

3 Proposed Approach

Motivated from related literature that in place of any sole process, the hybrid methods are most suitable for forecasting of diseases and utilization of only the most relative attributes from an entire attribute list of opted dataset is a worthy decision that significantly improves an act of build procedures, the anticipated approach mix the muscle of feed-forward neural network MLP (Multilayer Perceptron) and SMO into a single form with the exploitation of correlation-based feature selection and breadth-first search mechanism to discover and integrate only most relative features of opted medical dataset for the forecasting process of diseases. Typically, MLP is a form of ANN in which build network is connected with single or added hidden layers from input to output layer. At initial phase, proposed model preprocessed the opted dataset, remove incomplete data entries, and discover the most optimal attributes set for forecasting process of diseases. Subsequent to preprocessing process, the build model breaks the dataset into training and testing phase and starts the disease forecasting process with SMO approach. The approach trains the build model with a part of opted dataset and performs a testing phase with untouched part, utilizes 10-fold cross-validation process in which build model utilizes one part for training and other exploit for the testing phase. Once the SMO approach completes its task the build mechanism delivers the misclassified data to MLP method that significantly improves the forecasting accuracy of diseases. Additionally, the MLP technique uses 3 hidden layer to enhance an amount of accurate forecasting, following Fig. 1 shows the structure of MLP method with selected optimal features set for forecasting of diabetes, among total 9 features the proposed approach only utilizes 4 features as shown in Fig. 1.

Where Ds stand for opted dataset of diseases, collected from benchmarked collection library UCI. PDs are preprocessed data set, after removing incomplete data and selection of an optimal feature set (OFs). TRDs and TDs denote the training and testing part of preprocessed dataset. MD is misclassified data (Fig. 2).

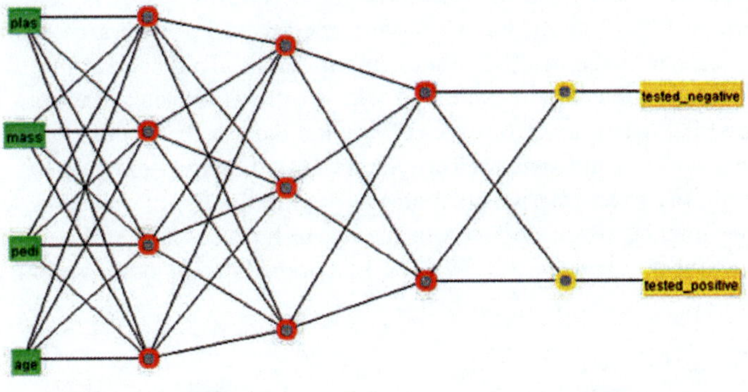

Fig. 1 Utilized multilayer perceptron model for forecasting of diabetes diseases

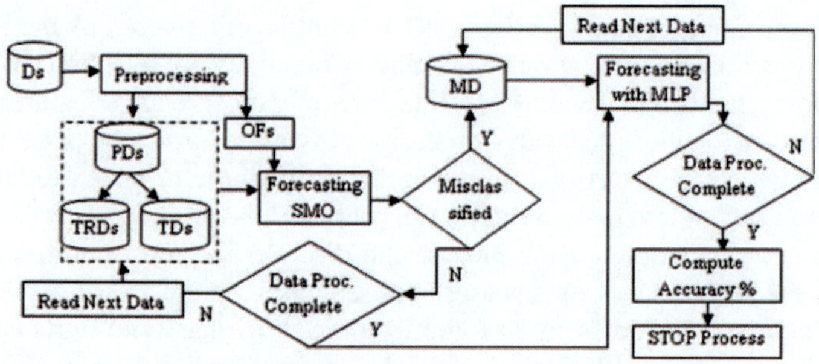

Fig. 2 Proposed methodology for diseases forecasting process

4 Result Analysis

To analyze an act of the build method two different medical dataset heart and diabetes taken into account have been collected from well-accepted library UCI. Fair evaluation simulation has been done with an identical Pima Indian diabetes data set that has been considered in [2, 7], enclose 768 diabetes patients record with total of 9 features. However, in place of 9 features, the build approach utilizes only 4 features for the forecasting process of diabetes. The following figure illustrates the significance and acceptability of the build method (Fig. 3).

The presented values in the above figure illustrate that proposed approach has attained high accuracy in comparison to other methods, presented in [2, 7]. However, difference between DIM and the presented method of this paper is not huge but in medical area a little amount has its significance therefore the proposed approach is most suitable in place of other obtainable methods. To discover an efficiency of proposed method it has again evaluated with a different disease dataset that has to consider by [4, 17–19], heart diseases Cleveland dataset that holds down reports of 303 patients with 14 attributes. However, as alike previous experiment, the build model opted only most relative attribute set among the list of attributes, elect only 7. The experimental result depicted in the figure has shown that the build method has attained the finest accuracy 97.02% in comparison to other recent proposed

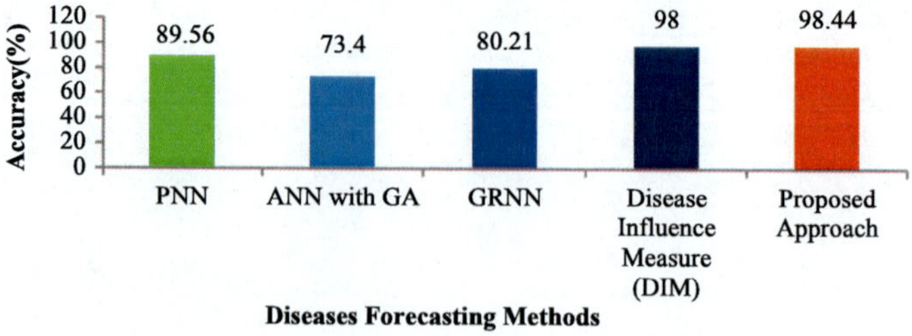

Fig. 3 Comparative Analysis over diabetes diseases dataset

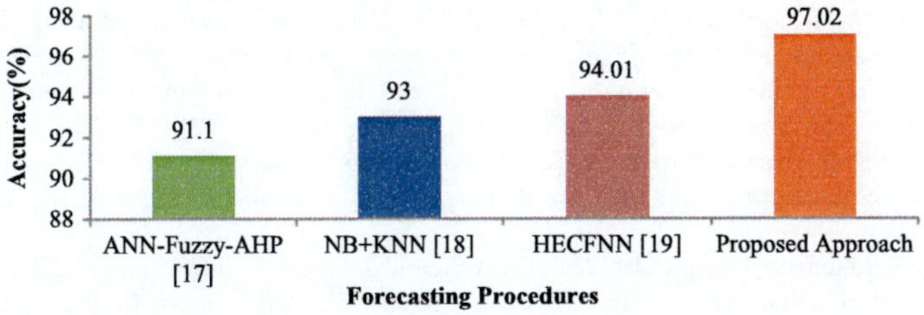

Fig. 4 Comparative analysis over heart diseases dataset

approaches. The both of above comparative fallouts of the build method demonstrate its efficiency over the other obtainable methods.

As the experimental result depicted in the Fig. 4, the build method has attained finest accuracy 97.02% in comparison to other recent proposed approaches. The both of the above comparative fallouts of the build method demonstrate its efficiency over the other obtainable methods.

5 Conclusion

Over past decades, heart and diabetes diseases are the two main causes of early mortality of human life. However, many investigators exploit different methodologies to aid medical professionals in the direction to save life but each and every obtainable method struggle due to the intricacy structure of the heart and human body. Recently machine learning-based method attained huge fame in medical field to aid related field professionals with quick responses to make life-saving decisions. This paper intends a naïve hybrid forecasting procedure of heart and diabetes diseases on the base of SMO and ANN methods. The proposed approach significantly improves the forecasting accuracy of diabetes and heart diseases with attaining 98.44% and 97.02%. Experimental evaluation demonstrates that the proposed approach has attained high accuracy in comparison to other modern algorithms and is more suitable for forecasting procedure. Future work will examine an act of the build method with other dataset and with inclusion of different feature selection procedures to enhance an accuracy percentage.

References

1. Rahiml P, Jafarian A (2016) Prediction of diabetes by using artificial neural network, logistic regression statistical model and combination of them. Bulletin de la Société Royale des Sciences de Liège 85:1148–1164
2. Sujarani P, Kalaiselvi K (2018) Prediction of diabetes using artificial neural networks: a review. J Adv Res Dynam Control Syst 10(4):67–72

3. Dbritto R, Srinivasaraghavan A, Joseph V (2016) Comparative analysis of accuracy on heart disease prediction using classification methods. Int J Appl Inform Syst 2249–0868 (2016)
4. Nagamani T, Logeswari S, Gomathy B (2019) Heart disease prediction using data mining with mapreduce algorithm. Int J Innov Technol Explor Eng (IJITEE) 8(3), 137–140
5. Srivastava S, Sharma L, Sharma V, Kumar A, Darbari H (2019) Prediction of diabetes using artificial neural network approach. In: ICoEVCI. Lecture Notes in Electrical Engineering, pp 679–687
6. Sharma P, Saxena S, Sharma YM (2018) An efficient decision support model based on ensemble framework of data mining features assortment & classification process. In: Proceedings of the international conference on communication and electronics systems (ICCES 2018). IEEE, pp 487–491
7. Baiju BV, John Aravindhar D (2019) Disease influence measure based diabetic prediction with medical data set using data mining. In: 1st International conference on innovations in information and communication technology (ICIICT). IEEE, pp 1–6
8. Kahramanli H, Allahverdi N (2008) Design of a hybrid system for the diabetes and heart diseases. Expert Syst Appl, Elsevier 35:82–89
9. Haq AU, Li JP, Memon MH, Nazir S, Sun R (2018) A hybrid intelligent system framework for the prediction of heart disease using machine learning algorithms. Hindawi, Mobile Information Systems, Article ID 3860146, pp 1–22
10. Beyene C, Kamat P (2018) Survey on prediction and analysis the occurrence of heart disease using data mining techniques. Int J Pure Appl Math 118:165–174
11. Singh P, Singh S, Pandi-Jain GS (2018) Effective heart disease prediction system using data mining techniques. Int J Nanomed:121–124
12. Liu X, Wang X, Su Q, Zhang M, Zhu Y, Wang Q, Wang Q (2017) A hybrid classification system for heart disease diagnosis based on the RFRS method. Hindawi Comput Math Method Med:1–11
13. Priscila SS, Hemalatha M (2017) Diagnosis of heart disease with particle bee-neural network. Special section: computational life sciences and smarter technological advancement. Biomed Res India, pp 1–7(2017)
14. Tayefia M, Tajfard M, Saffar S, Hanachi P, Amirabadizadeh AR, Esmaeily H, Taghipour A, Ferns GA, Moohebati M, Ghayour-Mobarhan M (2017) hs-CRP is strongly associated with coronary heart disease (CHD): a data mining approach using decision tree algorithm. Comput Methods Program Biomed, ELSEVIER, 141(2017):105–109
15. Wijaya SH, Pamungkas GT, Sulthan MB (2018) Improving classifier performance using particle swarm optimization on heart disease detection. In: IEEE international seminar on application for technology of information and communication (iSemantic), pp 603–608
16. Poornima V, Gladis D (2018) A novel approach for diagnosing heart disease with hybrid classifier. Biomed Res 29(11): 2274–2280
17. Samuel OW et al (2017) An integrated decision support system based on ANN and Fuzzy_AHP for heart failure risk prediction. Expert Syst Appl 68:163–172
18. Malav A, Kadam K, Kamat P (2017) Prediction Of heart disease using k-means and artificial neural network as hybrid approach to improve accuracy. Int J Eng Technol 9(4):3081–3085
19. Alkhasawneh MSh (2019) Hybrid cascade forward neural network with elman neural network for disease prediction. Arab J Sci Eng, Springer:1–12

Chapter 14
Power Quality Assessment of Solar PV Standalone System Using Various DC-DC Converters

Surbhi Shringi, Santosh Kumar Sharma, and Utkarsh Gupta

1 Introduction

Today the increasing energy demands with urbanization, industrialization, and increasing population have led to the need to find substitutes for conventional energy sources. Many renewable sources have lead to replacing conventional sources with time [1, 2].

The most used types of renewable power sources include solar and wind. Both of these sources use various types of power electronics devices in order to get synchronized with the grid [3].

The main issues that are created with the power electronics devices are the ripples and the Total Harmonic distortion which are introduced in the power supply. Hence in order to reduce these above-discussed problems, THD analysis of various DC-DC converters is done in order to find the suitable and best converter amongst the three [4, 5].

S. Shringi (✉) · S. K. Sharma · U. Gupta
Rajasthan Technical University, Kota, Rajasthan, India
e-mail: surbhishringi96@gmail.com

S. K. Sharma
e-mail: santoshelect@yahoo.com

U. Gupta
e-mail: utkarshgupta3095@gmail.com

M. Shorif Uddin et al. (eds.), *Intelligent Energy Management Technologies*, Algorithms for Intelligent Systems,
https://doi.org/10.1007/978-981-15-8820-4_14

2 System Description with Simulation Models and Results

The system consists of a PV array which acts as a power source keeping the irradiation and the temperature to be fixed which is further given to the DC-DC converter followed by the P&O MPPT technique to extract maximum power from the PV array this power is collected and measured at the DC link as it is considered as one of the parameters for power quality assessment. This DC voltage is converted into AC using the Voltage source converter and then this is connected with the Power grid or with the Load (Fig. 1).

2.1 System with Boost Converter

The simulation model of the system with Boost converter is shown in Fig. 2.

Figure 3 shows the output voltage at DC link of the Boost converter.

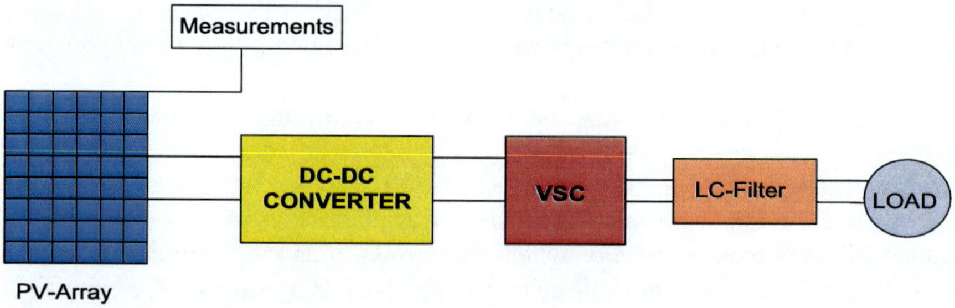

Fig. 1 Block diagram of the system

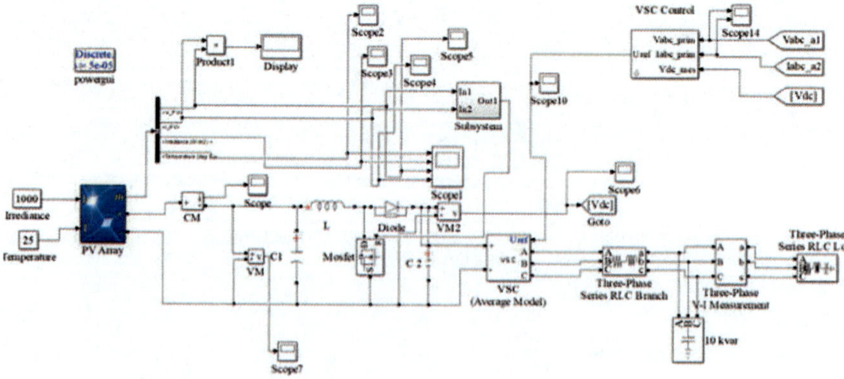

Fig. 2 Simulation diagram of the system using Boost converter

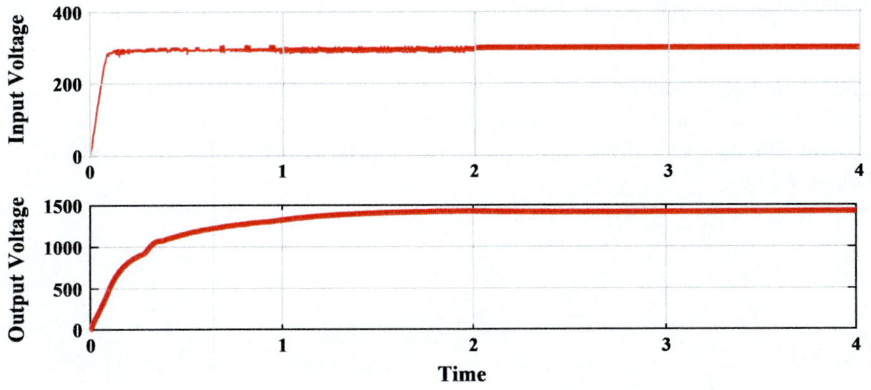

Fig. 3 Output Voltage of the DC link with Boost converter

The THD analysis of the VSC output waveforms is done whose results are shown below in Fig. 4.

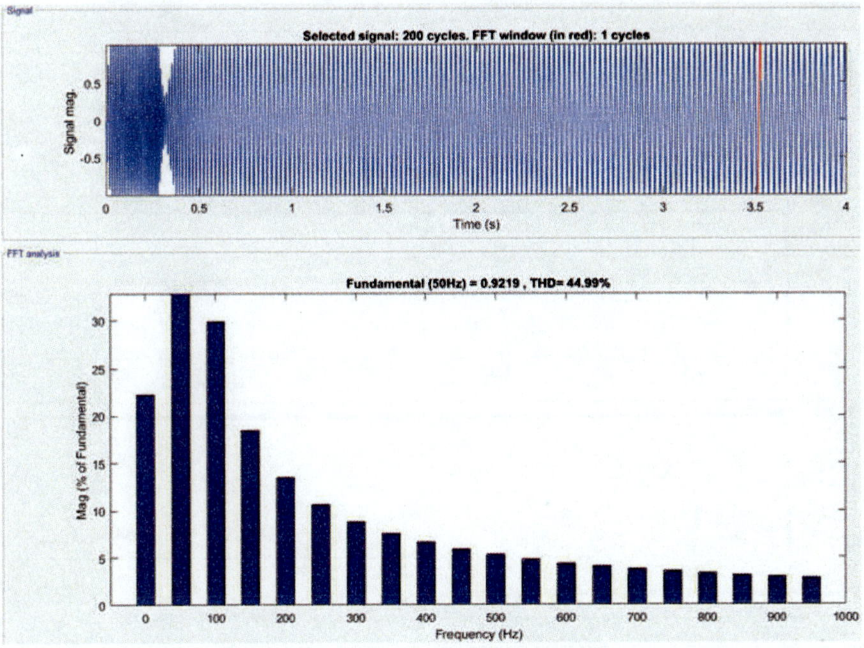

Fig. 4 THD analysis of VSC waveforms with Boost converter

2.2 System with Buck Converter

The simulation model of the system using Buck converter is shown in Fig. 5.

Figure 6 shows the output voltage at the DC link with the Buck converter.

The THD analysis of the VSC output waveforms is done whose results are shown below in Fig. 7.

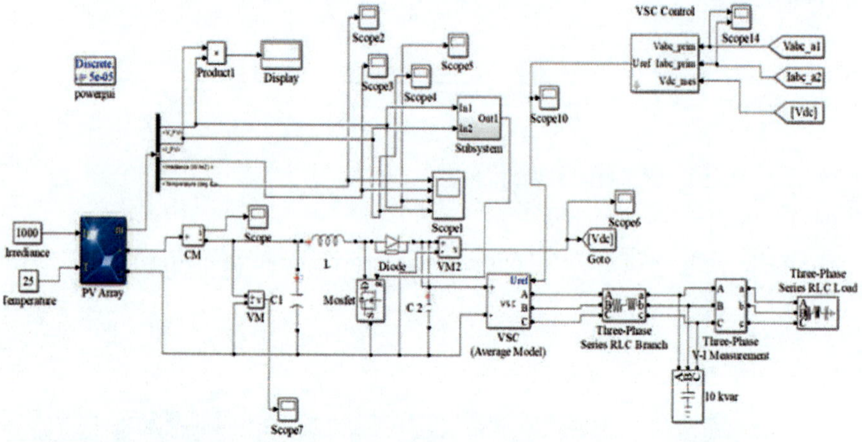

Fig. 5 Simulation model of the system with Buck converter

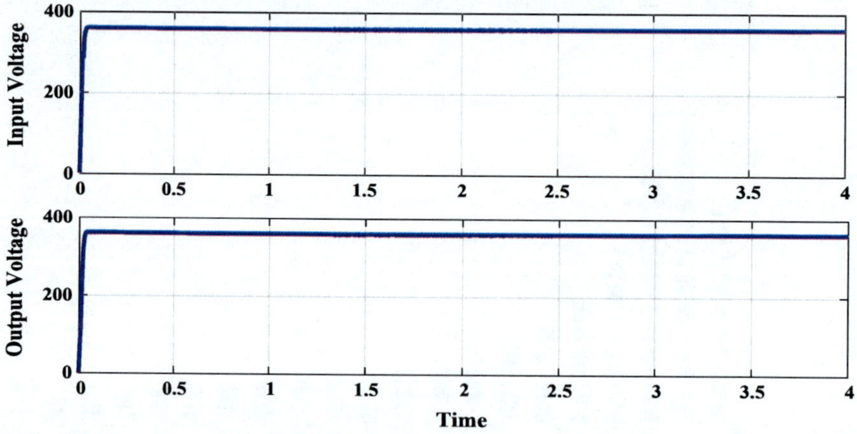

Fig. 6 Output voltage of the DC link with Buck converter

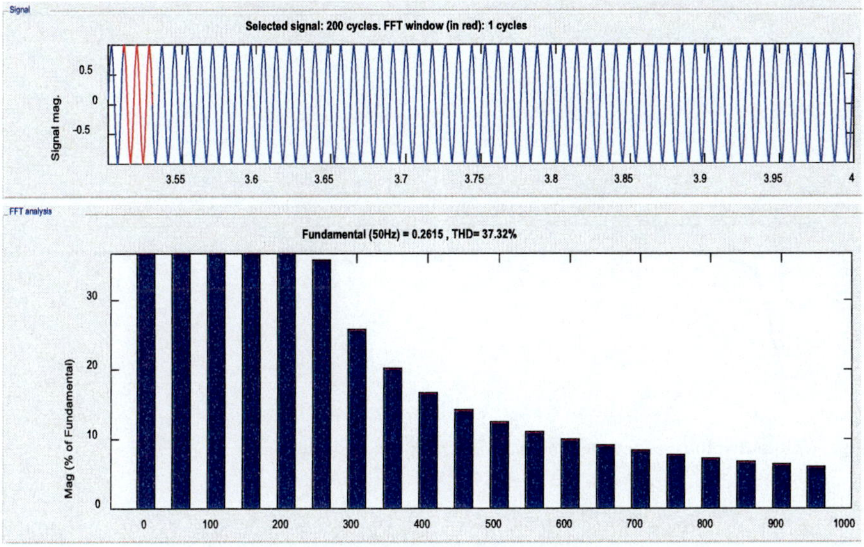

Fig. 7 THD analysis of VSC waveforms with Buck converter

2.3 System with Buck-Boost Converter

The simulation model of the system using Buck-Boost converter is shown in Fig. 8.

Figure 9 shows the output voltage at the DC link with the Buck-Boost converter.

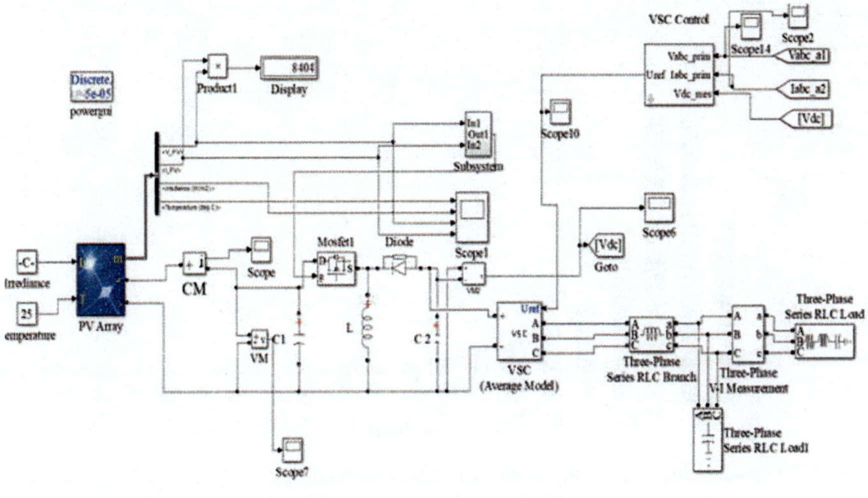

Fig. 8 Simulation model of the system with Buck-Boost converter

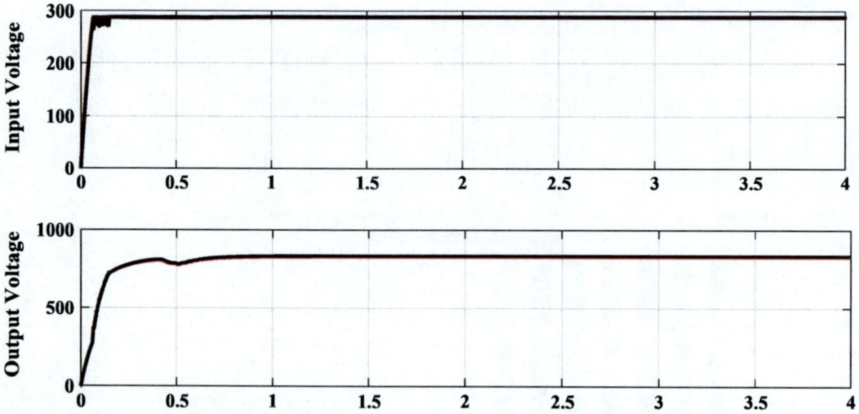

Fig. 9 Output voltage of the DC link with Buck-Boost converter

The THD analysis of the VSC output waveforms is done whose results are shown in Fig. 10.

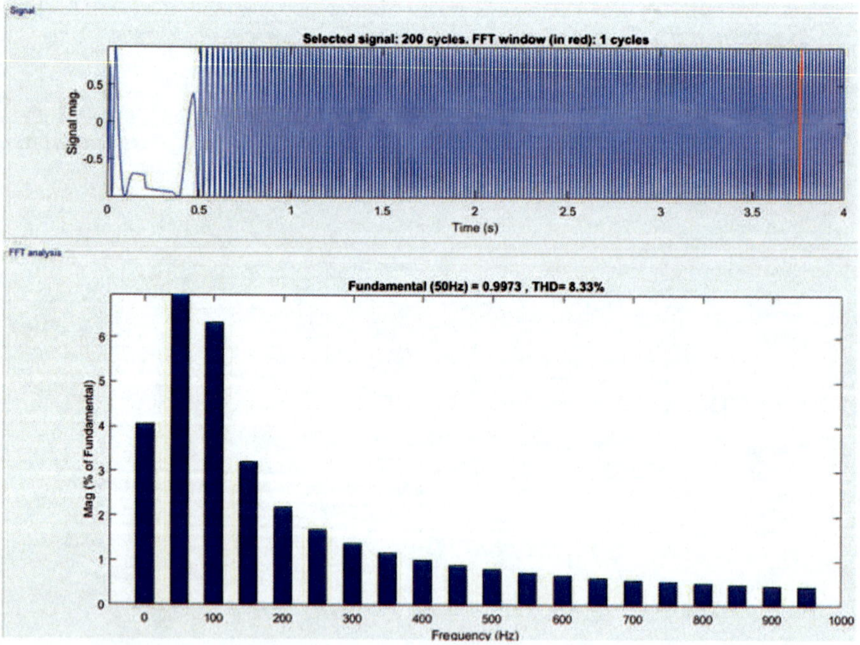

Fig. 10 THD analysis of VSC waveforms with Buck-Boost converter

2.4 System with Cuk Converter

The simulation model of the system using Cuk converter is shown in Fig. 11.

Figure 12 shows the output voltage at the DC link with the Cuk converter.

The THD analysis of the VSC output waveforms is done whose results are shown in Fig. 13.

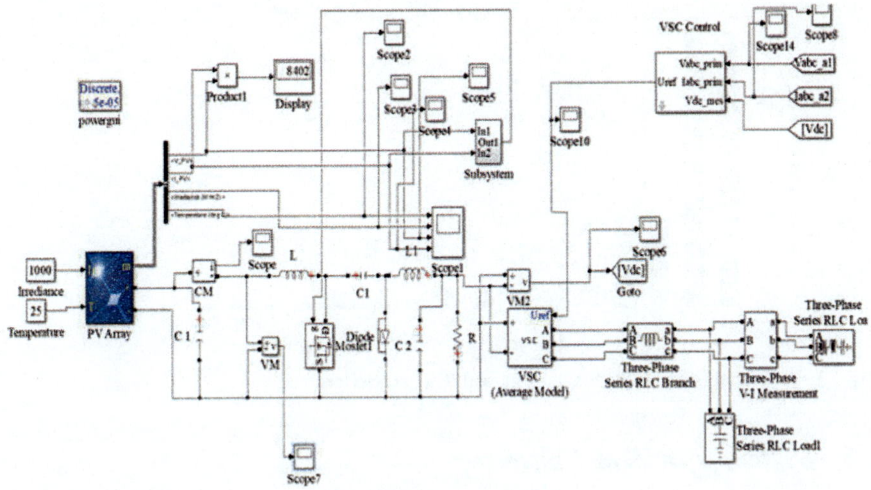

Fig. 11 Simulation model of the system with Cuk converter

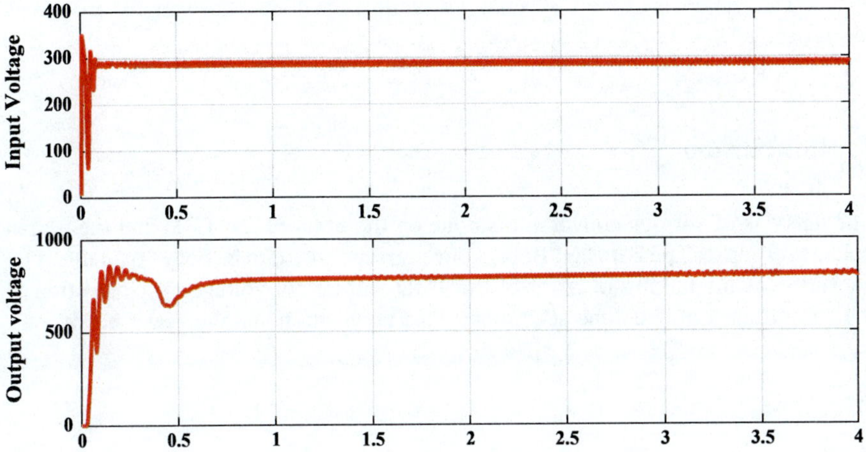

Fig. 12 Output voltage of the DC link with Cuk converter

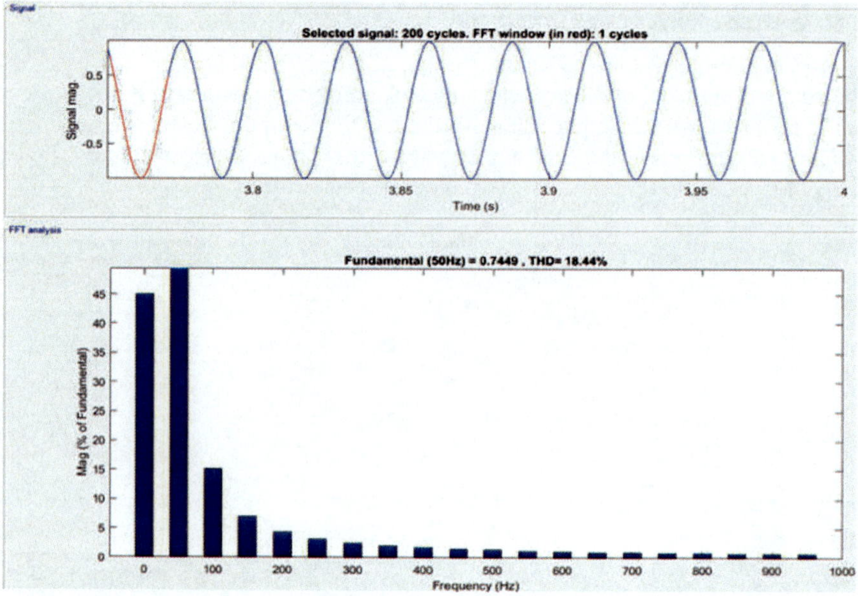

Fig. 13 THD analysis of VSC waveforms with Cuk converter

2.5 *System with Zeta Converter*

The simulation model of the system using Zeta converter is shown in Fig. 14.

Figure 15 shows the output voltage at the DC link with the Zeta converter.

The THD analysis of the VSC output waveforms is done whose results are shown in Fig. 16.

3 Conclusion

The analysis of various converters is done on the basis of the THD and the output voltage response. The Zeta and Boost converters are found to be better than the other converters while talking in terms of the THD and output voltage. The selection of the converters could be done according to the application and the user's needs.

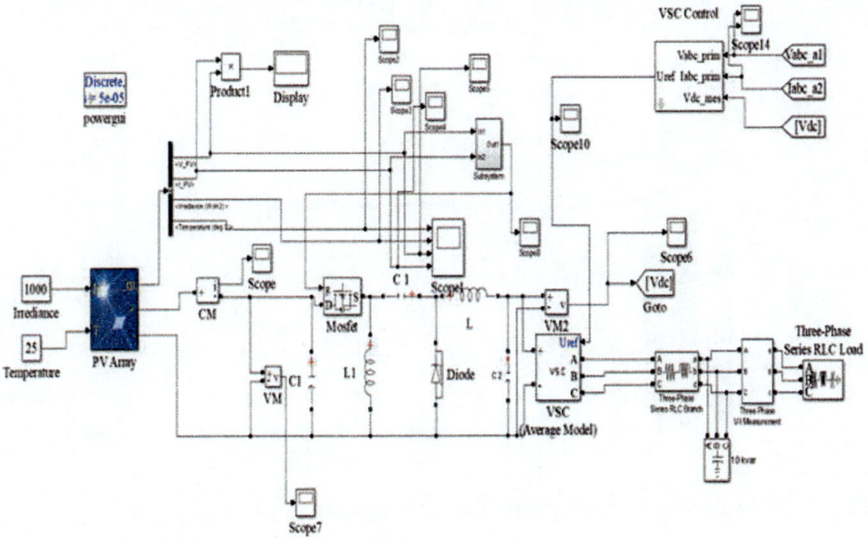

Fig. 14 Simulation model of the system with Zeta converter

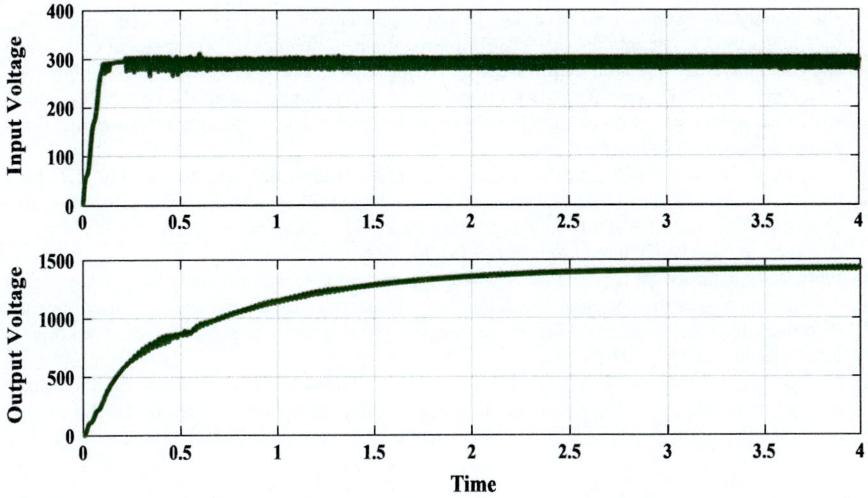

Fig. 15 Output voltage of the DC link with Zeta converter

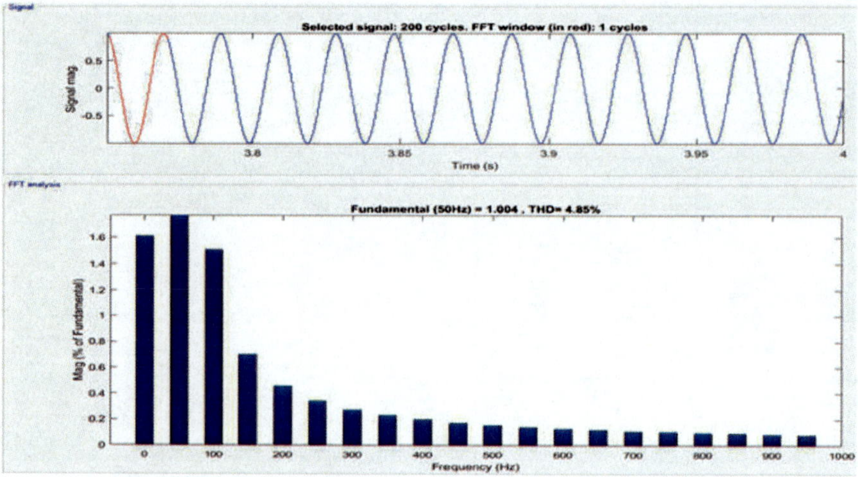

Fig. 16 THD analysis of VSC waveforms with Zeta converter

References

1. Bhattacharyya S, Samanta S (2018) DC link voltage control based power management scheme for standalone PV systems. In: 2018 IEEE international conference on power electronics, drives and energy systems (PEDES). https://doi.org/10.1109/pedes.2018.8707904
2. Fu J, Zhang B, Qiu D, Xiao W (2014) A novel single-switch cascaded DC-DC converter of Boost and Buck-boost converters. In: 16th European conference on power electronics and applications. https://doi.org/10.1109/epe.2014.6910723
3. Soheli SN, Sarowar G, Hoque MA, Hasan MS (2018) Design and analysis of a DC -DC buck boost converter to achieve high efficiency and low voltage gain by using buck boost topology into buck topology. In: 2018 International conference on advancement in electrical and electronic engineering (ICAEEE). https://doi.org/10.1109/icaeee.2018.8643001
4. Plotnikov I, Braslavsky I, Ishmatov Z (2016) The mathematical simulation of DC-DC coverter in the frequency-controlled electric drive with ultracapacitors. In: 2016 International symposium on power electronics, electrical drives, automation and motion (SPEEDAM). https://doi.org/10.1109/speedam.2016.7525910
5. Maroti PK, Padmanaban S, Wheeler P, Blaabjerg F, Rivera M (2017) Modified high voltage conversion inverting cuk DC-DC converter for renewable energy application. In: IEEE southern power electronics conference (SPEC). https://doi.org/10.1109/spec.2017.8333675

Chapter 15
Coordinated Control of UPFC-Based Damping Controller with PID for Power System

Amit Singhal and Ankit Tandon

1 Introduction

Flexible AC Transmission System Technology introduced in 1988 by Hingorani is an enabling technology and provides added flexibility and can enable a line to transfer power to the thermal rating. Unified Power Flow Controller is the best FACTS device which can control the various power system parameters like terminal voltage, phase angle, line impedance, etc. Therefore, it can be used not only for power flow but also for the power stabilizing control. The unified power flow controller (UPFC) is one of the most commonly used FACTS devices that provides the most important performance in damping small frequency oscillation in the power system [1, 2].

A comprehensive and analytical idea for modeling of UPFC for linearised and steady-state dynamic stability has been proposed. However, in some operating conditions, the PSS may fail to stabilize the power system, especially in low-frequency oscillations [3, 4].

2 SMIB System Without UPFC Considering Fault

A SMIB system installed without UPFC is considered (Fig. 1).

$$\Delta\dot{\delta} = \omega_0 \Delta\omega$$

$$\Delta\dot{\omega} = \left(-\frac{k_1}{M}\right)\Delta\delta + \left(-\frac{k_2}{M}\right)\Delta E_q' + \left(-\frac{D}{M}\right)\Delta\omega$$

A. Singhal (✉) · A. Tandon
Jodhpur Group of Institution, Jodhpur, Rajasthan, India
e-mail: amitsinghal.2891@gmail.com

M. Shorif Uddin et al. (eds.), *Intelligent Energy Management Technologies*,
Algorithms for Intelligent Systems,
https://doi.org/10.1007/978-981-15-8820-4_15

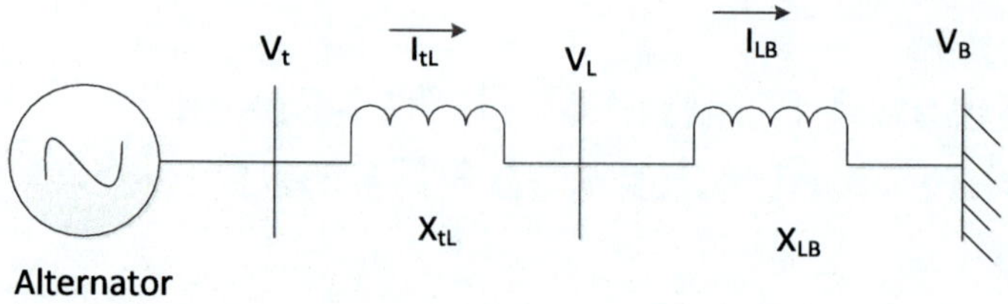

Fig. 1 A single machine infinite bus system (SMIB)

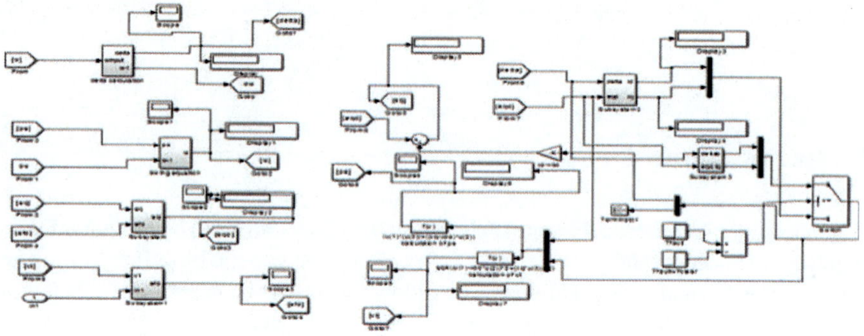

Fig. 2 Simulation model of SMIB system without UPFC

$$\Delta \dot{E'}_q = \left(-\frac{k_3}{T'_{d0}}\right)\Delta\delta + \left(-\frac{k_4}{T'_{d0}}\right)\Delta E'_q + \left(\frac{1}{T'_{d0}}\right)\Delta E_{fd}$$

$$\Delta \dot{E}_{fd} = \left(-\frac{k_5 k_a}{T_a}\right)\Delta\delta + \frac{k_a}{T_a}\Delta\omega + \left(-\frac{k_6 k_a}{T_a}\right)\Delta E'_q + \left(-\frac{1}{T_a}\right)\Delta E_{fd} + \left(\frac{k_a}{T_a}\right)\Delta V_{ref}$$

With the help of these linearised equations of SMIB, we obtained a simulation model of SMIB without UPFC considering fault (Fig. 2).

The output of this simulation is taken as angle deviation, electrical terminal voltage and power. Simulation process is being done for duration of 30 s (Figs. 3 and 4).

3 SMIB System with UPFC

A SMIB system installed with UPFC is shown in Fig. 5. Shunt converter is connected in parallel with the power system through an exciting transformer(ET), and second converter is attached in series with the system through a boosting transformer(BT). Both transformers are connected via a dc link.

The diagram of damping controller UPFC based is shown in Fig. 6.

By nonlinear model has been linearized, a dynamic linear model is obtained.

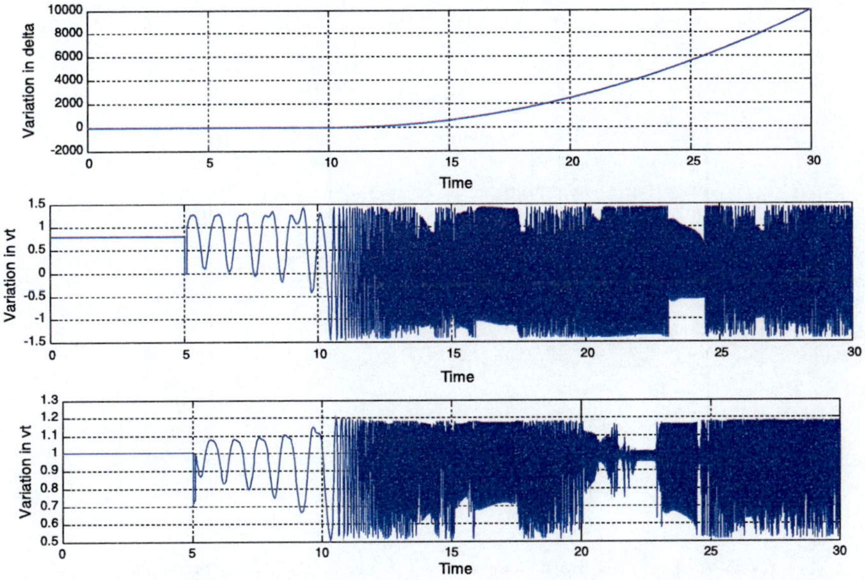

Fig. 3 Simulation results of SMIB system in variation of load angle, electrical power, terminal voltage at 80% loading considering fault

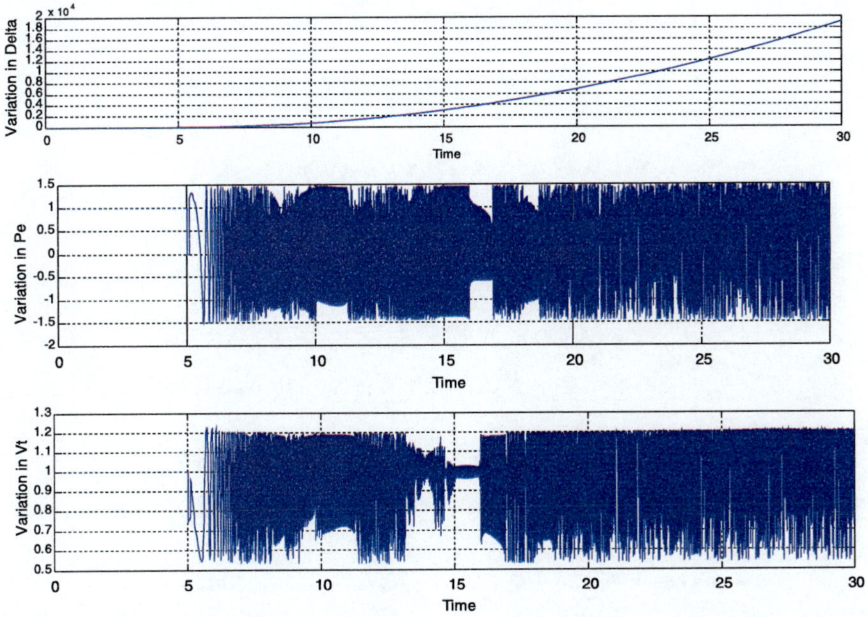

Fig. 4 Simulation results of SMIB system in variation of load angle, electrical power, terminal voltage at 100% loading considering fault

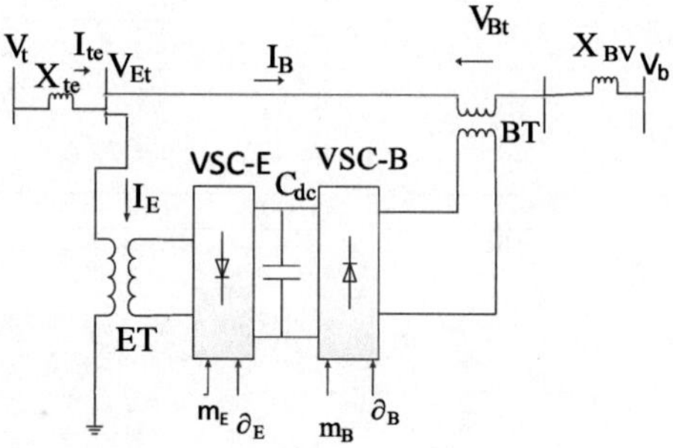

Fig. 5 SMIB system with UPFC

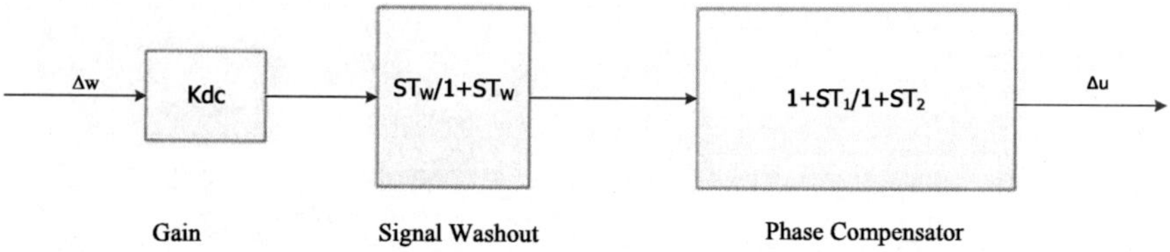

Fig. 6 Structure of UPFC-based damping controller

$$\Delta\dot{\delta} = w_0 \Delta w$$

$$\Delta\dot{w} = \frac{-\Delta P_e - D\Delta w}{M}$$

$$\Delta E' = \left(-\Delta E_q + \Delta E_{fd}\right)/T'_{d0}$$

$$\Delta\dot{E}_{fd} = -\frac{\Delta E_{fd}}{T_a} + \frac{K_a}{T_a}\left(\Delta V_{ref} - \Delta V_t\right)$$

$$\Delta E_q^{'} = K_4\Delta\delta + K_3\Delta E_q^{'} + K_{vd}\Delta V_{dc} + K_{qe}\Delta m_E + K_{q\delta e}\Delta\delta_E + K_{qb}\Delta m_B + K_{q\delta b}\Delta\delta_B$$

$$\Delta V_t = K_5\Delta\delta + K_6\Delta E_q^{'} + K_{vd}\Delta V_{dc} + K_{ve}\Delta m_E + K_{v\delta e}\Delta\delta_E + K_{vb}\Delta m_B + K_{v\delta b}\Delta\delta_B$$

$$\Delta V_{\dot{d}c} = K_7\Delta\delta + K_8\Delta E_q^{'} - K_9\Delta V_{dc} + K_{ce}\Delta m_E + K_{c\delta e}\Delta\delta_E + K_{cb}\Delta m_B + K_{c\delta b}\Delta\delta_B$$

With the help of these linearised equations of SMIB and UPFC, we obtained a simulation model of SMIB with UPFC considering fault (Figs. 7, 8 and 9).

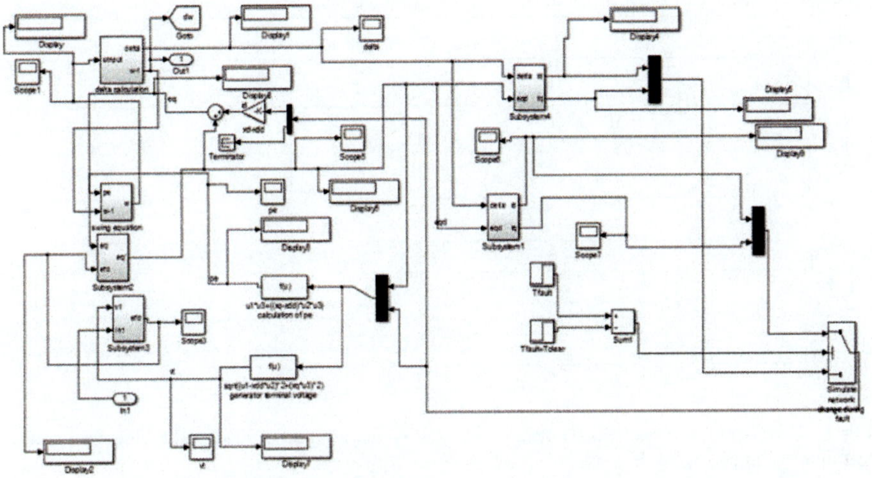

Fig. 7 Simulation diagram of SMIB with UPFC

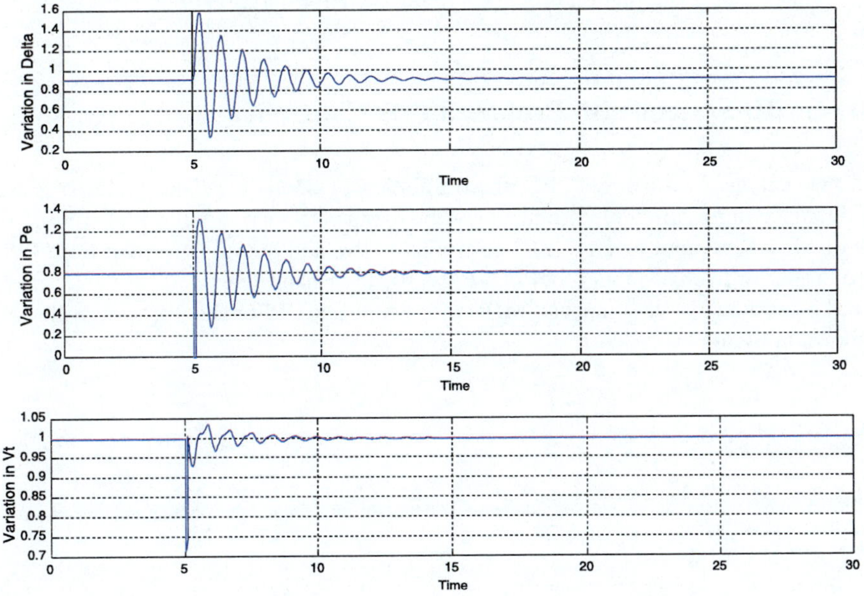

Fig. 8 Simulation results of SMIB system with UPFC in variation of load angle, electrical power, terminal voltage at 80% loading

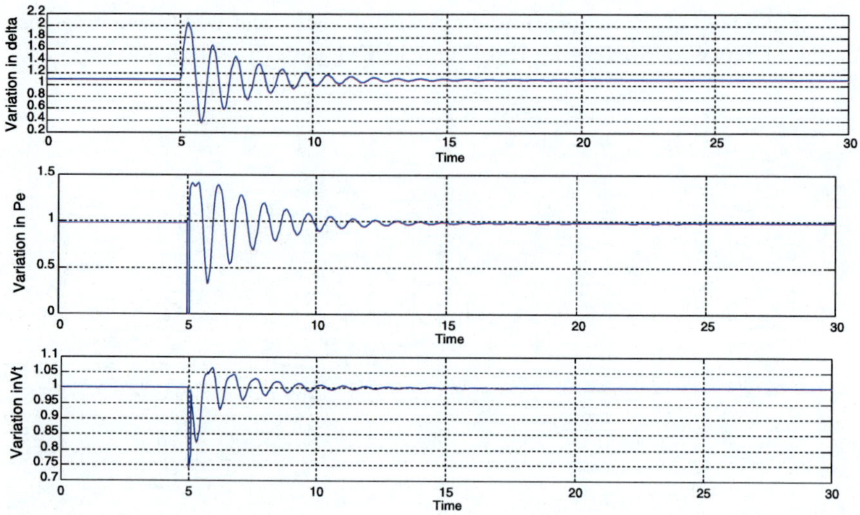

Fig. 9 Simulation results of SMIB system with UPFC in variation of load angle, electrical power, terminal voltage at 100% loading

The output of this simulation is taken as angle deviation, electrical power and terminal voltage. Simulation process is being done for duration of 30 s.

4 SMIB System with Proposed PID Controller

As we can see that using UPFC, the oscillation of SMIB is reduced. But still the system having oscillations which should be damped. So we install a PID controller on UPFC. The input given to PID controller is dw, and output is formed from PID controller is given as input to UPFC (Figs. 10, 11, 12 and 13).

The simulation diagram of SMIB with proposed UPFC and PID controller as shown in figure.

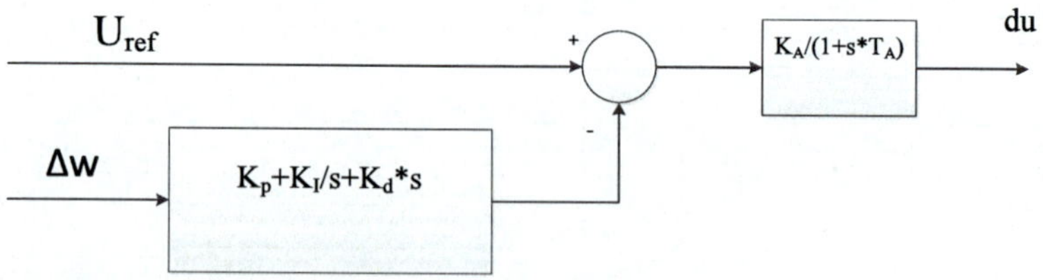

Fig. 10 Diagram of damping controller UPFC based with PID

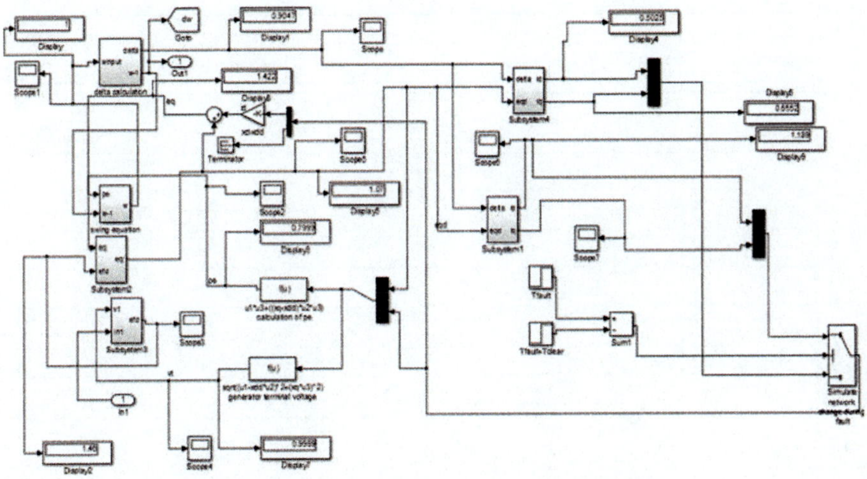

Fig. 11 Simulation model of SMIB with proposed UPFC with PID controller

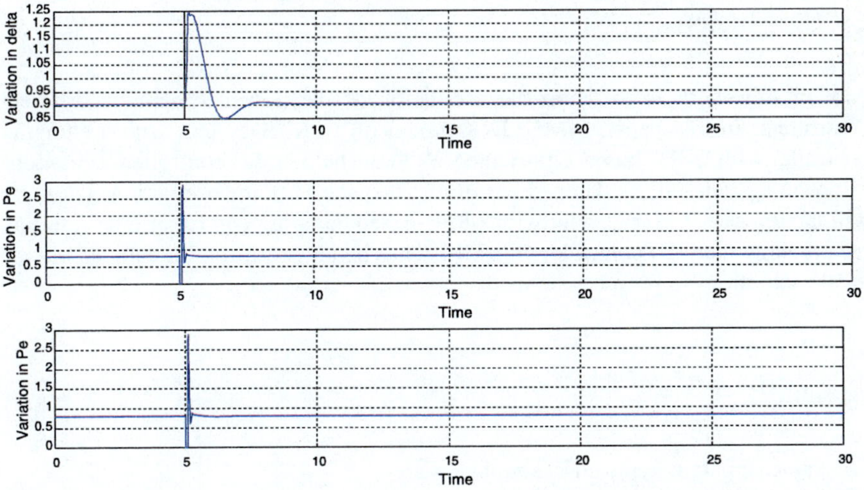

Fig. 12 Simulation results of SMIB system with proposed UPFC with PID controller in variation of load angle, electrical power, terminal voltage at 80% loading

The output of this simulation is taken as angle deviation, electrical terminal voltage and power. Simulation process is being done for duration of 30 s.

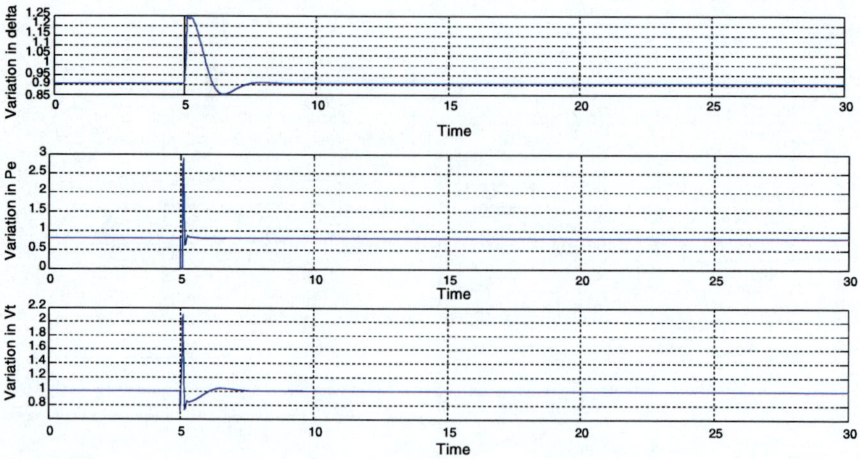

Fig. 13 Simulation results of SMIB system with proposed UPFC with PID controller in variation of load angle, electrical power, terminal voltage at 100% loading

5 Conclusion

Aim of this work is to damp the oscillation of the power system by different controllers. In this paper, SIMULINK model of a SMIB system with a damping controller with UPFC based is presented. With the help of this controller oscillations of power system will be damped out. But still system has the oscillation. Then, we will incorporate PID controller with UPFC-based damping controller. The simulink results show that PID controller with UPFC has better execution for damped out the local oscillations in power system.

Appendix

The value of parameter used in simulation are

Generator:

$M = 2H = 8.0$, $D = 0$, $T_{do'} = 5.044$, $Xd = 1.0$, $Xq = 0.6$, $Xd' = 0.3$

Excitation System:

$K_A = 100$, $T_A = 0.01$

Transformer:

$X_{tE} = 0.1$ p.u., $X_E = 0.1$ p.u., $X_B = 0.1$ p.u.

Transmission Line:

$X_{Bv} = 0.3$, $Xe = 0.5$

Operating conditions:

$Pe = 0.8$, $Vt = 1.0$ p.u.

$Vb = 1.0$ p.u.

UPFC parameter:

$m_E = 0.4013$, $m_B = 0.0789$

$\delta_E = -85.3478^{0,}$ $\delta_B = -78.2174^0$.

References

1. Larsen EV, Sanchez JJ, Chow JH (2005) Concept for design of FACTS controllers to damped power swings. IEEE Trans PWRS 10(2):948–955
2. Noroozian M, Angquist L, Ghandari M, Anderson G (2008) Use of UPFC for optimal power flow control. IEEE Trans Power Syst 12(4):1629–1634
3. Smith KS, Ran L, Penman J (2012) Dynamic modeling of a unified power flow controller. IEE Proc Gener Transm Distrib 144(1):7
4. Wang HF (2004) Damping function of unified power flow controller. IEE Proc 146(1):81

Chapter 16
Computational Neuroscience and Its Applications: A Review

Aisha Jangid, Laxmi Chaudhary, and Komal Sharma

1 Introduction

1.1 Central Nervous System (CNS):The Human Brain

Central nervous system (CNS): The central nervous system (CNS) is your body's centre of decision and communication. It consists of the brain and the spinal cord, and the peripheral nervous system (PNS) is made of nerves. They control and regulate every part of daily life from breathing and blinking to help you memorize facts for a test. The human brain is the command centre for the human nervous system. Nerves reach from your brain to your face, ears, eyes, nose and spinal cord and from the spinal cord to the rest of your body [1]. Sensory nerves collect the information from the environment which sends it to spinal cord, which then sends the message to the brain. The brain then makes sense of that message and fires off a response. Motor neurons distribute the instructions from the brain to the rest of your body.

A. Jangid · L. Chaudhary (✉) · K. Sharma
JIET, Jodhpur, India
e-mail: laxmichaudhary1911@gmail.com

A. Jangid
e-mail: aisha.jangid@jietjodhpur.ac.in

K. Sharma
e-mail: komal.sharmaece@jietjodhpur.ac.in

M. Shorif Uddin et al. (eds.), *Intelligent Energy Management Technologies*, Algorithms for Intelligent Systems,
https://doi.org/10.1007/978-981-15-8820-4_16

1.2 Single-Neuron Modelling

The neuron is the basic working unit of the brain, a specialized cell designed to transmit information to other nerve cells, muscle or gland cells. Neurons are cells within the nervous system that transmit information to other nerve cells, muscle or gland cells. Most neurons have a cell body, an axon and dendrites. In single-neuron modelling, we study about the behaviour and simulation of a single neuron [2].

For modelling single neurons, two types of complexity are usually associated. Firstly, the intrinsic properties of the cell membrane that make neuronal dynamics are so rich. Secondly, the morphology allows neurons to receive and integrate thousands of synaptic inputs from other cells.

1.3 Development, Axonal Patterning and Guidance

Axon guidance is a subfield of neural development concerning the process by which neurons send out axons to reach the correct targets. Axons generally follow very precise paths in the nervous system, and finding the way so accurately is a part of research [3].

1.4 Sensory Processing

Sensor Processing also called "sensory integration dysfunction" (SID) is the method in which the nervous system gets the message from the sense organs and convert those messages into suitable motor and responses for communication purpose. The sensorimotor system includes all of the sensor, motor, and central integration and processing components involved and also maintains joint homeostasis during bodily movements (functional joint stability). So, we can say that functional joint stability is basically a difficult and complex physiologic process [1].

1.5 Memory

Limbic system is the region of the brain which is deep inside the medial temporal lobe, which includes the hippocampus, the amygdala, the cingulate gyrus, the thalamus, the hypothalamus, the epithalamus, the mammillary body and other organs, many of which are of particular relevance to the processing of memory. They still don't fully understand exactly how we remember or what occurs during recall. The process of memory begins with encoding, then proceeds to storage and, eventually, retrieval. Encoding is the crucial first step to creating a new memory. It allows the perceived item of interest to be converted into a construct that can be stored within the brain, and then recalled later from short-term or long-term memory. Encoding is a biological event beginning with perception through the senses [3]. First, our brain consciously registers the memory, this is called encoding. The best way to improve our memory is to keep remembering the same thing, over and over.

1.6 Behaviours of Networks

This is called generalization. Highly complex pattern recognition can be achieved by using a network of neurons hence the name neural networks. The networks normally used for pattern recognition are called feed forward because they have no feedback. They simply associate inputs with output.

1.7 Cognition, Discrimination and Learning

1.7.1 Cognition

The study drawn on the basis of aspects of psychology, linguistics, philosophy and computer modelling gives the basis of thought, learning and mental organization. The use of the computer as a tool for thinking how the human mind handles information is known as the computer analogy. The computer gave cognitive psychologists an analogy, to which they could compare human mental processing.

1.7.2 Discrimination

Discrimination, in psychology, is the ability to perceive and respond to differences among stimuli. It is considered as a more advanced form of learning than generalization, the ability to perceive similarities, although animals can be trained to discriminate as well as to generalize (Figs. 1 and 2).

1.7.3 Learning

Discrimination, in psychology, is the ability to perceive and respond to differences among stimuli. It is considered as a more advanced form of learning that is the ability to perceive similarities, although animals can be trained to discriminate as well as to generalize. Figure 3 shows Process of Cognition, discrimination and learning [1]. Learning rule or Learning process is a method or a mathematical logic which improves the artificial neural network's performance by repeatedly applying over the network.

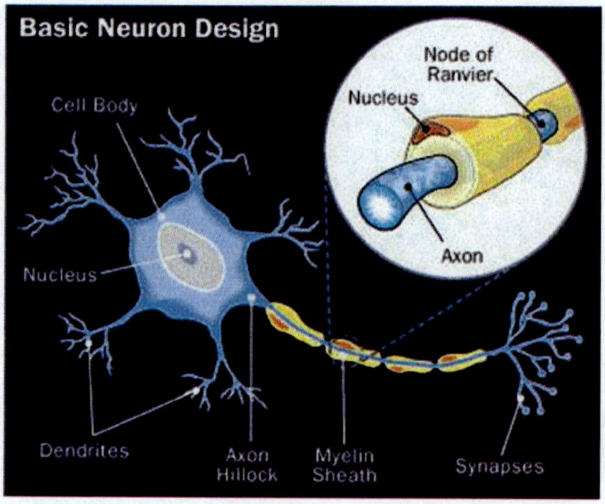

Fig. 1 Basic neuron design [1]

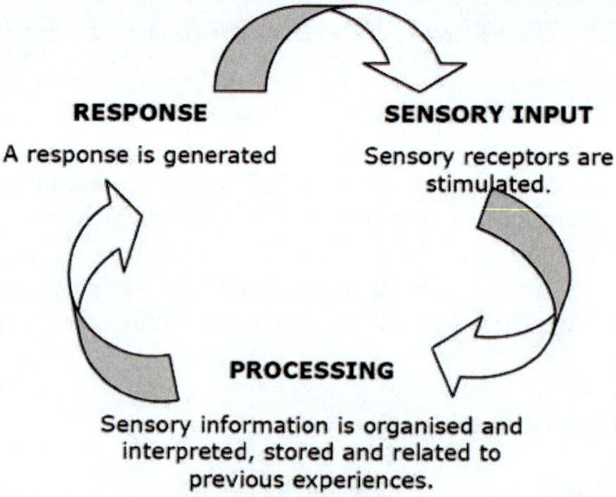

Fig. 2 Sensor processing

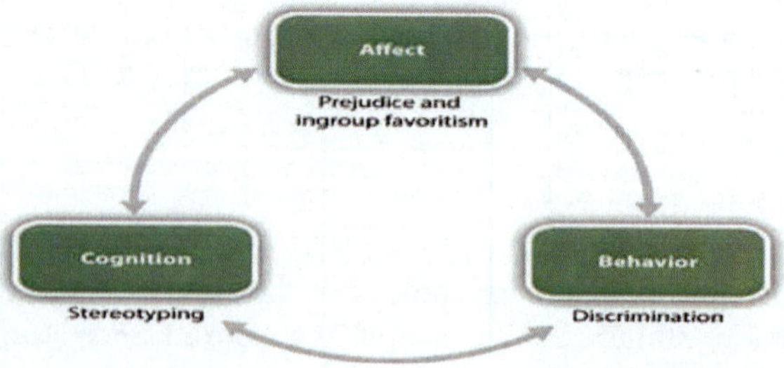

Fig. 3 Process of cognition, discrimination and learning [1]

2 Computational Neuroscience

2.1 Brain Data Recording

By the excitation of various neuron, the electric current flows in nanoamperes and EM waves is produced and that is recorded with the use of electrodes put on the head. The data obtained from this method is stored and processed in a special high-speed computer. After that these data are analyzed for the various reactions of the brain for different sensation towards body [4]. From this method, the scientist partially gets success to record the dreams of human. They succeeded to make the blurred image stored inside the brain which was shown to the human. After sometime they will also achieve the success to record the dreams of a person by this method.

2.2 MRI Magnetic Resonance Imaging

The other method to analyze the brain is MRI. MRI is acronym of magnetic resonance imaging. In this, the resonant property of hydrogen ion in brain is considered and makes an image of the brain at higher radio frequency oscillating magnetic field [5].

2.3 Quantum Computers

A quantum computer is a device for computation that distinctively uses quantum mechanical phenomena, such as superposition and entanglement to perform operations on data. An atom, not an electron, is the physical bit. An electron is 0 or 1. Quantum mechanics: at atom is 0, 1 or both "coherent superposition". The bit in quantum mechanics is a qu bit. The brain is a complex structure so we should require to store and process huge information at a time. The quantum computer is very fast computers and efficient storage schemes. So, the next generation is planning to use quantum computers for the analyzing of brain information [6]. It is a computer which makes use of the quantum states of subatomic particles to store information. Figure 4 shows that Quantum Computer is the best way to analyze the brain.

2.4 Neural Engineering

Neural engineering (also known as neuroengineering) is a discipline within biomedical engineering that uses engineering techniques to understand, repair, replace, enhance or otherwise exploit the properties of neural systems [7].

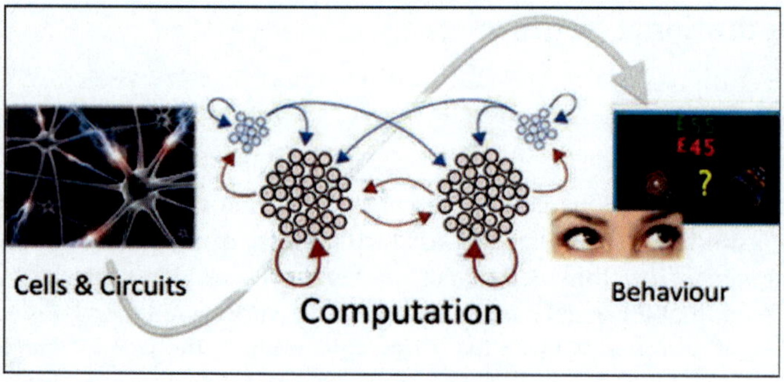

Fig. 4 Quantum computer is best way to analyze the brain

3 Neural System as Electronic Network

3.1 Hodgkin–Huxley Model

The Hodgkin–Huxley model, or conductance-based model, is a mathematical model that describes how action potentials in neurons are initiated and propagated. Figure 5 shows Hodgkin–Huxley Model.

Hodgkin–Huxley type models represent the biophysical characteristic of cell membranes. The lipid bilayer is represented as a capacitance (Cm). Voltage-gated and leak ion channels are represented by nonlinear Hodgkin–Huxley type models represent the biophysical characteristic of cell membranes. The lipid bilayer is represented as a capacitance (Cm). Voltage-gated and leak ion channels are represented by nonlinear (gn) and linear (gL) conductances, respectively. The electrochemical gradients driving the flow of ions are represented by batteries (E), and ion pumps, and exchangers are represented by current sources (Ip).

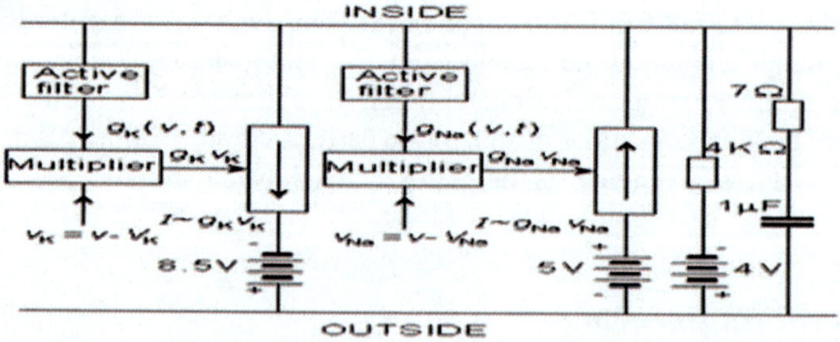

Fig. 5 Hodgkin–Huxley model

3.2 Neuron-Electronic Equivalent

We can make a neuron equivalent electronic circuits which works very similar to neuron so that the signal transmission can travel with the speed of electron. The neuron signal transmission speed is very slow as 12o m/s because it uses ion-exchange method. So the electronic equivalent would be fast (Figs. 6 and 7).

The electronic neuron model is Lewis Membrane Model which is based on Hodgkin–Huxley equation. The sodium and potassium conductance, synaptic connections and other functions of the model are represented by discrete transistors and associated components in form of parallel circuits connected between nodes representing the inside and outside of the membrane [8].

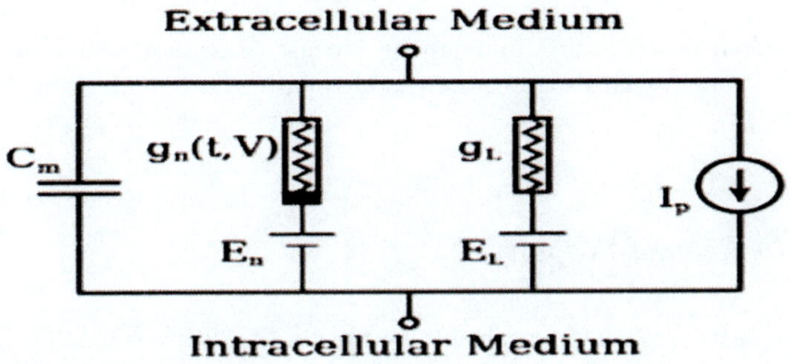

Fig. 6 Lewis membrane model [14]

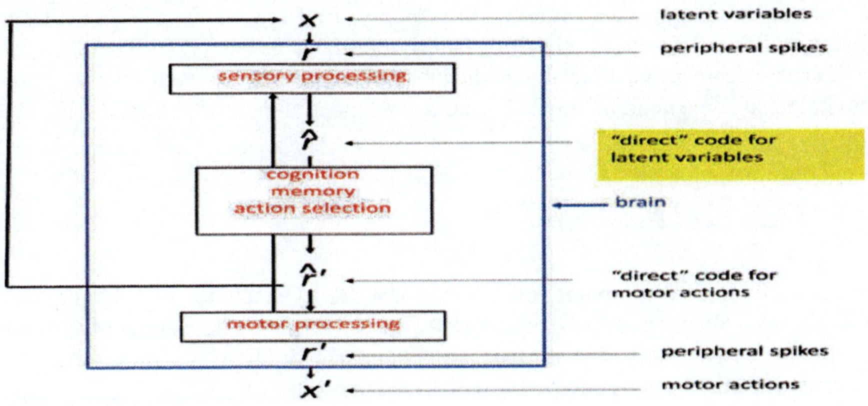

Fig. 7 Brain power processing [10]

3.3 Neural Networks

The interconnection of neurons inside the neural system is called neural networks. The fifth generation of computers uses neural networks. A neural network is referred as a series of algorithms that aims to understand relationships in a set of data through a process which behaves in the way the human brain operates.

4 Biological Properties of Neural System

4.1 Genetic Algorithm

GA is a search heuristic that imitates the process of natural evolution [9]. This heuristic is routinely used to generate useful solutions to optimization and search problems.

4.2 DNA Related Property

DNA is the basis of life and responsible for flow of life. It stores the information of body structure of a living being and can derive a living being correctly from a single cell called zygote. There are four types of base in DNA such as adinine, guanine, thymine and cytocine (A, G, T, C) [1]. A always combines with T by hydrogen bond. This one combination can be referred to as binary '1'. T always combines with C by hydrogen bond. This one combination can be referred to as binary '0'. There are 3.2 billions of such combinations inside the DNA of single cell of human. So, it can be possible that some neural features of parent can undergo their offspring.

4.3 Biological Computer

Brain is a biological computer which has neurons as its functional unit. Digital bits are only two voltages by which can store and process very less information per second as compared to neurons. It is an analogue system. The different neuron voltages act as several learning points and can be said bio–bits [8].

4.4 Section Division of Problems in Brain and CPU

The brain is divided into two sections as Left and Right brain which control the right and left side of the body control, respectively. The Left side of the body control includes the emotion, face recognition,creativity, Music/art awareness, imagination, subjectivity, etc., whereas the right side of the body control includes the math/science skills, written language, spoken language, objectivity, analytical, logic, reasoning, etc.

4.5 Brain Power Processing

The brain includes the sensory processing, cognition memory action selection and motor processing where the 'X' is the latent variables, 'r' is peripheral spikes which gives the motor actions [10]. Latent variables get encoded and are processed to give respective motor action using peripheral spikes.

5 Applications of Computational Neuroscience

5.1 Artificial Intelligence

It is the ability of computer system that makes it able to perform tasks which normally require human intelligence, such as visual perception, speech recognition, decision making and language translation [9, 11]. It includes knowledge representation issues, mapping and logic by rules, probabilistic reasoning, planning system, natural language processing, learning, artificial neural networks, Fuzzy logic & genetic algorithms, etc.

5.2 Artificial Neural Networks

Artificial neural networks (ANNs) or connectionist systems are a computational model used in computer science and other research disciplines, which is based on a large collection of simple neural units (artificial neurons), loosely analogous to the observed behaviour of a biological brain's axons [8, 12].

5.3 *Brain Like Working Machine*

Artificial brain (or artificial mind) is a term commonly used in the media to describe research that aims to develop software and hardware with cognitive abilities similar to those of the animal or human brain. Research investigating “artificial brains” and brain emulation plays three important roles in science. Neuroscientists are trying to understand how the human brain works, known as cognitive neuroscience. AI has ability to create a machine that can perofrm task similar to a human being. It is long-term project to create machines exhibiting behaviour comparable to those of animals with complex central nervous system such as mammals and humans, and this goal of creating a human mimicking machine and exhibiting human-like behaviour or intelligence is called as strong AI [11, 13].

5.4 *Immortality*

5.4.1 Brain Transplant (Limited Age of Brain)

A brain or whole-body transplant is a procedure in which the brain or organ of one organism is transplanted into the body of another [6]. It is different from head transplantation, which involves transferring the entire head to a new body unlike the brain only. Brain data transfer from old to child and few applications such as.

Brain data storage, Living machine (One kind of computer which uses brain as its CPU.), Living as robot-server, Cloud brain data storage.

Advanced display for blind people:
The blind people can’t see anything but due to improvements in these fields there may be displays that can be seen by blind people, and hence they can visualize the things through brain and take actions [5].

6 Conclusion

Computational neuroscience approaches further our understanding of brain function, and helps in translating the acquired knowledge into technological applications. From a scientific perspective discovering how the brain thinks is a major activity in the history of mankind. Be it cellular and synaptic dynamics or biophysical basis of neuronal computation or algorithms for the analysis of neuronal data, the field has provided analytical and computational skills for understanding the neuronal systems. The field of computational neuroscience explores the information processing strategies employed by neural circuits in the brain. Computation” and "information processing" are commonly used interchangeably, which presupposes that they are roughly synonymous terms. If that were true, computation would entail information processing, and information processing would entail computation. "Understanding the human mind in biological terms "has evolved as the challenging topic for science in the twenty-first century. There is need to understand the biological nature of perception, learning, memory, thought, consciousness and the limits of free will, and thus it is expansion of new science of mind not from extent of how we perceive, learn, remember, feel to a new perspective of ourselves in the context of biological evolution. The neural science explains the behaviour in terms of the activities of the brain.

References

1. Gazzaniga et al (2004) The cognitive neurosciences. MIT Press
2. Kriegeskorte N, Mok RM (2017) Building machines that adapt and compute like brains. Behav Brain Sci 40
3. Grill-Spector K, Weiner KS, Kay K, Gomez J (2017) The functional neuroanatomy of human face perception. Ann Rev Vision Sci 3:167–196
4. Mnih V et al (2015) Human-level control through deep reinforcement learning. Nature 518:529–533
5. Haynes J-D (2015) A primer on pattern-based approaches to fMRI: principles, pitfalls, and perspectives. Neuron 87:257–270
6. LeCun Y, Bengio Y, Hinton G (2015) Deep learning. Nature 521:436–444
7. Kriegeskorte N, Kievit RA (2013) Representational geometry: integrating cognition, computation, and the brain. Trends Cogn Sci 17:401–412
8. Thomas JI (2019) Current status of consciousness research from the neuroscience perspective. Acta Sci Neurol 2(2)
9. Dimitrov AG, Lazar AA, Victor JD (2011) Information theory in neuroscience. J Comput Neurosci, Springer 30:1–5. https://doi.org/10.1007/s10827-011-0314-3
10. Ward J (2015) The student's guide to cognitive neuroscience. Psychology Press
11. Frisoni GB et al (2012) N4U: expansion of neuGRID services and outreach to new user communities (poster). In: 9th e-Infrastructure concentration meeting of the european grid infrastructure, 22 Sept 2011. https://neugrid4you.eu/conferences
12. Dehaene S, Naccache L (2001) Towards a cognitive neuroscience of consciousness: basic evidence and a workspace framework. Elsevier
13. Redolfi A, McClatchey R et al (2009) Grid infrastructures for computational neuroscience: the neuGRID example. Future Neurol 4(6): 703–722. https://doi.org/10.2217/fnl.09.53
14. Sejnowski TJ, Jolla L, Chase C (2015) Computational neuroscience. In: International encyclopedia of the social &behavioral sciences, 2nd edn, pp 480–484, Elsevier

Chapter 17
Optimization of Band Notch Characteristic in Ultra-Wideband Microstrip Patch Antenna for Wireless Power Transfer

Ashish Mathur, Geetika Mathur, and Abhijit kulshrestha

1 Introduction

Federal communications commission (FCC) allocated a block of radio spectrum from 3.1 to 10.6 GHz for UWB operations [1]. UWB systems can support more than 500 Mbps data transmission within 10 m [1]. Compact size, low-cost printed antennas with Wideband and Ultra-wideband characteristic are desired in modern communications. The Ultra-wideband antennas can be classified as directional and omni-directional antennas [2]. A directional antenna has high gain and is relatively large in size. It has narrow field of view. Whereas the omni-directional antenna has low gain and relatively small in size. It has wide field of view as it radiates in all the directions [2]. The UWB antennas have broad band. There are many challenges in UWB antenna design. One of the challenges is to achieve wide impedance bandwidth. UWB antennas are typically required to attain a bandwidth, which reaches greater than 100% of the centre frequency to ensure a sufficient impedance match is attained throughout the band such that a power loss less than 10% due to reflections occurs at the antenna terminals. The bandwidth of the microstrip antenna can be enhanced by modifying the ground plane [3]. Many designers have tried various ways to improve the structure of the traditional circular antennas, and many valuable results have been obtained [4–8].

A. Mathur (✉) · A. kulshrestha
JIET Group of Institution-Jodhpur, Jodhpur, Rajasthan, India
e-mail: ashish.mathur@jietjodhpur.ac.in

A. kulshrestha
e-mail: abhijit.kulshrestha@jietjodhpur.ac.in

G. Mathur
Jagannath University Jaipur, Jaipur, Rajasthan, India
e-mail: geetika.mathur@jietjodhpur.ac.in

M. Shorif Uddin et al. (eds.), *Intelligent Energy Management Technologies*, Algorithms for Intelligent Systems,
https://doi.org/10.1007/978-981-15-8820-4_17

2 Antenna Configuration and Design

For the calculation of length and width of the patch antenna, we used basic formulas for length and width of patch. The ground plane is modified to enhance the bandwidth of the antenna. The proposed antenna designed on a FR4 substrate with dielectric constant $\varepsilon_r = 4.4$ and height of the substrate is $h = 1.6$ mm. The substrate has length $L = 50$ mm and width $W = 50$ mm. The substrate is mounted on ground of 18 mm length and 50 mm width (Figs. 1 and 2).

The proposed design is capable of passing four bands in the range of 2.1–3.25 GHz in the range of ISM (2.4–2.4835 GHz), Bluetooth (2.4–2.484 GHz) and rejection of Wi max IEEE 802.16 (3.3–3.7 GHz) band at absolute bandwidth in GHz below return loss of −10 dB is 2.1 to 3.25 GHz = 1.25 GHz, Second 5.2 to 6.2 GHz = 1.0 GHz, third 7.8 Ghz to 9.0 GHz and fourth 9.10 to 11.1 GHz. This antenna is resonant at four centre frequencies. First is 2.5 GHz with absolute bandwidth 1.0 GHz, second is 5.5 GHz with absolute bandwidth 2.0 GHz, third is 8.5 GHz and fourth is 10.5 GHz for UWB applications.

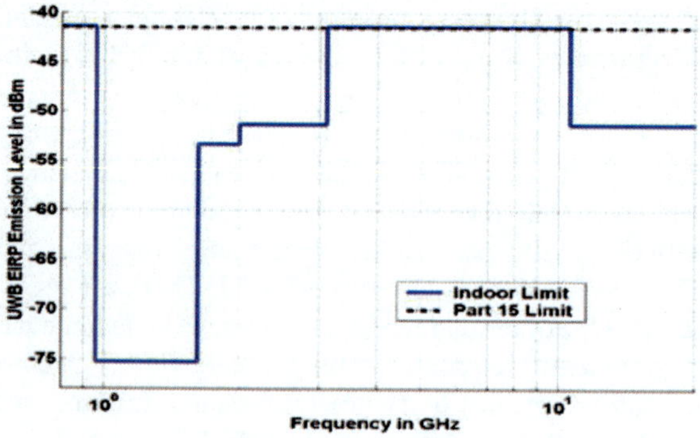

Fig. 1 UWB spectral mask per FCC (Modified) part 15 rules [1]

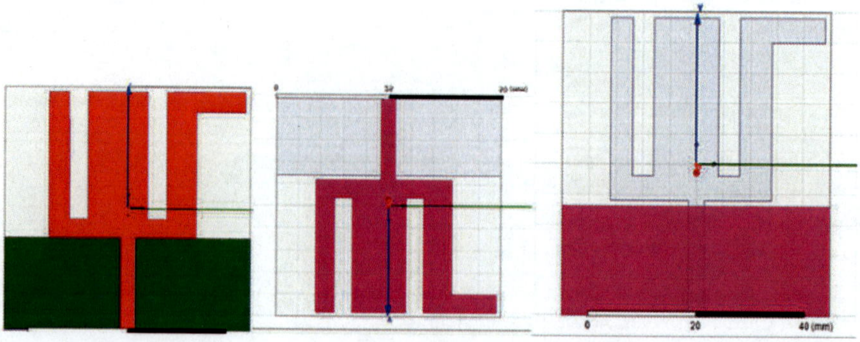

Fig. 2 Geometry of rectangular patch top view & bottom view

3 Simulation Results

Figures 3 and 4 show parametric study of S11 parameter for multiband patch antenna with optimized ground length Lg = 18 mm. This antenna is suitable for operating frequency of 2.1–3.25 GHz, 5.2–6.2 GHz, 7.8–9.0 GHz and 9.10–11.1 GHz in UWB. It is shown that return loss of the antennas is better than −10 dB. The VSWR obtained is less than 1.5 and the patch antenna is found to have the compact size. The return loss value of first, second, third and fourth band is −16 dB,−45 dB, −24 dB and −20.1 dB, respectively.

Figure 5 shows the relationship between VSWR with frequency for the proposed design. In this, the value of VSWR is ≤2 for four different centre frequencies. First is 2.5 GHz with absolute bandwidth 1.0 GHz and second is 5.5 GHz with absolute bandwidth 1.0 GHz and third is 8.5 GHz with absolute bandwidth 1.2 GHz and fourth is 10.5 GHz with absolute bandwidth 2.1 GHz for UWB applications.

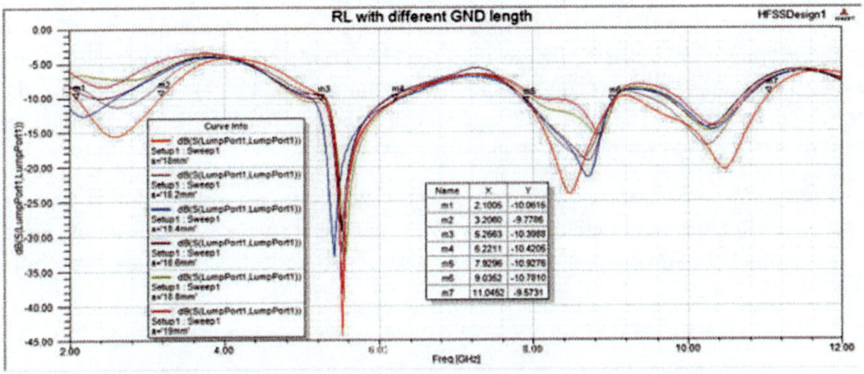

Fig. 3 S_{11} of Patch antenna with different ground plane effects Lg

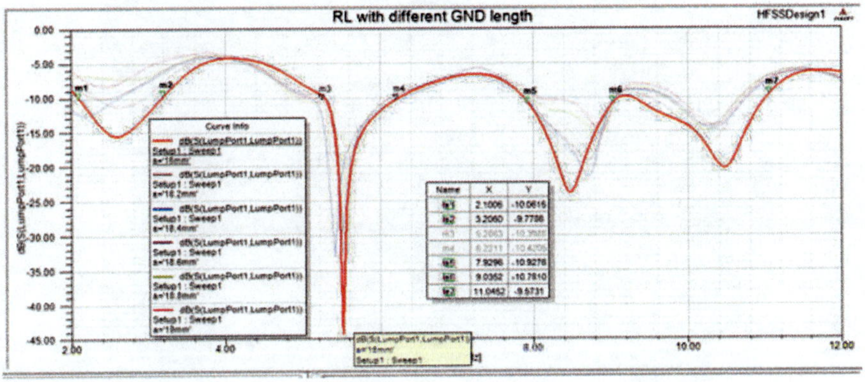

Fig. 4 S_{11} of patch antenna with Lg = 18 mm

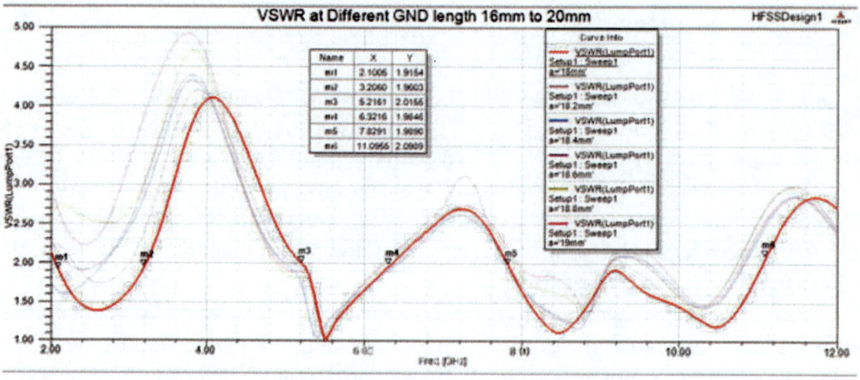

Fig. 5 VSWR of multiband notch patch antenna

The Plot curve of Directivity, Gain in 3D Polar are shown in Figs. 6 and 7, Radiation efficiency and radiation pattern are shown in Figs. 8 and 9, the current distribution of the proposed design is shown in Fig. 10. The simulated values of directivity are 3.65 dB with 2.64 dB antenna gain and 82% radiation efficiency is calculated for the proposed geometry. The uniformly current distribution and bidirectional radiation pattern are obtained for proposed geometry at 7 GHz.

Ground—Plane/Substrate—Related effect: The three important points can be observed after the parametric analysis of ground plane with FR4 substract. First, it is seen that the impedance matching is very sensitive to the feed gap, especially

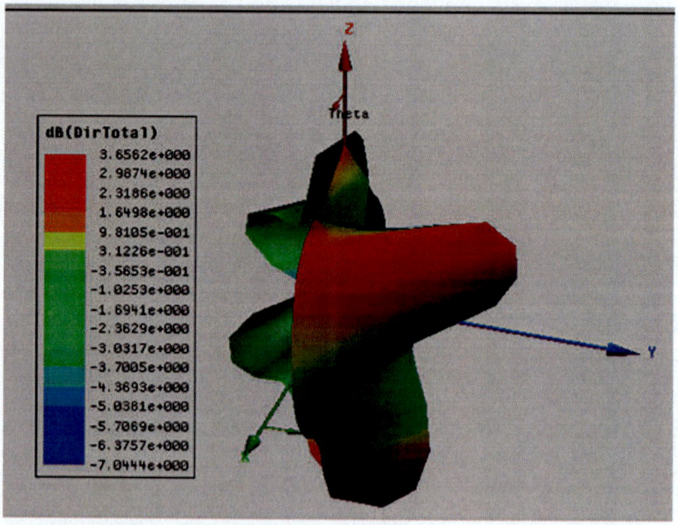

Fig. 6 Directivity of patch antenna at 7 GHz

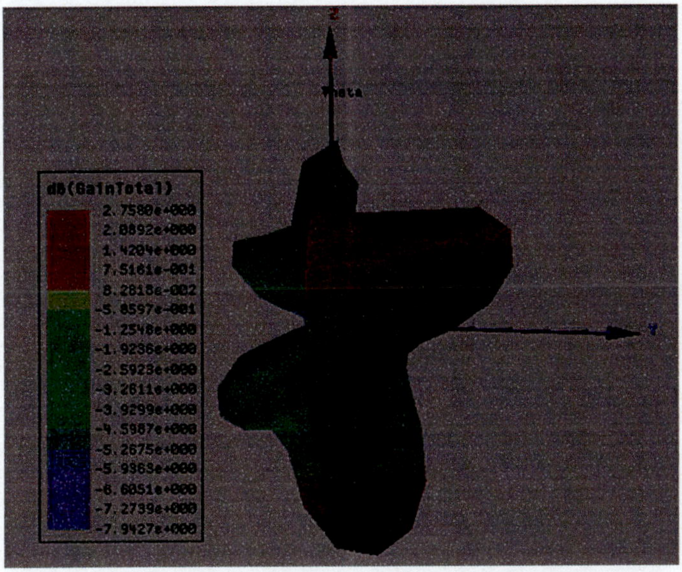

Fig. 7 3D Polar gain of patch antenna at 7 GHz

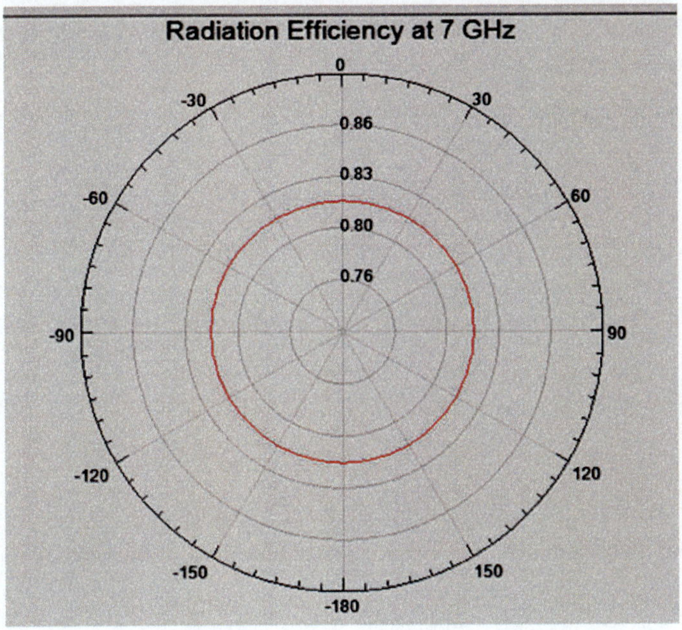

Fig. 8 Radiation efficiency at 7 GHz

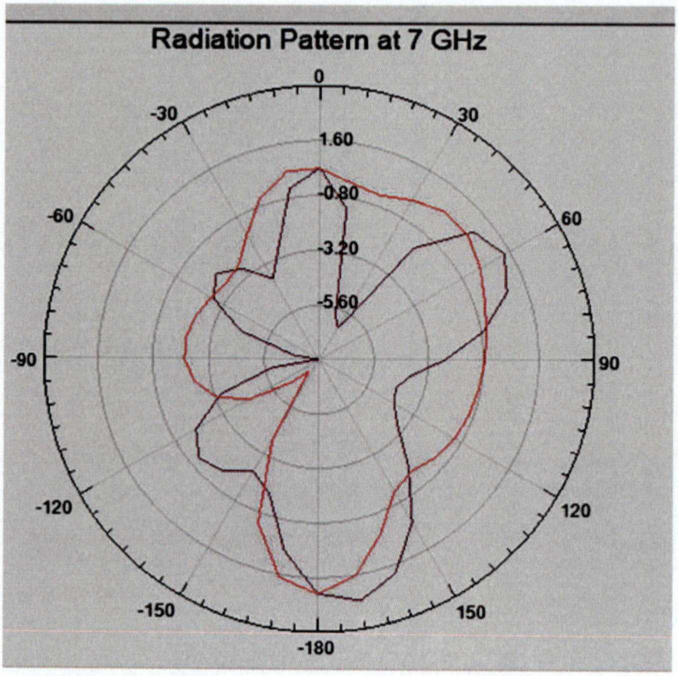

Fig. 9 Radiation patten at 7 GHz

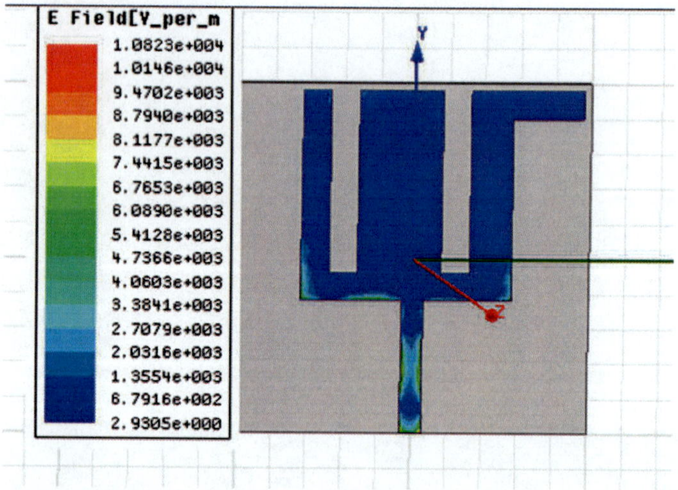

Fig. 10 Measurement of E field distribution at 7 GHz

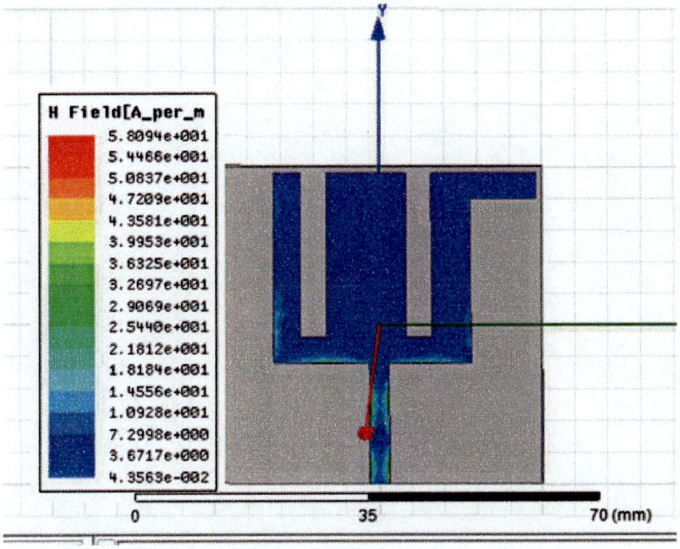

Fig. 11 Measurement of H field distribution at 7 GHz

at higher frequencies. Second, the length of the ground plane affects the impedance matching more significantly at higher frequencies than at lower frequencies [9]. This finding is consistent with the current distributions where more current is concentrated on the ground plane at the higher frequencies than at lower frequencies [10]. Last, the impedance response is also affected by the dielectric constant [11]. In this study, a change in the dielectric constant leads to a shift in the characteristic impedance of the feeding strip from 50 (Fig. 11).

4 Fabrication, Measurement Setup and Testing

The antenna structure is fabricated on FR 4 substrate using Photolithography technique. The proposed design is tested on vector network analyzer. The top view and measurement setup of fabricated antenna are shown in Figs. 12 and 13 (Table 1).

The measured result of S_{11} for the proposed design is calculated by vector network analyzer and on the basis of measured results, we conclude that this antenna is suitable for frequency band of 7.4–10.2 GHz with resonant frequency at 8.25 GHz with good return loss and VSWR values (Figs. 14 and 15).

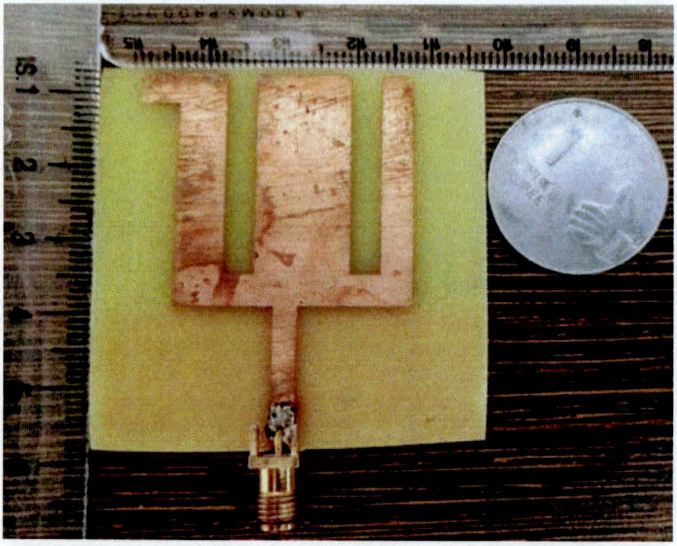

Fig. 12 Fabricated design of proposed antenna

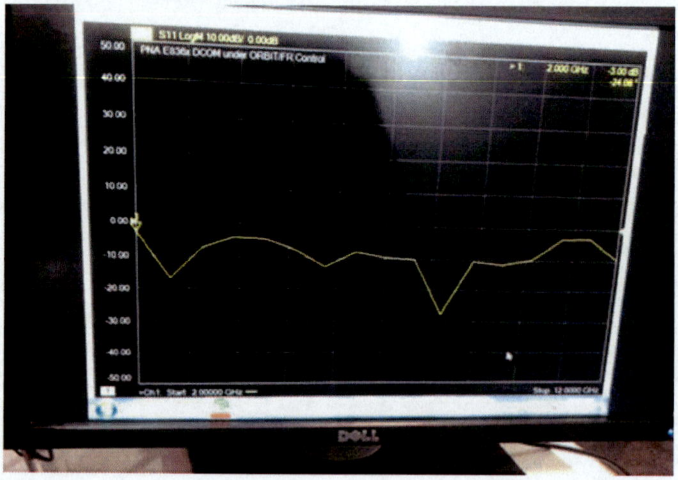

Fig. 13 Measurement setup of S_{11} at 7 GHz

Table 1 Antenna designing parameters

W sub	L sub	Wg	Lg	Ws	Ls	Wp	Lp
50	50	50	18	12	4	30	30

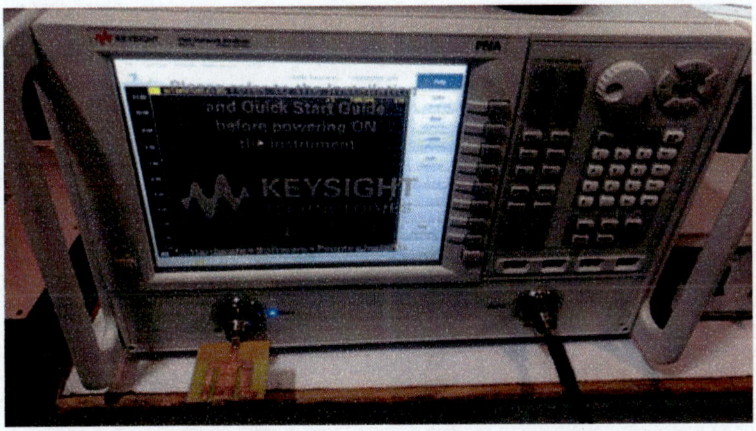

Fig. 14 Measurement setup of VSWR at 7 GHz

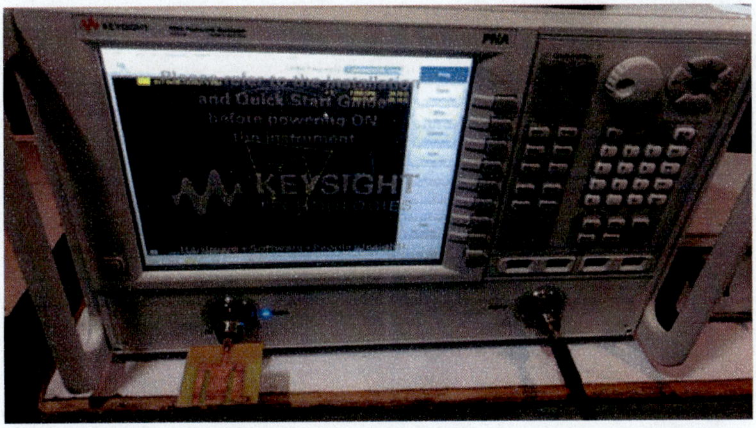

Fig. 15 Measurement setup of Input Impedance at 7 GHz

5 Conclusion

In this paper, multiband patch antenna with band Notch Characteristic in UWB is simulated using HFSS-13. The proposed antenna has advantages of small size, easy fabrication and simple construction. The simulated results of the proposed antenna show that return loss is less than −10 dB and VSWR is less than 1.5. The measured results of this antenna show that the antennas can be good candidates for the four operating frequency of 2.3–3.2 GHz, 5.4–6.2 GHz, 7.4–9.0 GHz and 9.1–10.2 GHz with four resonant frequencies and good return loss values. The gain of antenna is 2.64 dB and radiation efficiency 82% calculated. Microstrip line feeding is used for transmission of EM wave.

Acknowledgements The authors thank Prof. (Dr.) J. P. Agarwal Department of Electronics & Communication Engineering GIT Jaipur, India and Prof. R.K. Malviya Sectary ATMS for providing antenna design, fabrication and characterization facilities.

References

1. Ashtankar PS, Dethe CG (2012) Design and modification of circular monopole UWB antenna for WPAN application. Comput Eng Intell Syst ISSN 2222-1719 (Paper) ISSN 2222-2863 (Online) 3(5)
2. Barba Qinrino S (2010) UWB circular slot antenna provided with an inverted-l notch filter for the 5Ghz WLAN band. Progr Electromag Res PIER 104, 1–13
3. Chang CC, Watanable F, Inamura H (2006) Potential of UWB technology for the next generation wireless communications. In: IEEE ninth international symposium on spread spectrum techniques and applications, pp 422–429
4. Behdad N, Sarabandi K (2005) A compact antenna for Ultrawide- Band applications. IEEE Trans Antennas Propag 53(7)
5. Balanis C (1997) Antenna theory: analysis and design. John Wiley & Sons Inc, New York
6. Hassanien MA, Hamad EKI (2010) Compact rectangular u-shaped slot microstrip patch antenna for UWB applications. In: IEEE APS, middle east conference on antennas and propagation (MECAP), Cairo, Egypt
7. Bhomia Y, Kajla A, Yadav D (2010) V-slotted triangular microstrip patch antenna. Int J Electron Eng 2(1): 21–23
8. Kaur N, Sharma N, Singh N (2017) A study of different feeding mechanisms in microstrip patch antenna. Int J Microwave Appl 6(1)
9. Chen ZN, See SP, Qing X (2007) Small printed ultrawidwband antenna with reduced ground plane effect. IEEE Trans Antennas Propag 55(2)
10. Xi D, Wen LH, Yin YZ, Zang Z, Mo YN (2010) A compact dual inverted c-shaped slots antenna for WLAN applications. Progr Electromag Res Lett 17:115–123
11. Prombutr N, Kirawanich P, Akkaraekthal P (2009) Bandwidth enhancement of UWB micro strip antenna with a modified ground plane. Int J Microwave Sci Technol

Chapter 18
Experimental Investigation for Energy-Conscious Welding Based on Artificial Neural Network

Sudeep Kumar Singh, Suvam Sourav Swain, Amit Kumar, Prashanjeet Patra, Nitesh Kumar, and A. M. Mohanty

1 Introduction

Shielded Metal Arc Welding (SMAW) of Mild Steel (MS) finds wide application in structural frames, pipelines, visually aesthetic designs, and repair due to its high ductility and weldability properties [1–3]. Welding remains the most widely adopted joining process in the industry despite its high energy-intensive property. The selection of proper welding parameters is very important in a multi-input multi-output process like welding [4, 5]. The mechanical properties of welded joints largely depend on process parameters used in the manufacturing process [6]. The welder generally focuses on the quality aspects of the produced joints and pays lesser attention to the process parameters. In practice, the welding process parameters selected imparts a large influence on the resources consumed like joint quality, percentage of rework/rejection, and energy consumed [7]. Power consumption is one among many factors responsible for the negative environmental effects generated from welding operation, raising the need for characterization of the SMAW process considering sustainability aspects [3, 8]. Thus, the present study intends to draw a relationship between the four influential input parameters and the four output parameters adopted for investigation.

S. K. Singh (✉) · S. S. Swain · A. Kumar · P. Patra · N. Kumar · A. M. Mohanty
Centurion University of Technology and Management, Odisha Bhubaneswar, India
e-mail: sudeepuce008@gmail.com

M. Shorif Uddin et al. (eds.), *Intelligent Energy Management Technologies*, Algorithms for Intelligent Systems,
https://doi.org/10.1007/978-981-15-8820-4_18

2 Literature Survey

Adnan et al. [4] carried out Pareto Analysis to find uncontrollable input parameters of the GMAW welding process. They developed three different ANN models for input, output parameter prediction, and classifying products. ANN was also employed for investigating the effects of process parameters in laser welding of AA5754 aluminum alloy [9]. Two parameters welding speed and shielding gas were varied, and the optimization process was implemented using an Excel add-in named Neural Tools. In yet another study, authors developed two different ANN models one for classification of defective products and other for prediction of input parameters [5]. Welding processes have a poor environmental image for which optimization of key welding parameters is very crucial. A hybrid approach involving neural network and fuzzy logic is used for optimizing SMAW process parameters from the sustainability point of view [8]. Current, voltage and welding speed are considered for analysis. Welding of dissimilar metals involving Al alloy and stainless steel has been studied using the laser-arc welding technique [10]. Taguchi is used for studying the effect of various welding parameters to get optimum parameters of angular distortion in SMAW [11]. TIG welding parameter has been optimized using Response Surface Methodology (RSM), central composite design on mild steel [12], and grey wolf optimizer [13] on high strength low alloy 15CDV6 steel. RSM has also been adopted for optimizing GMAW parameters for welding mild steel IS:2062 [14]. Authors [15] have developed model for prediction of mechanical and microstructural properties of copper plate welding using Friction Stir Welding. RPLNN and GA have been used involving three inputs and two response parameters.

Weld quality considering tensile properties and microstructure were analyzed based on power distribution using an arc assisted fiber laser welding of Al–Mg alloy [16]. Tensile and impact properties in multi-pass SMAW have been investigated by Saxena et al. for determining the influence of welding consumables in Armox 500T alloy [17]. Mechanical properties and microstructure of MS welded parts under varying current were analyzed using the E7016 electrode [1]. The highest tensile strength was obtained at 75A with minor welding defects. Sheets of different thicknesses welded using SMAW and GMAW were investigated for finding a new set of welding parameters for structural grade steel welding [6].

The main aim of the current research work is to study the influence of varying input parameters on the output quality of the joint.

The arrangement of the paper is as follows. The experimental methodology is explained in Sect. 3. The next section discusses the outcomes of the experimental and test results. The fifth section discusses the application of ANN for welding parameter selection. The sixth section presents conclusions obtained from the analysis and also provides directions for future scope.

3 Methodology

The strategy followed in the current work can be divided into different sections of which, arc welding, testing for obtaining output data, and selection of input parameters to the welding process based on influential responses of welding are important. Arc welding of mild steel considering energy consumption has been considered in the present investigation. The strategy followed in the current investigation is presented pictorially in Fig. 1.

Mild steel plates of different thicknesses 3, 5, and 10 mm (three levels) were utilized in the welding process. The welding parameters, current, joint gap, and face width were also varied during the experiment. The input parameters considered in the investigation include the welding parameters and the plate thicknesses. The output parameters considered are Ultimate Tensile Strength (UTS), impact energy (Izod), Rockwell hardness, and energy consumption. The input parameters (factors) involved in the study are presented in Table 1.

Mild steel procured in flat form was first cut to a rectangular shape with length 200 mm and width 100 mm. One longitudinal edge of each plate was beveled to produce a double V-groove butt joint. The including angle of the V-shaped joint is 60° for all the plates used. The chemical composition of the plates was tested using XRF spectrometer, and the obtained values are tabulated in Table 2.

The data presented in the table displays close conformance in terms of composition for both the workpiece and filler metals. The filler rod used in the welding process is 3.15 mm in diameter Superweld E6013 manufactured by ESAB. The XRF samples for both material types were prepared by grinding on a surface grinder.

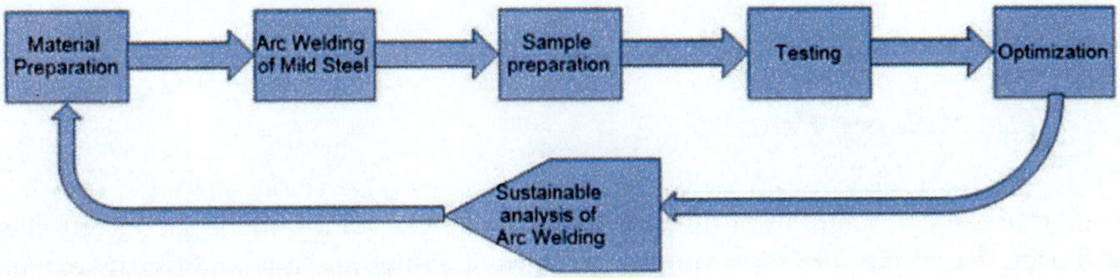

Fig. 1 Experimental methodology

Table 1 Different values for the input variables

Sl. no	Factors	Level 1	Level 2	Level 3
1	Current	100A	110A	120A
2	Plate thickness	3 mm	5 mm	10 mm
3	Root gap	0 mm	1 mm	2 mm
4	Face width	0 mm	1 mm	2 mm

Table 2 Material composition

Sl. no.	Base material	Si	Mn	S	P	Fe
1	Mild steel plate	0.720	0.709	0.132	0.029	96.840
2	Electrode E6013	1.451	0.437	0.125	0.034	96.115

Table 3 Experimental values in the investigation

Sl. no.	Current (A)	Plate thickness (mm)	Root gap (mm)	Face width (mm)	Power (kW)	UTS (MPa)	Hardness (HRB)	Impact energy (J)
1	100	3	0	0	4.73	481	76.4	60
2	100	5	1	1	4.52	411	77.25	62
3	100	10	2	2	5.32	305	83.9	74
4	110	3	1	2	4.59	295	78.1	50
5	110	5	2	0	5.52	501	78.6	52
6	110	10	0	1	5.14	406	85.55	160
7	120	3	2	1	5.88	458	80.6	52
8	120	5	0	2	6.59	362	84.05	112
9	120	10	1	0	5.89	329	82.65	110

The plates were cleaned properly using solvent to remove all dirt, rust present on the surface of the material to be welded. It is followed by welding the plates using process parameters obtained from TAGUCHI orthogonal array design presented in Table 3.

3.1 The Welding Process

Similar to the raw material of three different thickness values, the input current has also been varied into the same number of current values and adopted for the experiments: 100, 110, and 120 amperes. The remaining two input variables adopted are root gap and face width. Three different values were considered for both the variables as 0, 1, and 2 mm. All the varying parameters are taken together, including the plate thickness values, make the total number of factors involved in the experimental design as four. The number of levels for each factor is three. Thus, if the full factorial design of experiments was to be considered, the total number of experiments would become 27. To reduce the number of experiments, Taguchi Design of Experiment (DoE) method was adopted. Using L9 Taguchi orthogonal array design adopting a four-factor and three-level experimental approach, the total number of experimental runs was reduced to 9. The experimental design adopted for the experiments is presented in Table 3.

The welding process was carried out by using RS400 a Thyristorised MMA welding machine manufactured by ESAB India Ltd. The machine is equipped with 50 Hz 3-phase power supply with an input voltage of 415 volts and 27-ampere current. The welding runs were carried out using the AC power supply.

A 3-phase power analyser, model no DPATT-3Bi, manufactured by Uma Electronics Enterprises, Jaipur, India, was used for measuring the instantaneous power consumption values during the arc welding process. A three-phase four-wire connection was used in the process of measurement.

Table 3 presents the four factors and the values of the three levels of process parameters adopted in the experimental runs. It displays the values of different process parameters used in the welding process. Four different parameters welding current, plate thickness, root gap, and face width are used for designing nine number of experiments in total.

The welding speed was considered constant throughout the experiment. The plates of 3 mm thickness were welded using a single pass of welding, but multiple runs were necessary for plates with 5 mm and 10 mm thickness. The former was welded with two passes, and for the latter three number of welding, passes were used. In total, nine numbers of welding joints were produced and processed further for preparing test samples for tensile. Rockwell and Izod impact tests to be conducted further. The details of the test procedure and results have been explained in the next section.

4 Post-weld Testing

The welded steel plates were cleaned to remove the slag deposited during welding by using a chipping hammer and wire brush. Tensile, hardness, and Izod test specimens were extracted from the welded plates of different thicknesses with the respective dimensions, presented in Fig. 2.

Welding beads were removed by grinding operation from the welded surface for both the tests.

The tensile test was conducted on a Universal Testing machine manufactured by Blue Star Engineering & Electronics Ltd., having a maximum capacity of 1000 kN. The test specimens were made to undergo the tensile testing procedure, and the Ultimate Tensile Strength values for each test specimen were noted. The average value of HRB was calculated after measuring hardness values at two different points on the weld bead surface. The samples prepared for the Izod test were carried out using Impact test machine, and values of energy absorbed before failure for individual specimen were recorded. The values of UTS, HRB, and energy absorbed have been presented in Table 3 under respective columns.

Figure 3 displays phases of sample preparation for different tests after conducting the tensile, Rockwell, and Izod tests. The Fig. 4a displays the Impact testing machine, and 4 (b) depicts the Rockwell hardness testing machine used for the experimentation.

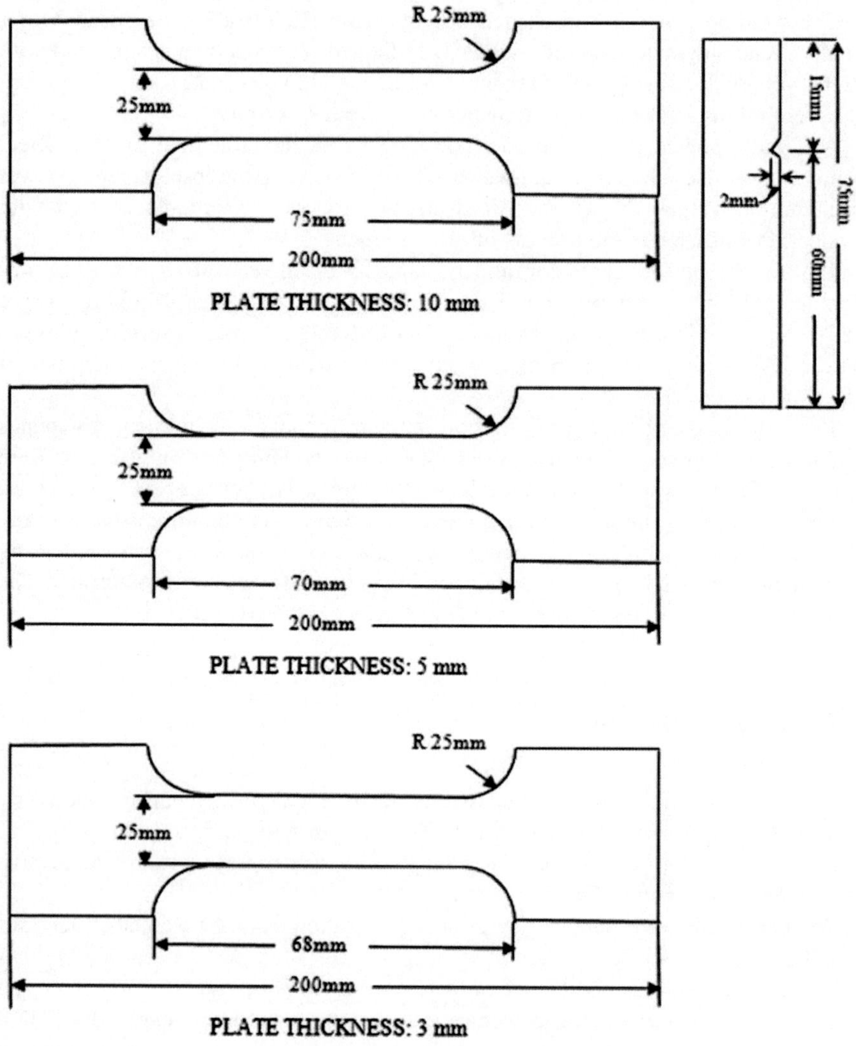

Fig. 2 Schematic diagrams of **a** Tensile Test specimen; **b** Izod specimen

5 Parameter Selection Using ANN

Neural networks find a wide application and recognized as efficient solvers of non-linear problems. Successful applications have been reported in literature containing real-world problems. Thus, ANN has been selected for finding optimum input parameters for SMAW in the present study. The architecture for the employed neural net is presented in Fig. 5.

Fig. 3 Pictorial representations of **a** Tensile Test; **b** Hardness; **c** Izod Specimens

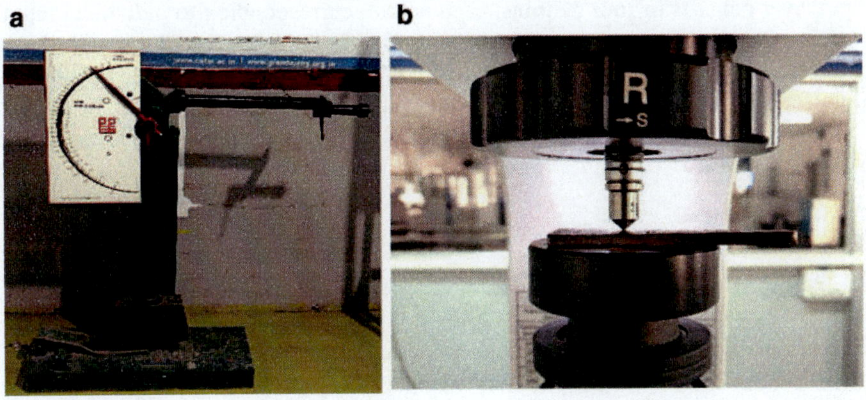

Fig. 4 **a** Impact testing machine; **b** Rockwell hardness testing machine

An Artificial Neural Network was modeled for training using the data collected from the conducted experiments. The Bayesian Regularization backpropagation method is used for the construction of the network. This method is generally used for difficult, small, and noisy datasets (Fig. 6).

In the current construction, the data set is small and prone to noise in the measured value; thus, the application of Bayesian Regularization fits our requirement. '*trainbr*' learning function is used in the Matlab R2019a platform. The network takes 70% of

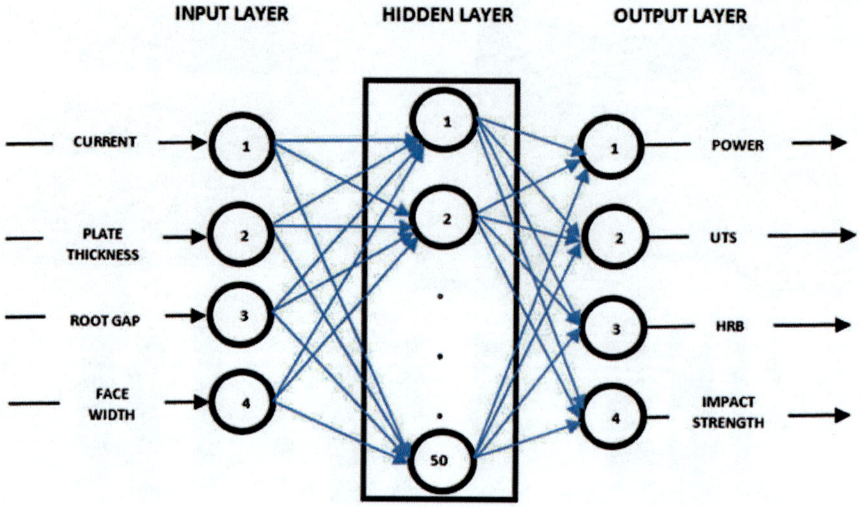

Fig. 5 ANN architecture

data for training, 15% for validation, and 15% for testing. The ANN model developed in this study involves an input layer, one hidden layer, and one output layer. The input layer consists of four neurons; each neuron corresponding to individual input parameters and the output layer containing four neurons, representing one output parameter each. The hidden layer employs 50 neurons. The most promising network architecture is based on the trial and error method for which many trials have been conducted to arrive at the best combination. The performance of the network has been discussed in detail in the conclusion section.

6 Conclusion

The current work involves four input and four output variables for SMAW welding of structural grade mild steel. The quality of the welding has been tested by measuring UTS of the welded joint by applying load in the transverse direction, measuring the impact energy absorbed by the joint before failure, hardness on the bead surface, and also the power consumed for joint preparation. The input and output were fed into an ANN network suitably designed for the purpose. The modeled network is capable of selecting all the four types of input parameters considered in the present work based on desired values of output parameters like energy consumed, UTS hardness, and impact energy. This work can be extended to other welding methods. Other crucial variables not considered in the present work may be considered as future research scope.

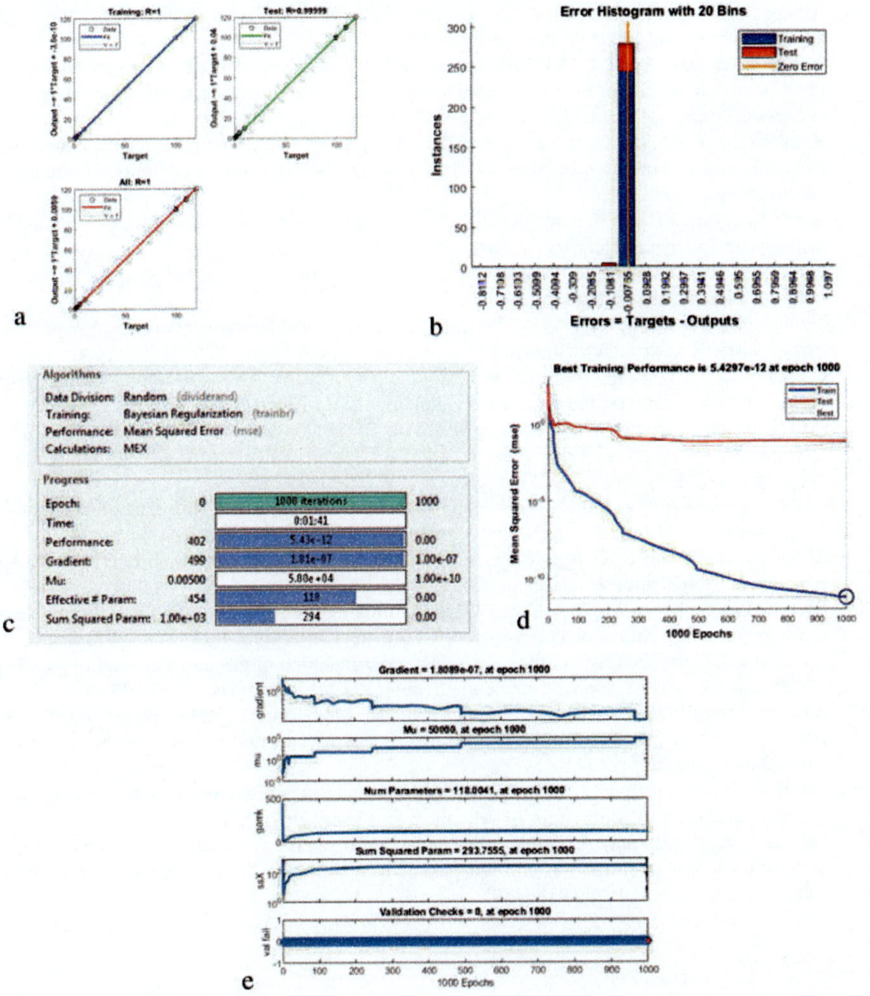

Fig. 6 **a** Regression; **b** error histogram; **c** performance; **d** training progress; **e** training state

References

1. Faqih IA, Ma'arif S, Sukarjo H (2019) The effect of current variation on mma welding to mechanical properties and microstructure of mild steel
2. Ahmed AN, Noor CM, Allawi MF, El-Shafie A (2018) RBF-NN-based model for prediction of weld bead geometry in Shielded Metal Arc Welding (SMAW). Neural Comput Appl 29(3):889–899
3. Alkahla I, Pervaiz S (2017) Sustainability assessment of shielded metal arc welding (SMAW) process. In: IOP conference series: materials science and engineering, IOP Publishing, 244(1):012001

4. Aktepe A, Ersöz S, Lüy M (2014) Welding process optimization with artificial neural network applications. Neural Netw World 24(6):655–670
5. Aktepe A, Ersöz S, Lüy M (2012) Backpropagation neural network applications for a welding process control problem. In: International conference on engineering applications of neural networks. Springer, Berlin, Heidelberg, pp 172–182
6. Khamari BK, Dash SS, Karak SK, Biswal BB (2019) Effect of welding parameters on mechanical and microstructural properties of GMAW and SMAW mild steel joints. Ironmaking Steelmaking:1–8
7. Singh SK, Samal BK, Pradhan SR, Ojha SR, Saffin MD, Mohanty AM (2019) Sustainable analysis of TIG parameters for welding aluminum alloy considering joint gap and welding current. In: International conference on application of robotics in industry using advanced mechanisms. Springer, Cham, pp 316–323
8. Vimal KEK, Vinodh S, Raja A (2017) Optimization of process parameters of SMAW process using NN-FGRA from the sustainability view point. J Intell Manuf 28(6):1459–1480
9. Casalino G, Facchini F, Mortello M, Mummolo G (2016) ANN modelling to optimize manufacturing processes: the case of laser welding. IFAC-PapersOnLine 49(12):378–383
10. Gao M, Chen C, Mei S, Wang L, Zeng X (2014) Parameter optimization and mechanism of laser–arc hybrid welding of dissimilar Al alloy and stainless steel. Int J Adv Manuf Technol 74(1–4):199–208
11. Arifin A, Gunawan AM, Yani I, Pratiwi DK, Yanis M, Sani KA (2019) Optimization of Angular Distortion on Weld Joints Using Taguchi Approach. Jurnal Kejuruteraan 31(1):19–23
12. Srivastava S, Kumar S, Garg RK (in press) A multi-objective optimisation of TIG welding parameters using response surface methodology
13. Skariya PD, Satheesh M, Dhas JER (2018) Optimizing parameters of TIG welding process using grey wolf optimization concerning 15CDV6 steel. Evol Intel 11(1–2):89–100
14. Srivastava S, Garg RK (2017) Process parameter optimization of gas metal arc welding on IS: 2062 mild steel using response surface methodology. J Manufact Process 25:296–305
15. Azizi A, Barenji A, Barenji R, Hashemipour M (2016) Modeling mechanical properties of FSW thick pure copper plates and optimizing it utilizing artificial intelligence techniques. Sensor Netw Data Commun 5(142):2
16. Leo P, Renna G, Casalino G, Olabi AG (2015) Effect of power distribution on the weld quality during hybrid laser welding of an Al–Mg alloy. Opt Laser Technol 73:118–126
17. Saxena A, Kumaraswamy A, Reddy GM, Madhu V (2018) Influence of welding consumables on tensile and impact properties of multi-pass SMAW Armox 500T steel joints vis-a-vis base metal. Defence Technol 14(3):188–195

Chapter 19
Detection of GSM Signal Using Energy Detection and Matched Filter-Based Techniques

Bablu Kumar Singh and Sanjay Bhandari

1 Introduction

The increase in demand of mobile communication and its reach to almost every person the bandwidth requirement has also increased, and current spectrum allocation technique is being over utilized in some areas and underutilized in other areas. To detect the presence of signal in GSM band, we have used energy detection and matched filter detection method, and using Cognitive Radio (CR) techniques the spectrum which are unused can be used to increase the spectrum efficiency. Cognitive Radio enables the user to determine the presence of primary user/signal, over available spectrum, if Primary User (PU) is not present then the available spectrum is considered as white spaces which can be made available for unlicensed users. To allocate the white spectrum for secondary users using spectrum division and spectrum mobility for spectrum utilization [1–3]. Spectrum Sensing is the most important and critical task to establish cognitive radio networks for spectrum utilization in GSM network [4–6].

B. K. Singh
MBM Engineering College, JNV University, Jodhpur, Rajasthan, India
e-mail: bablukumar.singh@gmail.com

S. Bhandari (✉)
Jodhpur Institute of Engineering and Technology, Jodhpur, Rajasthan, India
e-mail: sanjaybhandari41@gmail.com

M. Shorif Uddin et al. (eds.), *Intelligent Energy Management Technologies*,
Algorithms for Intelligent Systems,
https://doi.org/10.1007/978-981-15-8820-4_19

2 Spectrum Sensing

To enhance the detection probability in the available spectrum, Spectrum detection techniques are used. Sensing is done across Geographical Space, Frequency, Time, Phase, and code. The analysis of spectrum is based on several factors such as radio frequency spectrum used by mobile devices, network conditions, and communication protocol. The objective of Spectrum Sensing [6, 7] is to detect the presence of transmissions from primary users over an assigned channel in GSM network. If PU is absent, then how secondary users use the primary spectrum without disrupting the communication of primary user. This can only be possible if by some method we can detect the channel and identify signal is present on channel or not. To detect the presence of signal, we have considered two methods [8] in this paper. After detection and analysis of presence of signal characteristics, the channel characteristics of GSM can be determined which can be utilized for channel optimization using Store Forward Base Trans-receive System [9].

2.1 *Energy Detection Method*

Energy detection is a method which is used to detect the presence of signal over assigned spectrum. It is easier method because it does not require prior knowledge of type of signal available on channel. It is based on the power measurement of the received signal. The signal block for energy detection is shown in Fig. 1

In Fig. 1, the threshold decision may be made on two hypotheses for detection of primary and secondary user signals, first hypothesis is H_0 and the other hypothesis is H_1. H_0 is the case when signal is absent, and H_1 describes the case where signal is present. And by measuring the energy we can estimate hypothesis to decide whether H_0 or H_1 is correct. In energy detection, we have two general performance matrices which are used to evaluate the performance [8, 10, 11]. The hypothesis considered for analysis of presence/absence of primary users are H_0 and H_1.

H_0: Primary user is absent

$$y(n) = x(n)n = 1, 2, 3 \ldots N \quad (1)$$

H_1: Primary user present

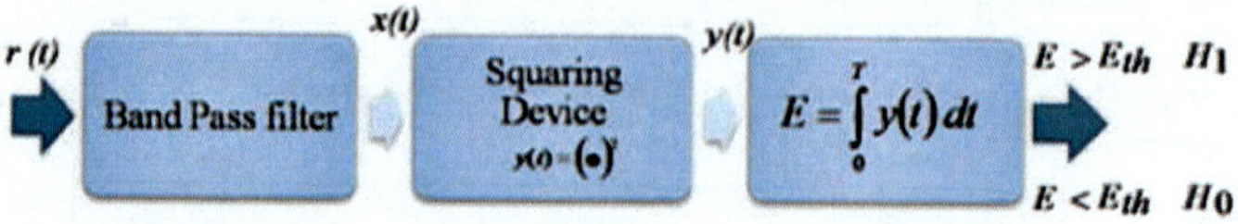

Fig. 1 Energy detector block

$$y(n) = x(n) + u(n)n = 1, 2, 3 \ldots N \tag{2}$$

where *u (n)* is noise and *x(n)* is the primary signal.

Sometimes the signal is detected and sometimes the noise acts like signal but it is very important to identify the signal which may be considered as probability of detection and probability of false alarm. The probability of detection P_d is defined by

$$P_d = P_r(Y > \lambda/H_1) \tag{3}$$

And the probability of false alarm P_f is defined by

$$P_f = P_r(Y > \lambda/H_0) \tag{4}$$

where λ is threshold value.

2.2 Matched Filter Detection Method

This method is a detection technique which performs a coherent detection of the primary signal [6]. Spectrum sensing using matched filter requires synchronization with the primary system and also able to demodulate the primary signal, i.e., the prior information about the primary system must be known to secondary sensing user like preamble signaling, channel estimation, modulation type of signal being transmitted over the GSM network. To detect signals with maximum Signal to Noise Ratio (SNR), matched filter detection techniques may be used at receiver. Matched filter is a linear filter that works on the phenomena to maximize the output signal to noise ratio and can be applied to cognitive radio user having evidence about the primary signal. Then, this detection method may be expressed as [8]

$$Y[n] = \sum_{K=-\infty}^{\infty} h[n-k]x[k] \tag{5}$$

where x is unknown signal convolved with h, the impulse response of matched filter. Detection by using matched filter is useful only when the information about the source signal is known.

3 Simulation and Results

Simulation provides interactive access to check the performance and comparative analysis for energy detection and matched filter detection method theoretically as well as practically by varying various parameters. The sample data is collected from USRP2920 in GSM band, and this data is being used for analysis. The detection techniques are applied on the received signal and processed data using MATLAB and are compared on the basis of probability of detection and probability of false alarm with AWGN channel. The results are shown with different SNR value. The change in observation of probability of detection and probability of false alarm is shown in Fig. 1.

The analytical and simulated performances of energy detection are shown in Fig. 2 with SNR with detection probability and varying probability of false alarm. The simulated result displays that with the increase in probability of false alarm from 10^{-5} to 10^{-1}, the probability of detection raises, approaching a similar tendency to the simulation. At high SNRs, the effect of probability of false alarm is negligible as the detector can achieve high detection probability. This is observed near 5 dB SNR, where the probability of finding the probability of false alarm values under consideration is at least 0.9. As the SNR declines, the effect of the various probability of false alarm values becomes easily distinguishable, i.e., at –5 dB SNR, the probability of detection is 0.4 for the probability of false alarm of 10^{-1} and 0.12, for the probability of false alarm of 10^{-2}.

Detection probability with varying SNR value can be observed in Fig. 3, which shows that as the SNR increases, the detection probability increases and the probability of false alarm decreases it means we should keep high SNR for detection of signal.

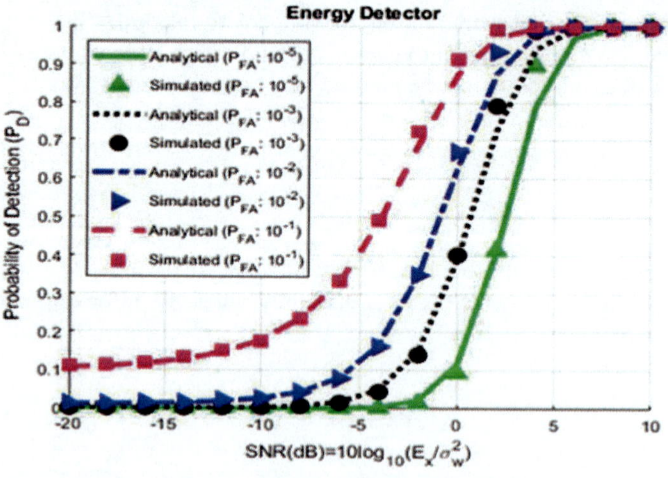

Fig. 2 Performance of energy detection (SNR, Vs, P_D under varying P_{fa})

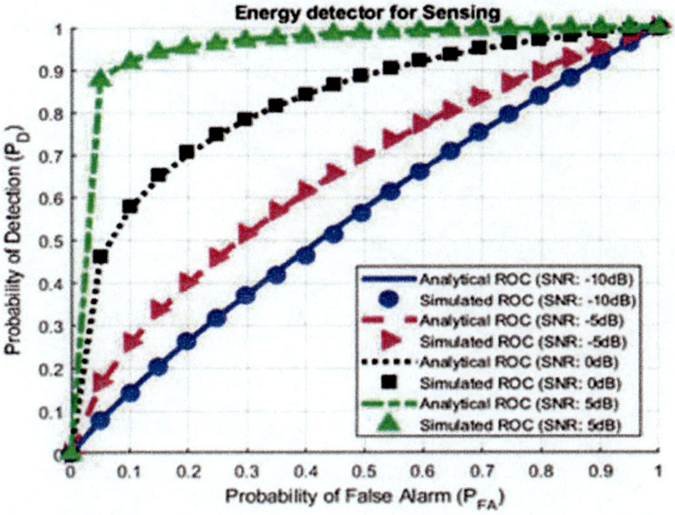

Fig. 3 Performance of energy detection (P_{FA}, Vs, P_D under varying SNR)

The analytical performance match filter in Fig. 4 shows that as we increase the probability of false alarm increases from 10^{-5} to 10^{-1}, the probability of detection increases, approaching an alike tendency to the simulation performance.

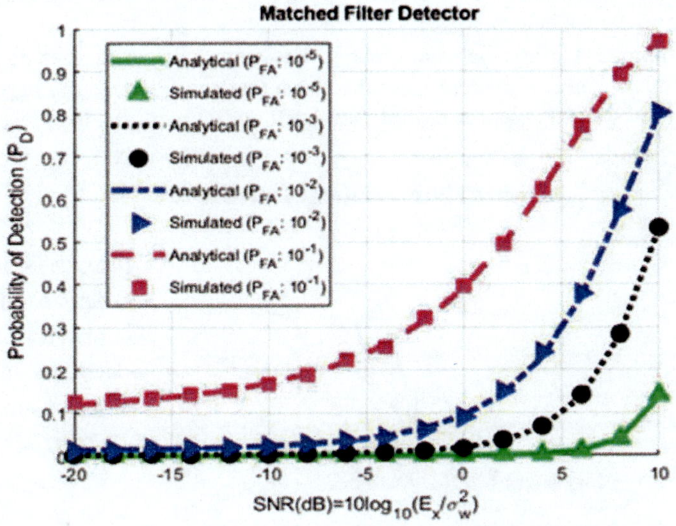

Fig. 4 Performance of matched filter detection (SNR, Vs, P_D under varying P_{fa})

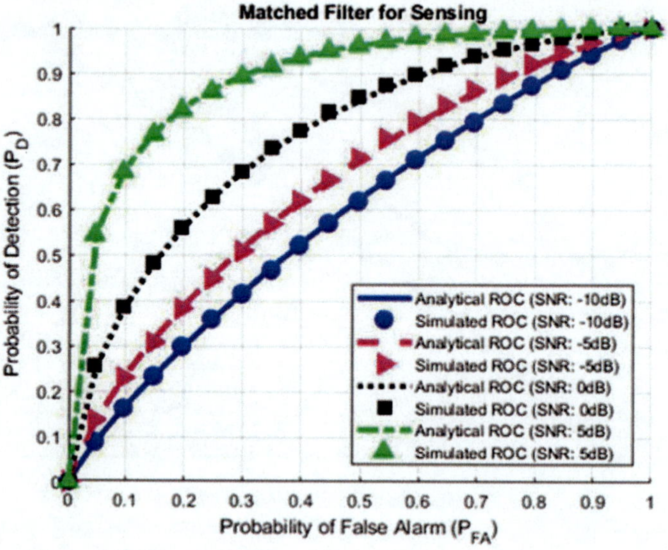

Fig. 5 Performance of matched filter detection (PFA, Vs, P_D under varying SNR)

The performance of match filter is shown in Fig. 5, at high SNR the effect of probability of false alarm is maximum. This can be observed at 5 dB SNR. The probability of detection for the probability of false alarm values under consideration is varying from 0.01 to 0.7. As the SNR declines, the effect of the various probabilities of false alarm values is lowest.

The comparative graph between energy detection and match filter detection is shown in Fig. 6, and it is evident from the graph that energy detection performance is poor at low SNR, it requires a minimum SNR for its working. The result shows

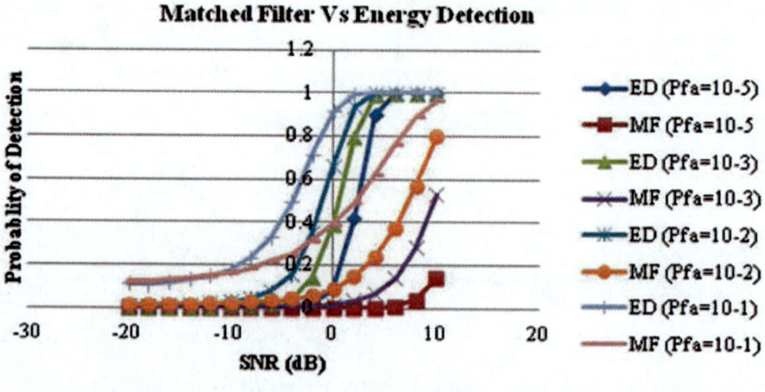

Fig. 6 Energy detection versus matched filter under varying P_{FA}

that energy detection starts working at –10 dB of SNR in our simulated result while matched filter detection performance is better under lower SNR values.

4 Conclusion

The performance of energy detection and matched filter increases with increasing SNR as the probability of detection of signal increases and correspondingly the probability of false detection decreases. At lower SNR, matched filter performance is better than energy detection (Below –5 dB). It is evidenced from Fig. 6 that with increase in SNR value the performance of energy detection improves over matched filter and for lower value of SNR matched filter method performance is improved compared to energy detection method.

References

1. Ali et al (2017) Advances on spectrum sensing for cognitive radio networks: theory and applications. IEEE Commun Surv Tutor 19(2):1277–1304
2. Dobariaa A, Sodhatarb S (2015) A literature survey on efficient spectrum utilization: cognitive radio technology. Int J Innov Emerg Res Eng 2(1)
3. Shahzad A et al (2010) Comparative analysis of primary transmitter detection based spectrum sensing techniques in cognitive radio systems", Aust J Basic Appl Sci 4522–4531. INS Inet Publication
4. Haykin S, Thomson DJ, Reed JH (2009) Spectrum sensing for cognitive radio. IEEE Proc 97(5):849–877
5. Arjoune Y, Mrabet ZE, Ghazi HE, Tamtaoui A (2018) Spectrum sensing: enhanced energy detection technique based on noise measurement. In: IEEE 8th annu comput commun workshop conf (CCWC), Las Vegas.
6. Abbas N, Nasser Y, Ahmad KE (2015) Recent advances on artificial intelligence and learning techniques in cognitive radio networks. EURASIP J Wirel Commun Network.
7. Masonta MT, Mzyece M, Ntlatlapa N (2013) Spectrum decision in cognitive radio networks: a survey. IEEE Commun Surv Tutor 15(3):1088–1107, 3rd Quart.
8. Salama U, Sarker PL, Chakrabarty A (2018) Enhanced energy detection using matched filter for spectrum sensing in cognitive radio networks. In: Joint 7th international conference on informatics, electronics & vision (ICIEV) and 2018 2nd international conference on imaging, vision & pattern recognition (icIVPR), Kitakyushu, Japan, pp 185–190
9. Singh BK, Bhadada R (2020) Enhancing spectral efficiency of mobile communication by store forward base transceiver system, 2020 7th Int Conf Signal Process Integrated Networks (SPIN), Noida, India, 2020, pp. 781–784, http://doi.org/10.1109/SPIN48934.2020.9071058
10. Talukdar J, Mehta B, Aggrawal K, Kamani M (2017) Implementation of SNR estimation based energy detection on USRP and GNU radio for cognitive radio networks. In: IEEE WiSPNET
11. Eduardo AF, Caballero RGG (2015) Experimental evaluation of performance for spectrum sensing: matched filter vs. energy detector. In: IEEE Colombian conference on communication and computing, Popayan, Colombia

Chapter 20
Simulation of Performance Characteristics of Different PV Materials

Harish Kumar Khyani and Jayashri Vajpai

1 Introduction

The research in the area of generation of solar electricity is motivated by both improvement of the performance for different Photovoltaic (PV) materials and by a decrease in the technological realization cost. Study and development of new photovoltaic materials is an emerging area of research. The variety of photovoltaic semiconductor materials, developed over the past two decades, possess different optical, electrical, and mechanical properties. However, the deployment of these new materials and their associated design technologies for manufacturing solar cell depend essentially on the availability of material and often impose an elevated price. These different materials are characterized by their typical properties, which can considerably influence the price and the performance of the variety of solar cells manufactured by employing them. The present stake in the growth of photovoltaic systems is to find a compromise between the cost and performance to guide the selection of photovoltaic material for designing modules for bulk energy generation. MATLAB/Simulink-based modeling of photovoltaic module has already been accomplished by the authors [1] in a previous work. This model can be adapted to represent different types of photovoltaic materials by changing certain parameters. The simulation of performance features of six commonly used different types of photovoltaic materials, and their comparative study is offered in this paper. This paper is divided into five sections. The next section discusses the PV module technologies being used presently. The third section presents MATLAB/Simulink-based model of PV system followed by the simulation of different PV materials. Finally,

H. K. Khyani (✉) · J. Vajpai
M.B.M Engineering College, JNVU, Jodhpur, India
e-mail: khyani.harish@gmail.com

J. Vajpai
e-mail: jvajpai@gmail.com

M. Shorif Uddin et al. (eds.), *Intelligent Energy Management Technologies*, Algorithms for Intelligent Systems,
https://doi.org/10.1007/978-981-15-8820-4_20

the comparison of the performance characteristics of these models has been carried out before concluding comments.

The research for mathematical modeling of PV cells was initiated by physicists in the year 1883 but engineering-oriented modeling began quite recently around the year 2000. With the emergence of computational modeling techniques along with increased interest of researchers toward simulation-based studies, the use of MATLAB software has become the ruling trend in this research area. Tsai et al. [2] pioneered the progress of generalized MATLAB/Simulink® model for solar PV cell to examine the consequence of solar irradiance and cell temperature. This was followed by a circuit-based simulation model designed by González-Longatt [3], in order to evaluate the electrical behavior of the cell with respect to changes in temperature and irradiance. Further, Bhatt and Thakker demonstrated MATLAB-based simulation of photovoltaic array at different temperatures to obtain their electrical characteristics as a function of temperature [4]. Yousef Mahmoud, Xiao, and Zeineldin have simplified the model and extended it for the PV modules while employing precise estimates of the model parameters directly from manufacturer datasheets [5]. Panwar and Saini have studied the problem of model parameter determination based on the four-parameter model using MATLAB/Simulink [6].

These researchers simulated MATLAB-based models of photovoltaic modules of crystalline silicon to examine the effect of temperature and irradiance on the performance characteristics. The authors of this paper have attempted to develop a simplified MATLAB/Simulink model that has been validated for performance characteristics by comparison with commercially available polycrystalline module from Easy Photovoltaic Private Limited, Ghaziabad, Uttar Pradesh. Then the developed model has been extended to simulate presently used commercial materials like Monocrystalline Silicon (Si-mono), Polycrystalline Silicon (Si-poly), Amorphous Silicon (a-Si) and the emerging like Cadmium Telluride (CdTe), Copper Indium Selenide (CIS), and Gallium Arsenide (GaAs).

2 PV Module Technologies

Several crystalline silicon based and thin-film PV technologies have been demonstrated commercially on a large scale in the past few years. In addition, several emerging PV technologies may be technically and economically competitive in the future. This subsection briefly describes some of these PV module technologies and compares their performance, particularly conversion efficiency. Efficiency is a significant characteristic for comparison of performance of different materials for design of modules. The efficiency of a PV cell or module is defined as the percentage of the solar energy striking the cell or module that is converted into electricity [7].

The PV materials considered here can be broadly classified as follows:

2.1 Crystalline Silicon PV Technologies

Crystalline Silicon (c-Si) technologies that constitute about 80% of the current PV market have reliable performance with working lifetime of more than 25 years. There are two types of crystalline silicon PV technologies: monocrystalline (Si-mono or Sc-Si) and multi-crystalline (Si-poly or mc-Si). The rated DC efficiencies of standard c-Si PV modules are about 14–16% [7].

2.2 Thin-Film PV Technologies

Thin-film PV cells have a few microns (μm) thick semiconductor layer which is about 100 times thinner than commercial c-Si cells. The most common thin-film semiconductor materials include Amorphous Silicon (a-Si), Cadmium Telluride (CdTe), Gallium Arsenide (GaAs), and alloys of Copper Indium Selenide (CIS). Thin-film modules have lower DC efficiencies than c-Si modules: about 9–12% for CdTe and 6–9% for a-Si. CdTe-based PV has proficient expressively higher market growth during the last decade than the other thin-film PV technologies [7].

3 Design of MATLAB-Simulink Model of PV System

The equivalent circuit and mathematical equations used to design MATLAB-Simulink model of the generalized PV cell have been described in this section. The model of [1] has been further extended for modeling a module and an array as described in Sect. 3.2 and simulated as described in Sect. 3.3. It is modified to suit the modeling of cells of different important materials in Sect. 3.4 to build a generalized model that is suitable for scaling at all levels of model, i.e., the PV cell, module, and an array.

The generalized PV model has been developed by using the following nonlinear voltage-current (V-I) characteristic equation of a solar cell [8]:

$$I = I_{PH} - I_S\left\{\exp^{\left[\frac{q(V+IR_S)}{AkT_C}\right]} - 1\right\} - \frac{V + IR_S}{R_{SH}} \tag{1}$$

where

I_{PH} Photocurrent,

I_S Saturation current,

q Charge on an electron, 1.6×10^{-19}C,

k Boltzmann constant $= 1.38 \times 10^{-23}$ J/K,

A Ideality factor,

T_C Working temperature,

R_S Series resistance,

R_{SH} shunt resistance.

The series loss and the leakage to ground is neglected in this model, i.e., $R_S = 0$ and $R_{SH} = \infty$.

The photocurrent I_{PH} is obtained by

$$I_{PH} = \left[I_{SC} + K_I\left(T_C - T_{Ref}\right)\right]G \tag{2}$$

where

I_{SC} Short-circuit current.
K_I Short-circuit current temperature coefficient.
T_{Ref} Reference temperature.
G Solar insolation.

The saturation current I_S is obtained by

$$I_S = I_{RS}\left(\frac{T_C}{T_{Ref}}\right)^3 \exp\left[\frac{qE_G\left(\frac{1}{T_{Ref}} - \frac{1}{T_C}\right)}{kA}\right] \tag{3}$$

where

I_{RS} Reverse saturation current.
E_G Energy bandgap of the semiconductor material used.

The reverse saturation current I_{RS} is represented by

$$I_{RS} = \frac{I_{SC}}{\exp\left(\frac{qV_{OC}}{kT_C A}\right) - 1} \tag{4}$$

Finally, the output current of the PV cell is given as follows:

$$I = I_{PH} - I_S\left[\exp\left(\frac{qV_{OC}}{kT_C A}\right) - 1\right] \tag{5}$$

These equations form the base for the development of model.

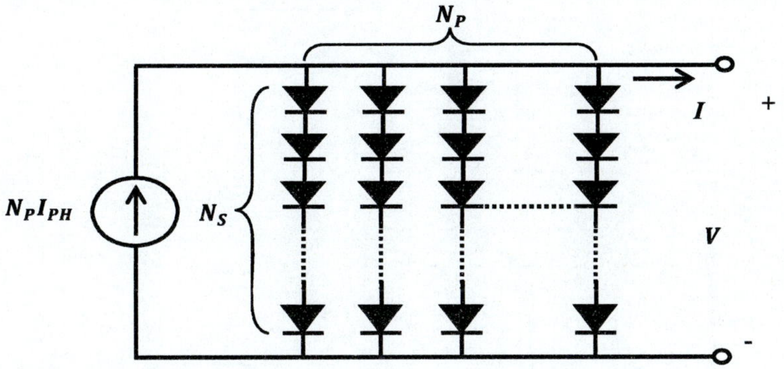

Fig. 1 Ideal model of PV module

3.1 Modeling of PV Module and Array

Since the PV cell produces less than 2 W at 0.5 V approximately, the model with a particular power rating requires that the number of cells must be connected in series–parallel configuration. Finally, PV array is a group of several PV modules connected in series and parallel circuits to generate the required current and voltage.

The Eqs. 4 and 5 can be modified by including these circuits to obtain the output current of a PV module. Hence, I_{RS} and I with Np parallel cells and Ns series cells are given as follows:

$$I_{RS} = \frac{I_{SC}}{\exp\left(\frac{qV_{OC}}{N_S k T_C A}\right) - 1} \tag{6}$$

$$I = N_P I_{PH} - N_P I_S \left[\exp\left(\frac{qV_{OC}}{N_S k T_C A}\right) - 1\right] \tag{7}$$

The circuit of an ideal PV module with Np parallel cells and Ns series cells is shown in Fig. 1 [2].

3.2 Simulation of PV Model

With the help of mathematical equations given in this section, simulation of model is carried out as explained in the following steps:

Step—1 Model I_{PH} using Eq. 2.
Step—2 Model I_S using Eq. 3.
Step—3 Model I_{RS} using Eq. 6.
Step—4 Model I using Eq. 7.

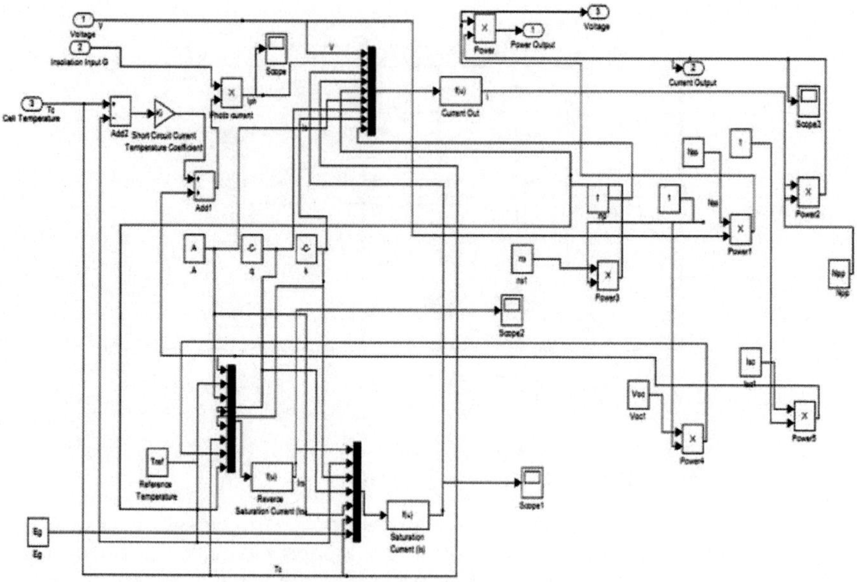

Fig. 2 MATLAB/simulink model of PV module

Step—5 The four models developed are finally interconnected as shown in Fig. 2.

A masked model is designed with dialog box, where the parameters of PV module can be configured in the same manner in case of standard Simulink block library. The generalized model was adapted to represent Si-mono PV module, and then it is simulated at Standard Test Conditions (STC) defined as temperature = 25 °C, Irradiance = 1 kW/m^2 and AM = 1.5, and further validated with respect to a reference model.

3.3 *Simulation of PV Model*

The reference module for the simulation is PSS1237, which is commercially available mono-crystalline module from Easy Photovoltaic Private Limited, Ghaziabad (UP). The technical parameters specified in manufacturer datasheet of this module are given in Table 1 [9].

Comparison of proposed model values with reference model values at remarkable points is summarized in Table 2.

This table depicts the I-V and PV simulation, and experimental results show a good agreement in terms of current at maximum point, voltage at maximum point, and maximum power. Hence, the designed model has been validated with a reasonable accuracy. However, Table 2 shows that the modeling error in power lies in the range

Table 1 Technical parameter specifications of PSS1237PV module

Characteristics	Specifications
Nominal voltage, ***V*** (volts)	12
Open circuit voltage, ***Voc*** (volts)	21
Voltage at maximum power, ***Vmax*** (volts)	16.8
Short circuit current, ***Isc*** (Amps)	2.55
Current at maximum power, ***Imax*** (Amps)	2.2
Maximum rated power, ***Pmax*** (watts)	37
No. of cells/module	36

The electrical specifications are under standard test conditions of irradiance of 1 kW/m^2, cell temperature of 25 °C, and AM of 1.5
Permissible range of error is ± 3%

Table 2 Comparison of proposed model values with reference model values (temperature = 25 °C) and irradiation G = 1 kW/m^2)

Remarkable points	Reference model value	Proposed model value
Imax	2.2	2.55
Vmax	16.8	22
Pmax	37	45.24

5–9%. The effect of series and shunt resistance has not been taken into account in the model for the sake of simplicity, i.e., $Rs = 0$ and $Rsh = \infty$. Hence, the error that has been introduced on this account is more than ± 3% which is admissible as per the limit specified on the manufacturer's datasheet.

3.4 Model for Comparative Study

The masked model developed is now available in the form of user-friendly interface that follows the style of standard MATLAB/Simulink block library components, and an attempt has been made in this paper to use the developed model for conducting simulative studies by considering the effect of change of material as shown in Fig. 3. This model is adapted by using suitable parameters in order to represent the following materials, which are commonly used for fabrication of PV modules:

- Monocrystalline Silicon (Si-mono)
- Polycrystalline Silicon (Si-poly)
- Amorphous Silicon (a-Si)
- Cadmium Telluride (CdTe)
- Copper Indium Selenide (CIS)
- Gallium Arsenide (GaAs).

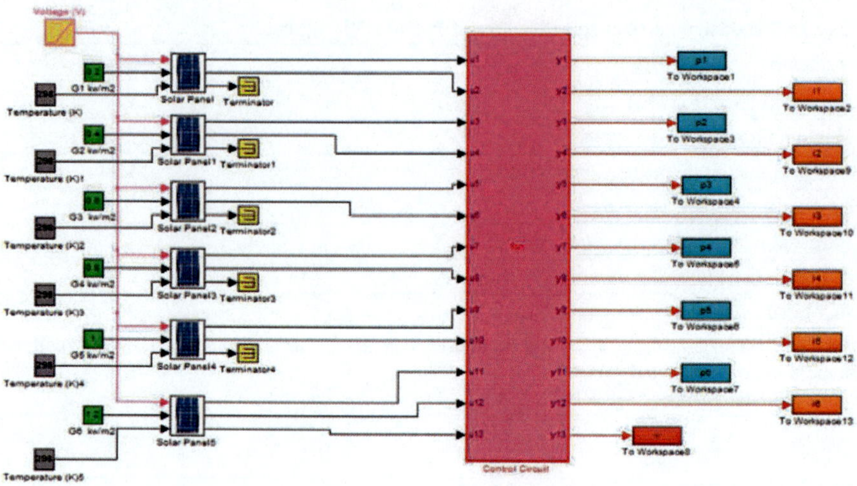

Fig. 3 Final model

The results of simulation for these different materials have been presented in the next section.

4 Simulation of Model for Different PV Materials

Six different models based on commonly used PV materials were simulated by incorporating their relevant parameters in the proposed model. These are as follows:

- Band-gap energy and
- Ideality factor.

The ideality factor and bang-gap energy are dependent on PV technology [9] and is listed in Table 3.

These six models hence developed were simulated at STC. The results are tabulated in Table 4.

Table 3 Ideality factor (A) and energy bandgap EG (in eV at 300 k) for PV technology [2]

Technology	Ideality factor	Energy bandgap
Si-mono	1.2	1.1
Si-Poly	1.3	1.1
a-Si	1.8	1.7
CdTe	1.5	1.49
CIS	1.5	2.42
GaAs	1.3	1.43

Table 4 Comparison of different PV modules at STC (temperature = 250 C) and irradiation G = 1 kW/m^2)

Name of material	Imax	Vmax	Pmax	Fill factor (Pmax/Voc*Isc)	%Efficiency (Pmax/G*Ac)*100)
Si-mono (sc-Si)	2.55	22	45.24	0.8448	11.51
Si-poly (mc-Si)	2.55	23.5	40.16	0.7947	10.82
a-Si	2.54	23.5	38.76	0.7499	10.21
CdTe	2.52	28	35.83	0.669	9.12
CIS	2.52	28	35.83	0.669	9.12
GaAs	2.55	23.5	40.16	0.7947	10.82

Table 4 shows that the efficiency of Si-mono is the highest among all the materials at standard test conditions closely followed by Si-poly. It can also be seen that the materials having same ideality factor having same range of output current, voltage, and power like Si-poly/GaAs and CdTe/CIS.

Although the development of new materials like CdTe, CIS, and GaAs has been received by the academic community with large fan force, they have not received favor of practical venture capitalist. Hence, Si-mono has emerged as the best choice among the six studied materials. The added advantage is that its cost is lowest among these materials and is available in abundance, which further justifies its widespread popularity.

5 Conclusion

A MATLAB/Simulink PV model has been developed for comparing the performance of six different PV materials in this paper. The proposed model was validated with PSS1237. The simulation results revealed a good agreement with the practical data from the manufacturer, Easy Photovoltaic Private Limited, Ghaziabad (Uttar Pradesh). The manufacturer's datasheet specifies that the rated power may vary by ±3%. However, Table 2 shows that the modeling error in power is somewhat higher than that of specified limit. The influence of series and shunt resistance has not been taken into account for the sake of simplicity, i.e., $Rs = 0$ and $Rsh = \infty$. Hence, the error that has been introduced on this account is more than ±3% admissible as limit specified on the manufacturer's datasheet. The validated model is used for carrying out study of PV modules manufactured from six commonly used PV materials. These were simulated by incorporating their relevant parameters in the proposed model. Their simulation for studying the behavior of different materials has depicted that the Monocrystalline Silicon shows the highest efficiency among all the materials at standard test conditions. It has also been found that all the materials having the same ideality factor exhibit similar behavior with respect to temperature and irradiation.

Finally, it can be concluded that Si-mono is the best PV material for commercial application.

References

1. Khyani HK, Vajpai J (2013) Mathematical modeling and experimental validation of performance characteristics of solar photovoltaic modules. Int J Appl Innov Eng Manag (IJAIEM) 2(11):295–301. ISSN 2319–4847
2. Tsai HL, Tu CS, Su YJ (2008) Development of generalized photovoltaic model using MATLAB/simulink. In: Proceedings of the World congress on engineering and computer science (WCECS '08), San Francisco (USA). https://www.iaeng.org/publication/WCECS2008/WCECS2008_pp846-851.pdf
3. Longatt FMG (2005) Model of photovoltaic module in matlab. In: 2nd international conference on Iberoamerican congress to electrical engineering students, electronics and computing (II CIBELEC: 2005), pp 1–5
4. Bhatt HG, Thakker RA (2011) Matlab based simulation of photovoltaic solar cell and its array at different temperature values. In: National conference on recent trends in engineering & technology, B.V.M. Engineering College, Gujarat, pp 1–4
5. Mahmoud Y, Xiao W, Zeineldin HH (2012) A simple approach to modeling and simulation of photovoltaic modules. IEEE Trans Sustain Energy 3(1):185–186
6. Panwar S, Saini RP (2012) Development and simulation of solar photovoltaic model using MATLAB/simulink and its parameter extraction. In: International conference on computing and control engineering (ICCCE'2012)
7. SunShot Vision Study (2012) A technical report on solar PV technology, Prepared by US Department of Energy. https://www1.eere.energy.gov/solar/pdfs/47927.pdf
8. Khyani HK (2013) Modeling and simulation of solar photovoltaic systems. M.E. thesis, Electrical Engineering Department, Faculty of Engineering (M.B.M. Engineering College) JNV University, Jodhpur, pp 27–75
9. Solar PV Module (2019). https://www.easyphotovoltech.com/solar-module.html,last. Accessed 12 Oct 2019

Chapter 21
Stabilization Analysis of Clay Soil by Using Waste Tile Particles

Saraswati Chand Dhariwal and Rajat Mangal

1 Introduction

Far reaching soils present huge geotechnical and auxiliary building difficulties the world over, with costs related with extensive conduct evaluated to run into a few billion yearly. Broad soils will be soils that experience critical volume change related with changes in water substance. These volume changes can either as swell or in the structure shrinkage and this is the reason, they are at some point known as swell/shrivel soils. Key points that need ID while overseeing expansive soils include soil properties, suction/water conditions, water content assortments transient and spatial, for instance created by trees, and the geometry/firmness of foundations and associated structures. So as to make an improved soil material which has the ideal designing properties, it is basic for the dirt to experienced adjustment for at least one properties. Either by doing the modification by mechanical or substance, this procedure is called as soil adjustment. The purpose of soil stabilization not only limited to enhance the load bearing of the soil capacity but also improve the shear strength, filter, drainage system, permeability, enhance soil resistance to the weathering process and traffic usage to meet specific engineering projects requirement [2, 6].

S. C. Dhariwal (✉)
M.B.M. Engineering College, J.N.V. University, Jodhpur, Rajasthan, India
e-mail: sonudhariwal14@gmail.com

R. Mangal
J.I.E.T Engineering College, Jodhpur, Rajasthan, India
e-mail: rajat.mangal0@gmail.com

M. Shorif Uddin et al. (eds.), *Intelligent Energy Management Technologies*, Algorithms for Intelligent Systems,
https://doi.org/10.1007/978-981-15-8820-4_21

2 Scope of Present Work

As a huge wastage of tiles waste is available in world so it can be used as admixture for stabilization of clay soil there is a great scope of stabilization of clayey soil with the admixture of tiles particles for construction of pavements, airfields and helipads. the aim of present work is the beneficial and economical utilization of such wastages for improving properties of clayey soil which is to be used as a base material for streets development and appropriate establishment material for different sorts of super structures [3].

Utilization of tiles particles for improvement of a soil is a sustainable and cost-effective technique. Since tiles particles can be easily available hence it was thought to be utilized as an admixture with which clay soil can be utilized as a low-cost building material. On the other hand, the problem of the disposal of tiles particles can be overcome by using it for stabilization of clay soil [1].

3 Materials Used in This Study

3.1 Clay Soil

Clay soils contain minerals that are capable of absorbing water. When water is added in the clay soils, they increase in volume. And again, more water added, the more their volume increases of the clay soils. this changing in volume can exerts enough forces on a building or other structure this cause damage the structure. Broken establishments, floors, and storm cellar dividers are common kinds of harm done by expanding soils. Clay soils will also be shrinking when they remove the water content. The clay sample for in this study were collected from Mundwa city in Mundwa Tehsil in Nagaur District of Rajasthan State, India. It belongs to Ajmer Division. It is found 31 km towards South from District base camp Nagaur [4].

3.2 Waste Tile Particle

A waste ceramic tile particle is an inorganic, non-metallic solid prepared by the action of heat and subsequent cooling. Waste ceramic tile material may have crystalline or partly crystalline structure because most common ceramics are crystalline materials. Ceramic Tile waste is cheap and non-reusable material, it is shown in all construction area and easy to collect. By the using of ceramic Waste Tile Particles to reduce the waste materials in earth and economical. The Waste ceramic tiles used in this study work were of hindware company. The waste ceramic tiles were bought from the local tiles shop from Jodhpur city, Rajasthan (India) [5, 7] (Fig. 1).

Fig. 1 Tiles particles

4 Experimental Programme

The Following Tests Were Conducted to Determine the Engineering Properties of The Clay Soils Mixed with Tiles Particles:

- Light compaction Test (Standard Proctor Test) for determining different dry densities of above-mentioned clayey samples and optimum moisture content of them.
- CBR test to determine % CBR values for clayey mix composition with wastes tile particles.
- Direct shear test to determine shear strength of clayey mix composition with waste tile particles (Figs. 2 and 3).

Fig. 2 CBR test mix composition in mould

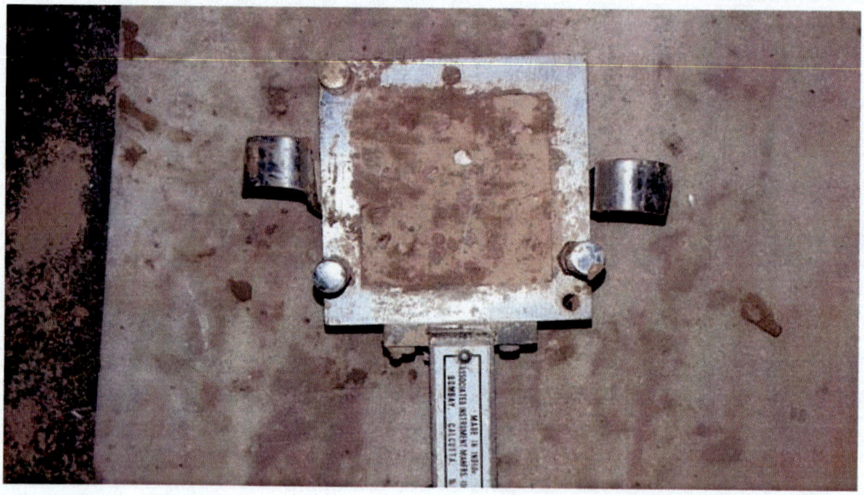

Fig. 3 Direct shear test apparatus

Table 1 Dry density variation with water content for clay soil

S. No.	% water added (by weight)	Dry density (gm/cc)
1	8	1.76
2	10	1.78
3	12	1.79
4	14	1.81
5	16	1.83
6	18	1.74
7	20	1.71
8	22	1.69
9	24	1.65
10	26	1.64

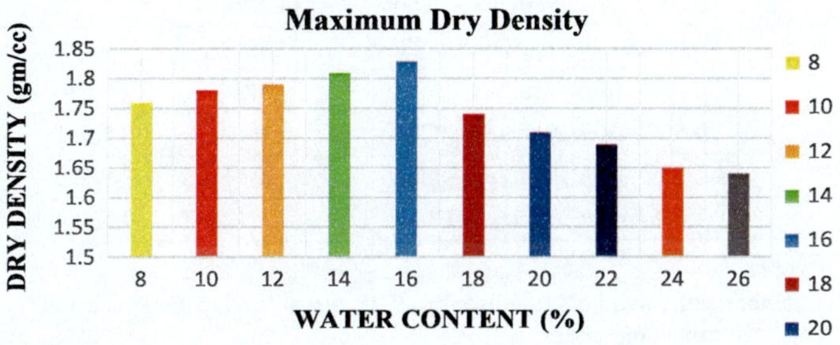

Fig. 4 Dry density variation of clay soil with water content

5 Test Results

5.1 Standard Proctor Compaction Test

The maximum dry density (MDD) of the clay soils without any admixture mixed is obtained as 1.83 gm/cc at the optimum moisture content (OMC) equal to 16%. The dry density variation with water content for clay soil is tabulated below in Table 1 and graphically shown below in Fig. 1 (Fig. 4).

5.2 California Bearing Ratio Test

The results of the unsoaked CBR tests conducted at MDD of 1.83, 1.78 and 1.69 gm/cc in Tables 2, 3 and 4.

The graphical variation of the percentage CBR values in unsoaked conditions for different mix compositions of admixture with Clay Soil is shown in Fig. 5.

Table 2 Mix compositions, symbols for unsoaked CBR test at MDD 1.83 gm/cc and % CBR value of clay soil with each mix composition

S. No.	Mix composition	Symbol	% CBR value
1	0.05% Tiles Particles + Clay	CB1	7.153
2	0.075% Tiles Particles + Clay	CB2	7.153
3	0.25% Tiles Particles + Clay	CB3	7.868
4	0.50% Tiles Particles + Clay	CB4	6.795
5	0.75% Tiles Particles + Clay	CB5	6.437
6	1.0% Tiles Particles + Clay	CB6	5.722

Table 3 Mix compositions, symbols for unsoaked CBR test at MDD 1.78 gm/cc and % CBR value of clay soil with each mix composition

S. No.	Mix composition	Symbol	% CBR value
1	0.05% Tiles Particles + Clay	CB7	5.722
2	0.075% Tiles Particles + Clay	CB8	6.437
3	0.25% Tiles Particles + Clay	CB9	6.676
4	0.50% Tiles Particles + Clay	CB10	5.961
5	0.75% Tiles Particles + Clay	CB11	5.722
6	1.0% Tiles Particles + Clay	CB12	5.484

Table 4 Mix compositions, symbols for unsoaked CBR test at MDD 1.69 gm/cc and % CBR value of clay soil with each mix composition

S. No.	Mix composition	Symbol	% CBR value
1	0.05% Tiles Particles + Clay	CB13	6.795
2	0.075% Tiles Particles + Clay	CB14	7.153
3	0.25% Tiles Particles + Clay	CB15	7.868
4	0.50% Tiles Particles + Clay	CB16	7.153
5	0.75% Tiles Particles + Clay	CB17	6.437
6	1.0% Tiles Particles + Clay	CB18	5.961

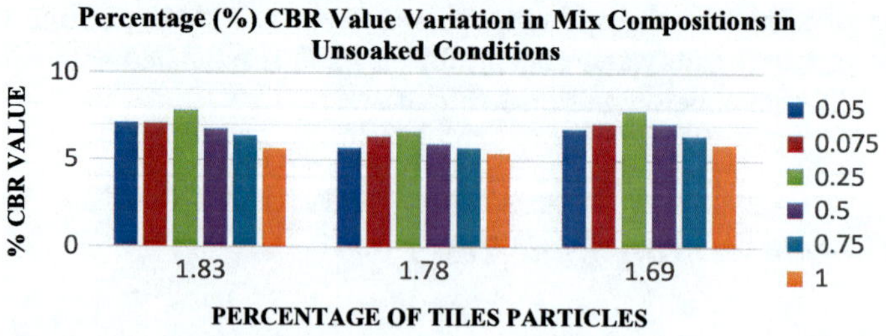

Fig. 5 Variation of % CBR values in unsoaked conditions for different mix compositions of admixture with Clay Soil of dry density 1.83, 1.78 and 1.69 gm/cc

From the above graph it is clear that among the three samples taken for the study, the maximum CBR value is obtained for dry density of Clay Soil of 1.69 gm/cc. It is also stated here that the maximum CBR value is obtained when the percentage by weight of admixture (Tiles Particles) mixed with the Clay Soil is 0.25%.

5.3 Direct Shear Test

The results of Direct Shear test at MDD 1.83 gm/cc of Clayey Soil mix with percentage of waste tiles particles are shown in Tables 5, 6 and 7 (Figs. 6, 7 and 8).

Table 5 Variation of φ with percentage of tiles particles at MDD 1.83 gm/cc

S. No.	Mix composition	Φ (Degree)
1	DB1	14.50°
2	DB2	17.76°
3	DB3	16.13°
4	DB4	14.40°
5	DB5	13.82°
6	DB6	15.47°

Table 6 Variation of C with percentage of tiles particles at MDD 1.83 gm/cc

S. No.	Mix composition	C (Cohesion)
1	DB1	0.338
2	DB2	0.345
3	DB3	0.369
4	DB4	0.348
5	DB5	0.387
6	DB6	0.408

Table 7 Variation of shear stress with normal stress for all mix composition at MDD 1.83 gm/cc

Shear Stress (kg/cm^2) for each mix composition at MDD 1.66 gm/cc	Normal stress (kg/cm^2)		
	0.5	1.0	1.5
0.05% Tiles Particles + Clay (DB1)	0.357	0.476	0.614
0.075% Tiles Particles + Clay (DB2)	0.366	0.531	0.687
0.25% Tiles Particles + Clay (DB3)	0.385	0.550	0.678
0.50% Tiles Particles + Clay (DB4)	0.366	0.495	0.623
0.75% Tiles Particles + Clay (DB5)	0.403	0.531	0.650
1.0% Tiles Particles + Clay (DB6)	0.421	0.595	0.705

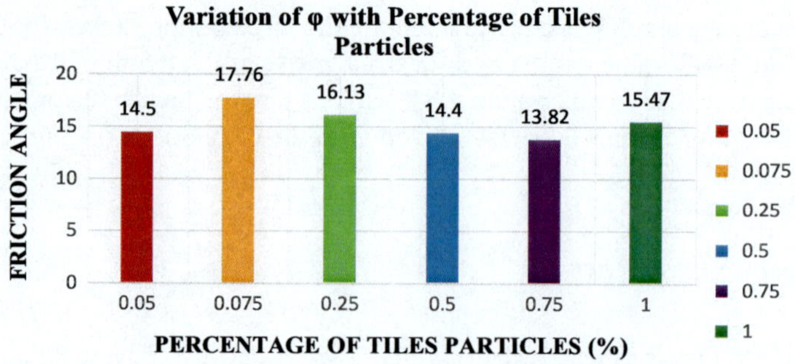

Fig. 6 Variation of φ with percentage of tiles particles at MDD 1.83 gm/cc

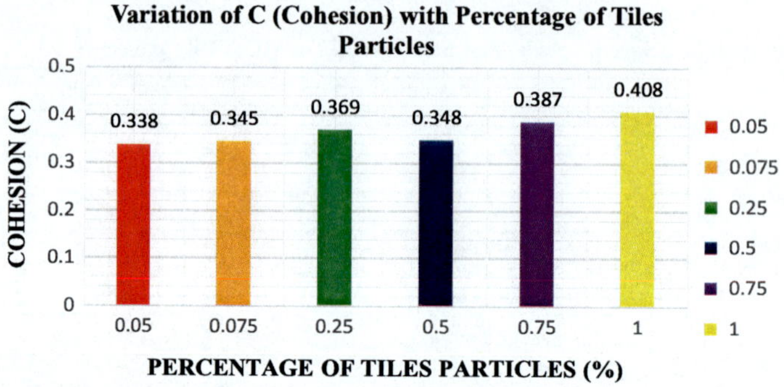

Fig. 7 Variation of C with percentage of tiles particles at MDD 1.83 gm/cc

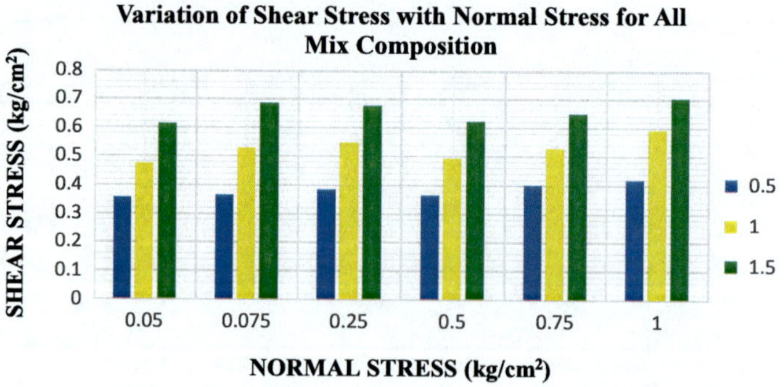

Fig. 8 Variation of shear stress with normal stress for all mix composition at MDD 1.83 gm/cc

6 Conclusions

The following conclusions can be made from the above study:

- It is observed that the Maximum Dry Density (MDD) of the clay soil is first Increased and after that it will be Decreased. In this study the Maximum Dry Density is Obtained 1.83gm/cc at the Optimum Moisture Content (OMC) to 16%.
- It is observed that the CBR values first increases when the percentage by weight of admixture (Tiles Particles) mixed with the clay soil increases up to 0.25% and then with the increase in percentage by weight of admixture the CBR values decreases.
- The above observation is same for all the three samples and thus it is concluded that the variation of CBR values with the percentage by weight of admixture is independent of the dry density of the clay soil.
- It is also seen here that the maximum CBR value is obtained when the percentage by weight of admixture mixed with the clay soil is 0.25% irrespective of the dry density of the clay soil.
- Among the three samples taken for the study, the maximum CBR value is obtained for dry density of clay soil equal to 1.69 gm/cc.
- It is observed that the maximum angle of friction is 17.76° and cohesion (C) is 0.408 is determined. And the Maximum shear stress value is obtained at 1.5 kg/cm^2 of normal stress in the Direct Shear Stress Test.
- Tiles Particles can be used as admixture and thus its recycling helps in reducing environmental problem related to its dumping and/or disposal.

References

1. Arora KR (2016) Soil mechanics and foundation engineering (geotechnical engineering)
2. Bhuvaneshwari S, Robinson RG, Gandhi SR (2005) Stabilization of expansive soils using fly ash. In: Fly ash utilization programme, FAUP, TIFAC, DST, vol 8, pp 5.1–5.9
3. Cokca E (2001) Use of class C fly ashes for the stabilization—of an expansive soil. J Geotech Geo Environ Eng 127:568–573
4. Purohit DGM, Ametha NK, Wayal AS (2008) Characteristics, problems and remedies of expansive soils of Rajasthan, India. In: E.J.G.E, vol 13
5. Robert M, Brooks (2009) Soil stabilization with fly debris and rice husk ash. Int J Res Rev Appl Sci 1:209–217
6. Singh A (2009) Basic soil mechanics and foundation. CBS Publishers and distributors, India
7. Tiwari S, Tiwari N (2016) Soil stabilization using waste fiber materials. Int J Innov Technol Res 4(3):2927–2930

Chapter 22
Renewable Energy as Biofuel from Mirabilis Jalapa Seed Oil

Akleshwar Mathur and Harish Kumar Khyani

1 Introduction

The improvement in the living standards of a citizen is directly or indirectly related to the energy as a critical input toward the same. This fact is now realized by the Government so it has become one of the most important ambitious announcements. All villages should get electricity at 24 h for every day throughout the year by 2019, The 175 GW of renewable energy capacity by the year 2022 followed by reduction in intensity of energy emissions by 33–35% by the year 2030 with increase in sharing of non-fossil fuel-based capacity in the electricity blend at above 40% in the same year [1].

Fossil fuels are more commonly used as source of energy in developing countries like India. Since the natural energy resources are continuously harnessed and demand is increasing the price of fossil fuel are very high, moreover, these fuel induces climate change, environmental pollution and greenhouse effect so there is an urgent need to explore new renewable energy sources to overcome these problems. Biofuels are considered as a promising source of energy as the best natural substitute for diesel. The present paper belongs to the recent status of non-conventional energy sources in our country with special emphasis on biodiesel development, central and state government policies and the physicochemical properties of selected species [2].

The good quality of biofuels can be blended easily with gasoline and can be used as an individual fuel. The biofuels emit lesser amounts of polluted gases so are carbon neutral and safe as compared with conventional petroleum fuels. The estimated production of biofuel in India was 500–600 million liters in 2017–18.

A. Mathur (✉) · H. K. Khyani
JIET, Jodhpur, Rajasthan, India
e-mail: akleshwar.mathur@jietjodhpur.ac.in

H. K. Khyani
e-mail: harish.khyani@jietjodhpur.ac.in

M. Shorif Uddin et al. (eds.), *Intelligent Energy Management Technologies*, Algorithms for Intelligent Systems,
https://doi.org/10.1007/978-981-15-8820-4_22

India can fulfill their fuel demands totally by biofuels. Our country produces a large amount of cellulosic material, which can be converted into biofuels [3, 4].

1.1 Indian Energy Scenario

On 15 August 2015, Dr A. P. J. Abdul Kalam, President of India, quoted a remarkable commitment that, "By the year 2030, India should achieve energy Independence through solar power and other forms of renewable energy." This would be like to achievable through the fast-growing applications of new and novel energy resources. The targets of renewable energy sources till 2021–22, Indian Energy Scenario fuel wise and contribution of various renewable sources in total energy mix are summarized in Tables 1, 2 and 3, respectively (Tables 4 and 5).

Although biofuels have a small fraction as compared with other resources, Biofuels are being considered as important commercial fuels, resulting in tremendous growth of biodiesel industry. By observing the current fuel-wise sharing, 2.7% of total fuel is being sourced from biofuels [5–7].

Table 1 Target of different renewable sources by MNRE by the year 2021–22 [1]

S.No	Source	Target (GW)
1	Solar	100
2	Wind	60
3	Biopower	10
4	Small hydropower	5
Total		175

Table 2 Indian energy scenario fuel wise as on 30.09.2019

Fuel	MW	% share
Coal	196,894.50	54
Lignite	6260.00	1.7
Gas	24,937.22	6.8
Diesel	509.71	0.14
Hydro	45,399.22	12.5
Nuclear	6,780.00	1.86
RES** (MNRE)	82,588.95	23
Total	3,63,369.59	

Table 3 Contribution of various renewable sources in total energy share as on 30.09.2019

Fuel	MW	% share w.r.t. to RES (82,588.95)	% share w.r.t. to Overall (3,63,369.59)
SHP	4610.81	5.6	1.3
Wind	36,930.32	44.7	10.3
Solar	31,101.71	37.6	8.7
Biopower	9946.11	12.1	2.7
Total	82,588.95		

Table 4 Physicochemical properties of seeds and oil

Sr. no	Property of seeds/oil	Value
1	Moisture content	4.0%
2	Oil content	3.5%
3	Refractive Index at room temperature	1.466
4	Specific gravity	0.789
5	Acid value (mg NaOH/g of oil)	1.345
6	Saponification value (mg KOH/g of oil)	174
7	Iodine value (g I_2/100 g of oil)	83
8	Total saturated fatty acids	12.57%
9	Total unsaturated fatty acids	87.43%
10	Poly unsaturated fatty acids	94.39%
11	Category of oil	Non-drying

Table 5 Comparison of mirabilis jalapa seed oil with jatropha and karanj seed oils [16, 17]

Sr.	Component	Jatropha curcus	Pongammia pinneta	Mirabilis jalapa
1	Oil Content	35	38	3.5
2	Specific Gravity	0.918	0.934	0.789
3	Saponification Value	195	192	174
4	Iodine value	86	90	83
5	% Unsap matter	1	1	1
6	Myristic acid	0.6	NIL	NIL
7	Palmitic acid	14	5.5	3.41
8	Stearic acid	7.5	5.6	1.01
9	Oleic acid	49.9	56	1.20
10	Linolenic acid	–	**2.6**	**60.70**
11	Linoleic acid	28	14	33.69
12	Unusual component	Absent	Absent	Absent
13	PUFA	28.0	16.8	94.39

1.2 Biofuels in India

The gasoline blend rate in 2016 was remarkable as 3.3%, which results the production of 143 billion liters biofuels with an average growth of biofuels as 2.5% per annum in the next year [5]. The USA, Brazil and Germany are the first three chief producers of biodiesel in the world, and the 43.5% of global biofuel production is achieved by the USA [6]. As earlier reported that the use of biofuels results reduction in the dependence of the petroleum products, control on CO_2, SO_2, NO_2-like greenhouse gas emissions and other major problems, with additional advantages. India is a well-recognized biodiversity ecosystem and produces tones of biowaste from plants, so this resource becomes more productive compared with other countries. Sugarcane and sugar mother liquor, seed-producing crops like corn, soybean, sorghum and non-edible oils jatropha, Pongammia, wild sunflower verities, cellulosic materials, are considered as raw materials for biofuels [7, 8]. The current initiatives of the Government including the Make in India and Swatch Bharat Abhiyan and various offer great opportunity to catch the targets of increasing Farmer's Income, Import Reduction, Employment Generation, Waste to Wealth production, etc.

The utilization of dry and wastelands for the cultivation of wild or non-edible seed verities would result to generate wealth for the local population with more important effect in the sustainable development. The goal of the policy is to increase the blending percentage to provide biofuels in the market at low cost. According to the recent literature, ethanol blending percentage in petrol is around 2.0% and biodiesel blending percentage in diesel is less than 0.1%. Blending of ethanol in petrol to 20% and blending of biodiesel in diesel is 5% proposed by the year 2030. This target is achievable as India is the seventh largest ethanol-producing country in the world. In addition to this, the Indian government aims to reduce its 10% of crude oil imports by 2022 [9, 10].

1.3 Biodiesel Blending Program

Feedstock availability for biodiesel is a key factor related to the growth of this sector. The overall blending percentage of biodiesel in diesel is less than 0.5% and the biodiesel is manufactured from imported sources. Thus to ensure domestic raw material source is integral for long-term success of this program.

In addition to the raw materials, household resources also have great importance, used or waste cooking oil termed as UCO and WCO, respectively, offers a potential source of biodiesel production. The variable and cultural food products provide a rich amount of waste oils, which are not safe for reuse but are reused through various small eateries, vendors and traders resulting health issues. The entry of UCO in food stream and developing a suitable collection mechanism to augment its supply for biodiesel production could be run simultaneously.

1.4 Initiatives by the Government of Rajasthan

The biofuel rules 2019 were released on the eve of the World Biofuel Day, August 9, 2019, by the Rajasthan government and Rajasthan is the first state in the country to bring the policy for biofuel after the notification of the Centre [11].

1.5 Requirement of New Sources and Current Species

Although biodiesel has a number of advantages, the selection of raw material is a challenge to the producers so production of a cost-effective fuel is not achieved as edible oils are being used in biodiesel production. Already the use of edible oils in industries has resulted in the high cost, so one would not capable to increase again the consumption of edible oils in a non-edible sector. To make economical biodiesel, non-edible, low cost, easily available and low-grade oils with value-added by-products to be searched as alternative raw material. Figure 1 indicates the data about the Production and Consumption of Ethanol in India of the last 10 years and 2, the import of crude oil in India in the same time period (Fig. 2).

The demand and supply from other countries expanding huge amount of finance are not good for the economy of a developing country, especially India as a number of plant resources are still unexploited by us. So it the correct time to explore the new and novel resources of oils for the fuel industry.

In the above-mentioned scenario of oils, the phytochemical analysis of number of seed oils has been in progress since the last two decades. Mirabilis jalapa commonly known as 4 o'clock flower or the marvel of Peru is a member of botanical family Nyctaginaceae. It is a non-conventional, inexpensive, low-grade seed oil and unexploited for biodiesel production. Furthermore, the crude oil is rich in a variety of phytochemicals like flavonoids, triterpenes, proteins, steroids and alkaloids with Mirabilis antiviral proteins abbreviated as MAPs. Therefore, use of mirabilis jalapa

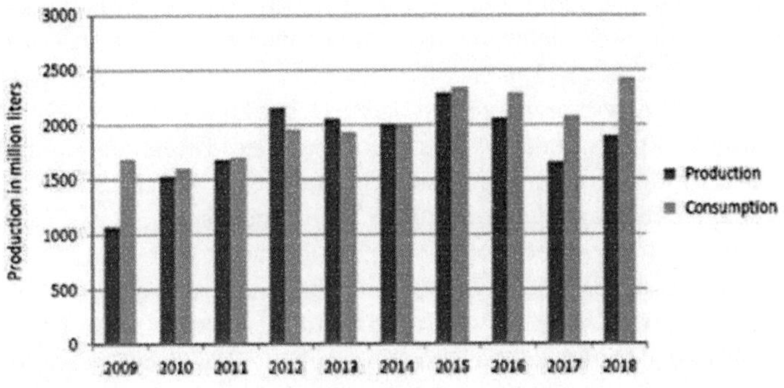

Fig. 1 Production and consumption of ethanol in India

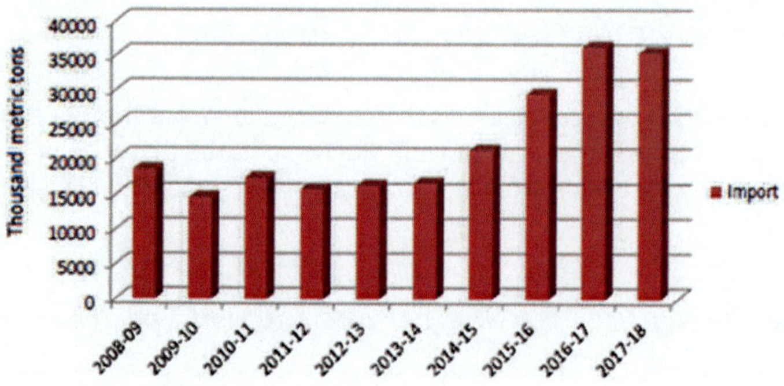

Fig. 2 Import of crude oil in India

seed oil as raw material for biodiesel production would not only make the process economical but also generate value-added phytochemicals or bioactive compounds. In the present investigation, physicochemical properties of this seed oil have been reported with fatty acid composition [12, 13].

As per literature available, this oil has not been a viable biodiesel feedstock due to its low oil content and less emphasis on fixed oils as compared with corn oil, canola oil, rice bran oil, etc., but as earlier discussed, it would be a good source of raw material [14].

2 Materials and Methods

The seeds were collected from cultivated plants after maturation by hand-picking methods. The dust were removed by washing seeds with water. Seeds were air dried at 60 °C for 7 h in oven. The seeds were crushed by Mortar and pestle into a coarse powder. Petroleum ether of 40–60° boiling point range was selected as extraction solvent for oil using Soxhlet apparatus. The solvent was recovered by Rota-evaporator under vacuum. Percentage of moisture, oil percentage and protein content of the seeds were determined recommended by American Oil Chemical Society as shown in Table 1 [15–19].

The analytical TLC was prepared by coating 0.25 mm layers of Silica gel G using methanol solvent on glass plates. A mixture of petroleum ether and diethyl ether in 80:20 ratio was applied as eluting solvent. The spots were visualized by exposure to Iodine vapor and then charring with Sulfuric dichromate spray.

The refractive index was recorded on Abbey Refractometer. GLC analyses were carried out with Varian Vista 6000 Gas Liquid Chromatograph using DEGS and SE-30 columns. Various TLC methods were used for further confirmation. The Refractive Index was determined by the Adobe Refractometer, at 30 °C. Density bottle was used for specific gravity. ISO 3657 (1988) indicator method was applied to determine the saponification value [20, 21]. Acid value was determined by as FFA/2.ISO 3961

(1989) method was used to determine the Iodine value. The GLC of methyl esters was done to analyses the fatty acid composition of the seed oil as usual methods [15–17].

3 Results and Discussion

The oil content in the collected sample was found as 3.5% which may be due to environmental and climatic conditions and also affected by the mode of extraction. In earlier reports, it has been found in the range of 3–5%. The oil is light, transparent, odorless, colorless and clear. The refractive index of oil was found as 1.466, specific gravity as 0.72, acid value as 1.345 (mg NaOH/g of oil), saponification values (mg KOH/g of oil) as 172. The Iodine value (g I2/100 g of oil) was found as 83. The oil is categorized as a non-drying oil. The physicochemical properties of seeds and oils determined are summarized in the table.

Total Saturated Fatty Acids: The total amount of saturated fatty acids was found as 12.57%. The fatty acids like Myristic, Palmitic and Behinic acids were absent in selected species.

Total Unsaturated Fatty Acids: The total amount of oleic acid, linoleic and linolenic acids was found as 87.43% as the major component of all the fatty acids. Other unsaturated acids were also found in small amount.

Poly Unsaturated Fatty Acids (PUFA): The combined content of linolic and linolenic acid was found as 94.39%.

Category of Oil: On the basis of PUFA content, the seed oil of *Mirabilis jalapa* is categorized as non-drying oil.

3.1 Comparison with Earlier Studies and Other Seed Oils

The comparison of present findings with earlier reports was done [22, 23]. The richness of Poly Unsaturated Fatty Acids was found similar except the presence of oxygenated fatty acid component due to environmental and species variation factors. The table indicates the comparative chemical composition of seed oils of Jatropha curcus and Pongammia pinneta seed oils with mirabilis Jalapa seed oil. These two plant species are commonly in practice for biodiesel preparation. It is clear that most of the properties resemble with jatropha and karanj oil.

4 Conclusion and Future Scope

The physicochemical properties of the *mirabilis jalapa* seed oil seem to have good scope and applications in industrial approaches. On the basis of unsaturated fatty acids, it is a non-drying oil so it can be used as an additive in biodiesel and lubricants also in manufacturing of medicated cosmetics, drug binder, etc. Further the yield of the oil is influenced by environmental conditions, variety and age of plant, collection timings, storage, etc. On the basis of the present results and previous research findings, more applications could be suggested. It is concluded that Mirabilis jalapa oil could be a substitute for preexisting seed oil. It can be cultivated as a medicinal crop-producing specific alkaloid. Since the oil is found in small amount hence it might be a part of research regarding oil improvement experiments to prove a promising source of Indian economy, especially in our desert zone. It is suggested that the cultivation of this plant should be granted under government projects.

References

1. http://www.cea.nic.inhttp://cea.nic.in/monthlyinstalledcapacity.html. Accessed 11 Nov 2019
2. Order no 30/80/2014–15/NSM, dated 1st July 2015
3. Ho DP, Ngo HH, Guo W (2014) A mini review on biofuel. Bio resource technology, pp 1–31
4. Kitani O (1999) CIGR book of agriculture engineering, energy and biomass engineering, vol V. ISBN 0-929355-97-0
5. Moon, Pakrasi HB (2013) Front Microbiol 4:1–14. Article 246
6. Genetically engineering cyanobacteria to convert CO, water and light into the long-chain hydrocarbon farnesene, Bioenergy and Biofuels. Appl Microbiol Biotechnol 98:9869–9877 (2014)
7. Alternative fuels data center—fuel properties comparison. www.afdc.energy.gov
8. Berla BM, Saha R, Immethun CM, Maranas CD, Seok T, Synthetic biology of cyanobacteria: unique challenges and opportunities
9. The Global Dynamics of Biofuels (2007) Brazil institute Special report (3):1–8
10. Ethanol industry outlook (2012) Renewable fuel assosiation America
11. https://www.biofuel.rajasthan.gov.in/about.aspx
12. Chisti Y (2008) Biodiesel from microalgae beats bioethanol. Trends Biotechnol 26(3):126–131
13. Miller CD, Bazore K, Bartow M (1955) Fruits of Ha-waii, 2nd edn. University of Hawaii Press, Hawaii
14. Barrett OW (1928) The tropical crops. The Macmillan Co., New York
15. Kennard W, Winters H (1960) Some fruits and nuts for the tropics. Misc. Pub. 801, Agricultural Research Service, US Dept. Agric, Washington
16. Deora MA, Jaiswal P, Mathur A, Sherwani MRK (2003) Fatty acid composition of some minor seed oils from arid zone of Rajasthan. JICS 80:141–142
17. Jaiswal P, Mathur A, Sherwani, Phytochemical studies of some seed oils from arid zone of Rajasthan, Oriental J Chem 19(2):487–488
18. Ahmad MS, Rauf A, Mustafa J, Sheikh MO (1984) An 8-hydroxyoctadeca-cis-11, 14-dienoic acid from Mirabilis Jalapa seed oil. Phytochem 23:2247–2249
19. Kyriakidis NB, Katsiloulis T (2000) Calculation of iodine value from measurements of fatty acid methyl esters of some oils: comparison with the relevant American oil chemists society method. J Am Oil Chem Soc 77(12):1235–1238
20. Standard methods for the analysis of oils, fats and derivatives 1st Supplement to the 7th Edition, International union of pure and applied chemistry commission on oils, fats and derivatives.
21. Endo Y (2013) Analytical methods to evaluate the quality of edible fats and oils: the JOCS standard methods for analysis of fats, oils and related materials; (2018) Advanced methods. J Oleo Sci 67(1):1–10
22. Mathur A (2017) A comparative physicochemical analysis of crude and refined jamun (Syzygium cumini) seed oil. Iconic Res Eng J 1(3)
23. Mathur A (2014) Renewable energy sources from Moringa Oleifera seed oil: a rich source of oil for bio diesel. Int J Comput Appl (0975–8887) National conference on innovations and recent trends in engineering and technology (NCIRET-2014)

Chapter 23
A Comparative Study of Various Approaches to Lossy Image Compression Process

Nitesh Agarwal and Rajendra Purohit

1 Introduction

1.1 Image

An image sensing device captures a real-world image and converts it into a digital image, which is represented in a two-dimensional (2D) discrete space and denoted by $f(x, y)$. The dimension of digital image $f(x, y)$ is represented by N rows and M columns in discrete space, where N, M are integer values and $f(x, y)$ represents the gray value at location (x, y). x, y are varied from 0 to N-1 and 0 to M-1, respectively [1] (Fig. 1).

1.2 Image Compression Model

Image compression model consists of two separate modules named as encoder and decoder as illustrated in Fig. 2. The encoder uses some data compression algorithm to compress input image, whereas decoder may be viewed as a reverse procedure of encoder to decompress the output of the encoder. In image compression, decompressed image can be or cannot be an exact replica of the input image; if decompressed image is an exact replica of input image then it is lossless model, else it is a lossy model. If the input image of resolution $M \times N$ is represented by $f(x, y)$ and decompressed image is represented by $\widehat{f}(x, y)$, then for lossless Image Compression can

N. Agarwal (✉) · R. Purohit
Jodhpur Institute of Engineering and Technology, Jodhpur, India
e-mail: niteshagarwal.234@rediffmail.com

R. Purohit
e-mail: Rajendra.purohit@jietjodhpur.ac.in

M. Shorif Uddin et al. (eds.), *Intelligent Energy Management Technologies*, Algorithms for Intelligent Systems,
https://doi.org/10.1007/978-981-15-8820-4_23

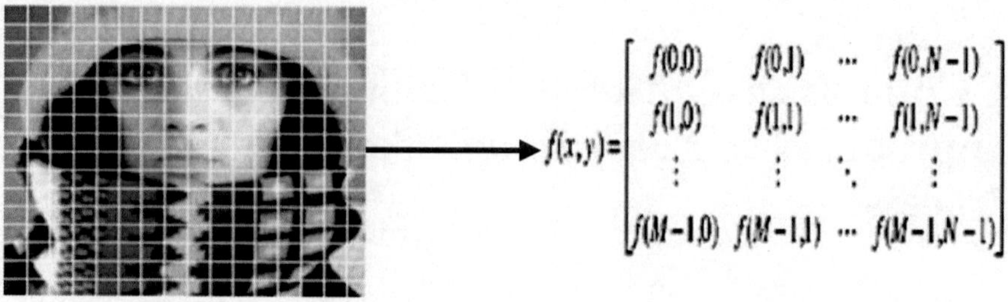

Fig. 1 Digitization of a continuous image [1]

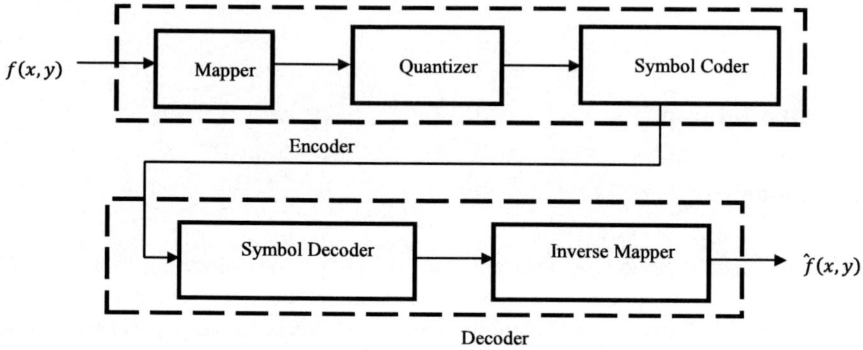

Fig. 2 Image compression model

be given as

$$f(x, y) = \hat{f}(x, y), \forall x \in [0, M-1], \forall y \in [0, N-1] \quad (1)$$

and for lossy Image Compression as

$$f(x, y) \neq \hat{f}(x, y), for\ any(x \in [0, M-1], y \in [0, N-1]) \quad (2)$$

2 Methodology

This research is based on lossy compression and embeds the n-Modulus Method in lossy compression and analyzes it with three different mathematical transformations, DCT, DST, DHT (Discrete Hartley Transform), and two different quantization tables.

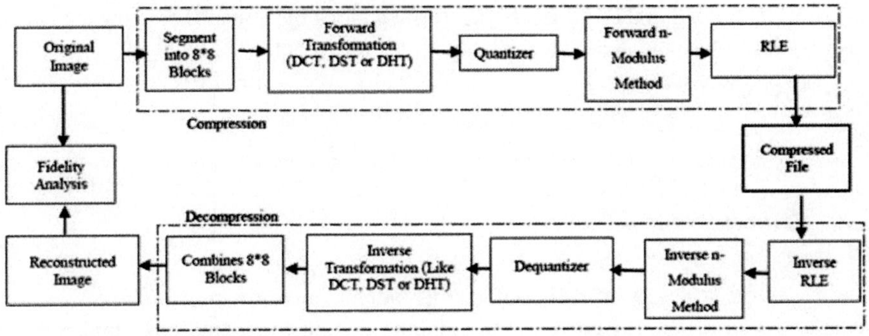

Fig. 3 Functional block diagram of proposed work

A functional block diagram of this research is shown in Fig. 3. Proposed process has three phases: first is compression, second is decompression, and the third is analysis.

2.1 Compression and Decompression

Compression starts with dividing the input image into blocks of size 8 × 8 each. The total number of blocks depends upon the resolution of input image, if the input image has resolution $M \times N$, then the total number of blocks represented by p can be defined as shown in (Fig. 4)

$$p = \frac{M \times N}{64} \tag{3}$$

After the decomposition of image into blocks, each block is transformed into another domain (frequency domain) using three basic mathematical transformation

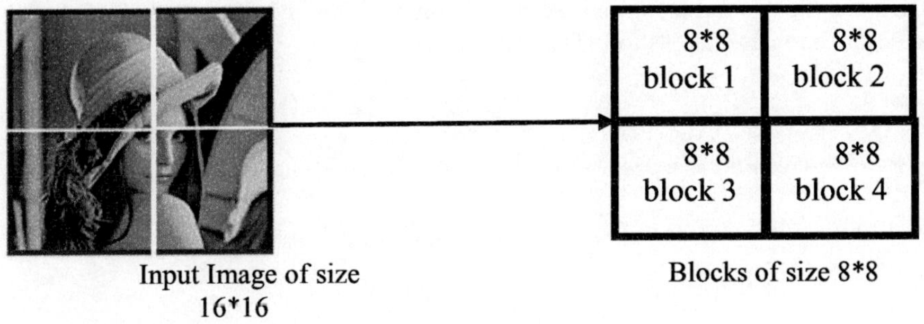

Fig. 4 Division of image into blocks

function DCT, DST, and DHT. Each transformed block then quantizes using standard quantization table and generates quantized blocks. Each quantized block is then operated by the n-modulus method and, finally, RLE performs the final compression phase and generates compressed files of input image [2, 3].

2.2 *Transformation*

It is based on a variety of different mathematical transformations. The choice of specific transformation depends on the amount of transformation error that can be tolerable and availability of computational resources.

A transformed block can be represented by $T(u, v)$ for input image $f(x, y)$ of resolution $M\ X\ N$ and can be expressed as

$$T(u, v) = \sum_{x=0}^{M-1} \sum_{y=0}^{N-1} f(x, y) r(x, y, u, v), \quad (4)$$

where $r(x, y, u, v)$ is termed as the forward transformation kernel and its inverse can be defined as

$$f(x, y) = \sum_{u=0}^{M-1} \sum_{v=0}^{N-1} T(u, v) s(x, y, u, v), \quad (5)$$

where $s(x, y, v)$ is termed as the inverse transformation kernel. The present work is processed by three basic transformation functions individually, named as DCT, DST, and DHT.

2.2.1 Discrete Cosine Transform (DCT)

DCT has the same transformed kernel in compression and decompression phases and can be represented as follows:

$$r(x, y, u, v) = s(x, y, u, v) = \alpha(u)\alpha(v) \cos\left[\frac{\pi(2x+1)u}{2N}\right] \cos\left[\frac{\pi(2y+1)v}{2N}\right] \quad (6)$$

$$\alpha(u) = \begin{cases} \sqrt{\frac{1}{N}} \text{ for } u = 0 \\ \sqrt{\frac{2}{N}} \text{ for } u \neq 0 \end{cases} \quad (7)$$

$$\alpha(v) = \begin{cases} \sqrt{\frac{1}{N}} \text{ for } v = 0 \\ \sqrt{\frac{2}{N}} \text{ for } v \neq 0 \end{cases} \tag{8}$$

$$t_k(u, v) = \alpha(u)\alpha(v) \sum_{x=0}^{7} \sum_{y=0}^{7} f_k(x, y) \cos\left[\frac{\pi(2x+1)u}{16}\right] \cos\left[\frac{\pi(2y+1)v}{16}\right] \tag{9}$$

for $u = 0, 1, 2 \ldots 7$ *and* $v = 0, 1, 2 \ldots 7$, the inverse transform IDCT of block $t_k(u, v)$ can be defined as

$$f_k(x, y) = \sum_{u=0}^{7} \sum_{v=0}^{7} \alpha(u)\alpha(v) t_k(u, v) \cos\left[\frac{\pi(2x+1)u}{16}\right] \cos\left[\frac{\pi(2y+1)v}{16}\right] \tag{10}$$

for $x = 0, 1, 2 \ldots 7$ *and* $y = 0, 1, 2 \ldots 7$. Here, k defines the block number that has been processed, $f_k(x, y)$ is the kth block of input image of size 8*8 and $t_k(u, v)$ is the kth transform block of block $f_k(x, y)$ [4, 5].

2.2.2 Discrete Sine Transform (DST)

Like DCT, DST is also a member of Discrete Trigonometry Transform (DTT), hence, it also can be used two as transform input blocks. Like DCT, it has same forward and inverse transform kernels as following:

$$\begin{aligned} r(x, y, u, v) &= s(x, y, u, v) \\ &= \alpha(u)\alpha(v) \sin\left[\frac{\pi(2x+1)(u+1)}{2N}\right] \sin\left[\frac{\pi(2y+1)(v+1)}{2N}\right] \end{aligned} \tag{11}$$

Using Eq. (11), forward transform FDST can be defined as

$$t_k(u, v) = \alpha(u)\alpha(v) \sum_{x=0}^{7} \sum_{y=0}^{7} f_k(x, y) \sin\left[\frac{\pi(2x+1)(u+1)}{16}\right] \sin\left[\frac{\pi(2y+1)(v+1)}{16}\right] \tag{12}$$

and its inverse transform IDST can be defined as

$$f_k(x, y) = \sum_{u=0}^{7} \sum_{v=0}^{7} \alpha(u)\alpha(v) t_k(u, v) \sin\left[\frac{\pi(2x+1)(u+1)}{16}\right] \sin\left[\frac{\pi(2y+1)(v+1)}{16}\right] \tag{13}$$

Here, terms $t_k(u, v)$, $f_k(x, y)$, $\alpha(\mathrm{u})$, $\alpha(\mathrm{v})$ are similar to DCT [6].

2.2.3 Discrete Hartley Transform (DHT)

DHT is also one of the important transformations in image processing, it looks like DFT but it does not generate complex numbers, hence, it is faster than DFT. For a one-dimensional (1D) function $f(x)$, DHT can be defined as [6, 7]

$$t(u) = \sum_{x=0}^{N-1} f(x)\mathrm{cas}\left(\frac{\pi(2x+1)u}{N}\right) \tag{14}$$

and its inverse as

$$f(u) = \frac{1}{N}\sum_{u=0}^{N-1} t(u)\mathrm{cas}\left(\frac{\pi(2x+1)u}{N}\right), \tag{15}$$

where $t(u)$ is the transform function, N is a positive integer, and casθ is defined as

$$\mathrm{cas}\theta = \sin\theta + \cos\theta \tag{16}$$

For 2D function, DHT has two variants of forward and inverse transform kernels and can be defined as

$$r(x, y, u, v) = \mathrm{cas}\left[\frac{\pi(2x+1)u}{M} + \frac{\pi(2y+1)v}{N}\right] \tag{17}$$

$$s(x, y, u, v) = \frac{1}{MN}\mathrm{cas}\left[\frac{\pi(2x+1)u}{M} + \frac{\pi(2y+1)v}{N}\right] \tag{18}$$

$$r(x, y, u, v) = \mathrm{cas}\left[\frac{\pi(2x+1)u}{M}\right]*\mathrm{cas}\left[\frac{\pi(2y+1)v}{N}\right] \tag{19}$$

$$s(x, y, u, v) = \frac{1}{MN}\mathrm{cas}\left[\frac{\pi(2x+1)u}{M}\right] * \mathrm{cas}\left[\frac{\pi(2y+1)v}{N}\right] \tag{20}$$

Equations (17) and (18) are known as normal forward and inverse transform kernel, respectively, whereas Eqs. (19) and (20) are termed as forward and inverse transforms for Separable 2D Discrete Hartley Transform (SDHT)n respectively. Using Eq. (17) forward transform FDHT on image block, $f_k(x, y)$ can be applied as

$$t_k(u, v) = \sum_{x=0}^{7}\sum_{y=0}^{7} f_k(x, y)\mathrm{cas}\left[\frac{\pi(2x+1)u}{8} + \frac{\pi(2y+1)v}{8}\right] \tag{21}$$

and its inverse transform IDHT can be defined as

$$f_k(x, y) = \frac{1}{64}\sum_{u=0}^{7}\sum_{v=0}^{7} t_k(u, v)\text{cas}\left[\frac{\pi(2x+1)u}{8} + \frac{\pi(2y+1)v}{8}\right]. \tag{22}$$

Similarly, forward SDHT (FSDHT) can be defined using forward kernel as described by Eq. (19)

$$t_k(u, v) = \sum_{x=0}^{7}\sum_{y=0}^{7} f_k(x, y)\text{cas}\left[\frac{\pi(2x+1)u}{M}\right]{*}\text{cas}\left[\frac{\pi(2y+1)v}{N}\right] \tag{23}$$

and its inverse transform ISDHT can be defined as [8]

$$f_k(x, y) = \frac{1}{64}\sum_{u=0}^{7}\sum_{v=0}^{7} t_k(u, v)\text{cas}\left[\frac{\pi(2x+1)u}{M}\right]{*}\text{cas}\left[\frac{\pi(2y+1)v}{N}\right] \tag{24}$$

All the blocks generated by these transform are acting as input to quantization process, where quantize blocks are formed for further processing.

2.3 *Quantization*

Quantization is a process of representing a large possible input set of value with a smaller set. Quantization is an irreversible and lossy process because there is no method by which input symbol can be reconstructed using quantized value.

In this phase, each transformed block is quantized using the following formula:

$$QTB(u, v) = ROUND\left(\frac{t(u, v)}{QT(u, v)}\right), \tag{25}$$

where QTB is termed as Quantized Transform Block, $u = 0, 1, 2\ldots 7$, $v = 0, 1, 2\ldots 7$. $t(u, v)$ is the transform block obtain from Eqs. (9), (12), (21) or from (23) and QT represents quantized matrix which can be either of following quantize tables as shown below.

Because quantization is a lossy process, hence block $t(u, v)$ cannot reconstruct in decompression process but its equivalent block $t^{'}(u, v)$ can be obtained by following a formula:

$$t'(u, v) = QTB(u, v) \times QT(u, v) \tag{26}$$

Table 1 3Quantize Table 1 [9]

16	11	10	16	24	40	51	61
12	12	14	19	26	58	60	55
14	13	16	24	40	57	69	56
14	17	22	29	51	87	80	62
18	22	37	56	68	109	103	77
24	35	55	64	81	104	113	92
49	64	78	87	103	121	120	101
72	92	95	98	112	100	103	99

Table 2 3Quantize Table 2 [10]

17	18	24	47	99	99	99	99
18	21	26	66	99	99	99	99
24	26	56	99	99	99	99	99
47	66	99	99	99	99	99	99
99	99	99	99	99	99	99	99
99	99	99	99	99	99	99	99
99	99	99	99	99	99	99	99
99	99	99	99	99	99	99	99

2.4 *n-Modulus Method*

The n-Modulus Method (shortly n-MM) transforms each input value into a number that is divisible by n, this transformation will not affect the Human Visual System (HVS).

n-MM reads each QTB block row by row and divides each input value by n and adds or subtracts the reminder from the original pixel to get repeated pixel values. The basic idea in n-MM is to check the whole pixel metrics and transform each pixel into a number divisible by n. The reason behind using this method before encoding is that an entropy encoding technique gives a high compression as the input values have repeated data sequentially [11, 12].

Mathematically, forward n-MM is given as

$$n{-}MM(u, v) = ROUND\frac{QTB(u, v)}{n} \tag{27}$$

and its inverse work as

$$QTB'(u, v) = n{-}MM(u, v) \times n \tag{28}$$

2.5 *Entropy Encoding*

Encoding techniques are the key to data compression techniques. However, a baseline or base value is required to identify how much amount of compression can be achieved, this base value is given by entropy. Entropy is a probability-based method to identify the average amount of information required representing each input symbol and its mathematical statement can be defined as

$$H = -\sum_{j=1}^{n} p(a_j) log p(a_j) \tag{29}$$

Here, H is the entropy of input symbols $a_1, a_2, a_3, \ldots\ldots\ldots\ldots, a_n$, $p(a_j)$ is the probability of jth symbol in the sequence of input symbols. In the present paper for final compression, RLE is used as an entropy encoding [13].

2.5.1 Run Length Encoding (RLE)

RLE is one of the simplest encoding techniques for compression; it is mainly used for compressing sequential data. In most of the digital image, gray values of pixels are repeated sequentially. Hence, RLE is suitable for such type of images. For every repeated gray value, it only stores the gray value with their frequency rather than storing that gray value every time [13].

2.6 *Fidelity Analysis*

The image compression model can be used for lossless compression as well as for lossy compression; if it is used for lossy, then there is need to identify some criteria, by which amount of information can be decided. There are two types of criteria for such assessment [10].

2.6.1 Objective Fidelity Criteria

The criteria are given by mathematical functions that determine the amount of error or noise arising in the image due to information loss. The terms to define these criteria are MSE, PSNR, CR, CC, and NCC. For an input image $f(x, y)$ and output image $\widehat{f}(x, y)$ with resolution $M \times N$, these criteria can be defined as

$$e(x, y) = \hat{f}(x, y) - f(x, y) \tag{30}$$

$$TE = \sum_{x=0}^{M-1} \sum_{y=0}^{N-1} \hat{f}(x, y) - f(x, y) \tag{31}$$

$$MSE = \frac{1}{M * N} \sum_{x=0}^{M-1} \sum_{y=0}^{N-1} \left[\hat{f}(x, y) - f(x, y)\right]^2 \tag{32}$$

$$e_{rms} = \sqrt{\frac{1}{M * N} \sum_{x=0}^{M-1} \sum_{y=0}^{N-1} \left[\hat{f}(x, y) - f(x, y)\right]^2} \tag{33}$$

$$SNR_{ms} = \frac{\sum_{x=0}^{M-1} \sum_{y=0}^{N-1} \left[\hat{f}(x, y)\right]^2}{\sum_{x=0}^{M-1} \sum_{y=0}^{N-1} \left[\hat{f}(x, y) - f(x, y)\right]^2} \tag{34}$$

$$\text{PSNR} = 20* \log_{10}(\text{MAX}) - 10*\log_{10}(\text{MSE}) \tag{35}$$

$$\text{CR} = \frac{\text{Memory Required to store Original Image}}{\text{Memory Required to store Encoded data}} \tag{36}$$

$$CC = \frac{\sum_{x=0}^{M-1} \sum_{y=0}^{N-1} (f(x, y) - i_{\text{mean}})(f'(x, y) - i'_{\text{mean}})}{\sqrt{\sum_{x=0}^{M-1} \sum_{y=0}^{N-1} (f(x, y) - i_{\text{mean}})^2} \sqrt{\sum_{x=0}^{M-1} \sum_{y=0}^{N-1} (f'(x, y) - i'_{\text{mean}})^2}} \tag{37}$$

$$NC = \frac{\sum_{x=0}^{M-1} \sum_{y=0}^{N-1} (f(x, y) \times f'(x, y))}{\sqrt{\sum_{x=0}^{M-1} \sum_{y=0}^{N-1} (f(x, y))^2}} \tag{38}$$

Here, $f(x, y)$ and $f'(x, y)$ are the pixel values at point (x, y) for original image and decompressed image, respectively, i_{mean} and i'_{mean} are the mean values for original image and decompressed image, respectively, $e(x, y)$ represents the error at (x, y). TE represents the Total Error, MSE represents the Mean Square Error, e_{rms} represents Root Mean Square Error, SNR_{ms} represents the Mean Square Signal to Noise Ratio, PSNR represents the Peak Signal to Noise Ratio, and MAX represents the maximum gray value of a pixel. For example, for an 8-bit image, MAX would be 255 [10].

2.6.2 Subjective Fidelity Criteria

Subjective fidelity criteria are purely based on the visual properties of output image $\widehat{f}(x, y)$ and can be rated as follows based on the recognition of HVS [10] (Table 3).

Table 3 Subjective fidelity criteria [10]

Value	Rating	Description
1	Excellent	An extremely high-quality image
2	Fine	A high-quality image, but information losses not noticeable
3	Passable	Quality of image is acceptable but degradation in image noticeable
4	Marginal	A poor quality image and losses can be seen easily
5	Inferior	A very poor image, but can be seen
6	Unusable	Due to extreme degradation-level image cannot be seen

3 Results

The results vary on the basis of the input image and the value of *n* (from 1 to 10) in n-MM algorithm. For each input image, results are categorized into two sections, one is objective fidelity and other is subjective fidelity. Objective fidelity gives variation in the objective parameter of input image, which are MSE, PSNR, Size, CR, CC, and NC. Subjective fidelity shows the variation in vision quality on the basis of HVS. To complete this research, the following platforms are used.

Hardware: Intel(R) Core(TM) i5-6200U CPU @ 2.30 Ghz with 8 GB RAM.

Programming Language: JAVA programming language, MATLAB.

Tools: NetBeans IDE 8.0.2 for Java, MATLAB R2010a for MATLAB, Microsoft Office 2013.

Hardware: Intel(R) Core(TM) i5-6200U CPU @ 2.30 Ghz with 8 GB RAM (Table 4).

3.1 Object Fidelity Results for Lena.bmp

3.1.1 Variation in MSE for Lena.bmp

See Fig. 5.

Table 4 Sample input images

S. No	Image name	Image type	Resolution	Image size (KB)
1	Lena.bmp	BMP	512*512	768
2	Pappers_grey.bmp	BMP	512*512	768
3	Baboon.bmp	BMP	512*512	768

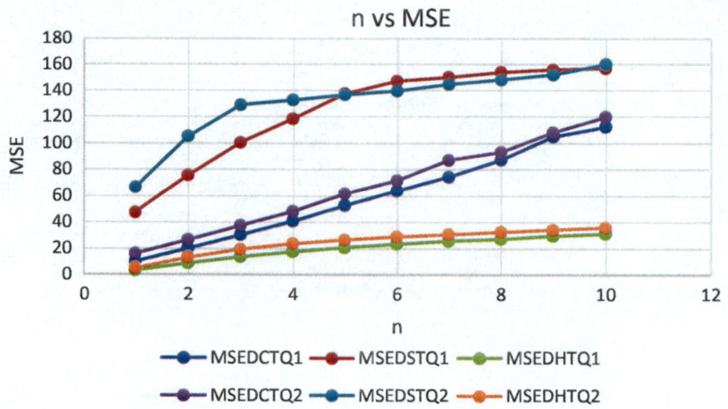

Fig. 5 Variation in MSE for Lena.bmp based on different transformation and quantization

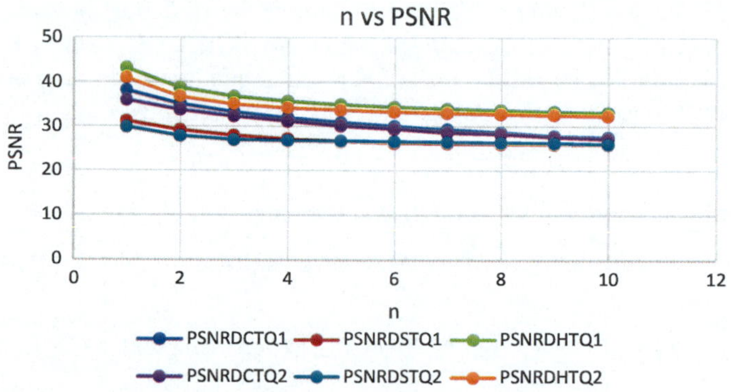

Fig. 6 Variation in PSNR for Lena.bmp based on different transformation and quantization

3.2 *Variation in PSNR for Lena.bmp*

See Figs. 6.

3.3 *Variation in CR for Lena.bmp*

See Fig. 7.

3.3.1 Variation in CC for Lena.bmp

See Fig. 8.

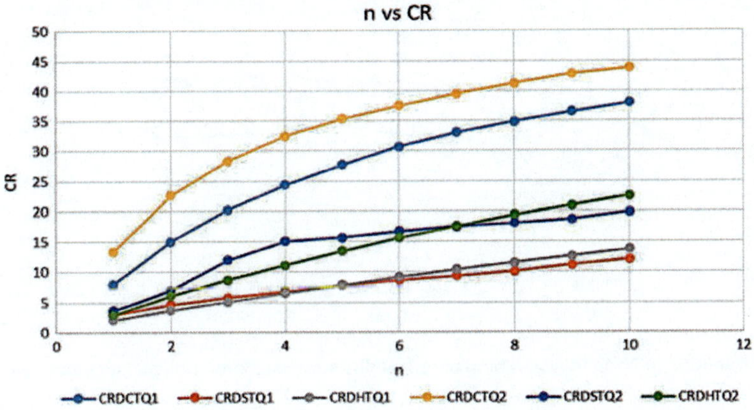

Fig. 7 Variation in CR for Lena.bmp based on different transformation and quantization

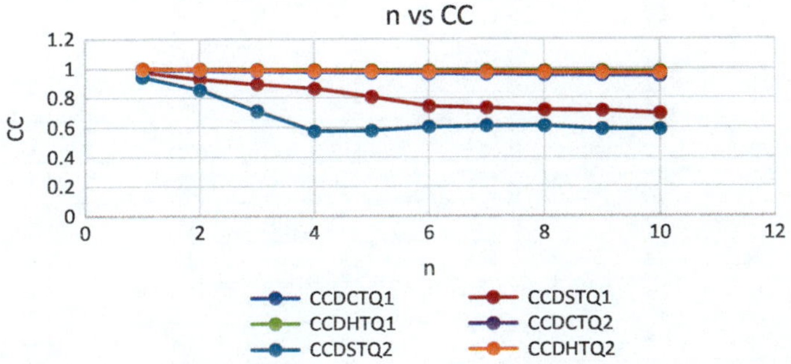

Fig. 8 Variation in CC for Lena.bmp based on different transformation and quantization

3.3.2 Variation in NCC for Lena.bmp

See Fig. 9.

Similar results for Pappers_grey.bmp and Baboon.bmp are obtained.

3.4 Subjective Fidelity Results for Lena.bmp

See Table 5.

Here, SFV refers to Subjective Fidelity Value as described in Table 3.

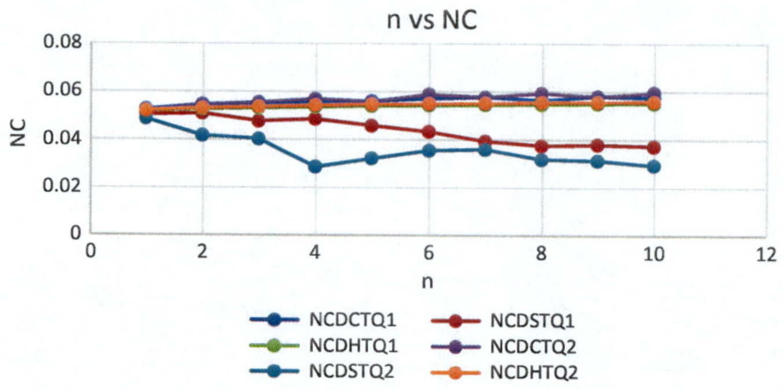

Fig. 9 Variation in NCC for Lena.bmp based on different transformation and quantization

Table 5 Variation in value of subjective fidelity, for Lena.bmp based on different transformation and quantization

	DCT + Q1	DST + Q1	DHT + Q1	DCT + Q2	DST + Q2	DHT + Q2
n	1	1	1	1	1	1
SFV	1	2	1	1	2	1
n	2	2	2	2	2	2
SFV	1	2	1	1	2	1
n	3	3	3	3	3	3
SFV	2	3	2	2	3	2
n	4	4	4	4	4	4
SFV	2	3	2	2	3	2
n	5	5	5	5	5	5
SFV	2	4	3	2	4	2
n	6	6	6	6	6	6
SFV	3	4	3	3	4	3
n	7	7	7	7	7	7
SFV	3	4	3	3	4	3
n	8	8	8	8	8	8
SFV	3	4	3	3	4	3
n	9	9	9	9	9	9
SFV	3	4	3	3	4	4
n	10	10	10	10	10	10
SFV	4	5	4	4	5	5

4 Conclusion and Future Scope

4.1 Conclusion

The present paper gives an analytical view of implementation and describes how the size of input image is effected at the cost of quality of input image. The results shown in this paper are categorized into two categories, objective fidelity and subjective fidelity. Based on results obtained for objective, following conclusion can be carried out.

- For all the input images taken in this research, DHT gives the best results with respect to quality as shown in the table, but when compression ratio has priority, then DCT is best among these three transformations as shown in the table. DCT and DHT both are better than DST in all aspect. Hence, where quality has higher priority, then order of good result is DHT > DCT > DST and for high compression, the order will be DCT > DST > DHT. For the higher value of n order becomes DCT > DHT > DST because DHT gives high CR than DST. However, if good quality of image and good compression both are required, then DCT is more suitable. It gives high compression at average degradation in input image (Tables 6 and 7).

- Mathematical transformation does not compress the input image without quantization and entropy encoding. In this research, two quantization tables Q1 and Q2 are used; results show that matrix Q2 nnot only gives high compression than Q1 but it also degrades input image in higher amount than Q1. Hence, if quality is considered, then Q1 is more suitable than Q2, otherwise Q2 can be used for high compression, as shown in Tables 8 and 9.

- The n-MM algorithm, used in this research, gives a range of degraded image from low degradation to high degradation. Level of degradation depends upon the value of n. As the value of n increases, the degradation level also increases and high compression is also obtained for higher value of n, as given inTables 10 and 11.

$$n \propto Compression \propto Degradation$$

- The results of subjective fidelity are purely based on human vision and results clearly show that input image is always visible and understandable even for higher value of n, i.e., for n > 5.
- The results also show that the compression and degradation in input image not only depends on algorithms used in the compression process but also depends on input image (Table 12).
- Based on the value obtained for CC, it is clear that as the value of n increases, deviation from original image also increases, but image remains visible for all input test cases (Table 13).

Table 6 Comparison of Mathematical transformation DCT, DST and DHT for all the sample input images with respect to average MSE value

Input image		MSE DCT + Q1	MSE DCT + Q2	MSE DST + Q1	MSE DST + Q2	MSE DHT + Q1	MSE DHT + Q2
Lena.bmp	Range	9.7–112.3	15.9–120	47.2–157.1	66.2–160.1	3.0–31.0	5–35.9
	Average	59.5	66.9	124.5	131.5	19.8	25
Baboon.bmp	Range	35.3–116.3	50.7–122.8	58.5–158.3	74.2–159	3.9–63.8	6.6–70.3
	Average	84.1	91.2	125.8	135.8	43.5	52.19
Pappers_grey.bmp	Range	14.6–113.4	19.3–120.0	47.5–157.0	65.5–159.9	3.8–34.2	6.0–38.9
	Average	62.7	69.9	122.2	129.2	22.9	28.3

Table 7 Comparison of Mathematical transformation DCT, DST and DHT for all the sample input images with respect to average CR value

Input image		CR DCT + Q1	CR DCT + Q2	CR DST + Q1	CR DST + Q2	CR DHT + Q1	CR DHT + Q2
Lena.bmp	Range	7.9–37.9	13.3–43.7	2.96–11.9	3.6–19.7	1.9–13.6	2.89–22.4
	Average	26.8	33.64	7.86	14.3	8.21	13.78
Baboon.bmp	Range	3.5–36.0	6.5–45.2	2.4–11.54	3.18–18.9	1.3–7.15	1.5–14.1
	Average	20.16	30.99	7.32	13.58	4.15	7.8
Pappers_grey.bmp	Range	7.7–39.5	13.8–44.64	2.9–12.7	3.7–21.9	1.6–14.05	2.24–23.14
	Average	27.8	34.4	8.2	15.1	8.02	14.07

Table 8 Comparison of quantization tables Q1 and Q2 for all the sample input images with respect to average MSE value

Input image		MSE DCT + Q1	MSE DCT + Q2	MSE DST + Q1	MSE DST + Q2	MSE DHT + Q1	MSE DHT + Q2
Lena.bmp	Range	9.7–112.3	15.9–120	47.2–157.1	66.2–160.1	3.0–31.0	5–35.9
	Average	59.5	66.9	124.5	131.5	19.8	25
Baboon.bmp	Range	35.3–116.3	50.7–122.8	58.5–158.3	74.2–159	3.9–63.8	6.6–70.3
	Average	84.1	91.2	125.8	135.8	43.5	52.19
Pappers_grey.bmp	Range	14.6–113.4	19.3–120.0	47.5–157.0	65.5–159.9	3.8–34.2	6.0–38.9
	Average	62.7	69.9	122.2	129.2	22.9	28.3

Table 9 Comparison of quantization tables Q1 and Q2 for all the sample input images with respect to average CR value

Input image		CR DCT + Q1	CR DCT + Q2	CR DST + Q1	CR DST + Q2	CR DHT + Q1	CR DHT + Q2
Lena.bmp	Range	7.9–37.9	13.3–43.7	2.96–11.9	3.6–19.7	1.9–13.6	2.89–22.4
	Average	26.8	33.64	7.86	14.3	8.21	13.78
Baboon.bmp	Range	3.5–36.0	6.5–45.2	2.4–11.54	3.18–18.9	1.3–7.15	1.5–14.1
	Average	20.16	30.99	7.32	13.58	4.15	7.8
Pappers_grey.bmp	Range	7.7–39.5	13.8–44.64	2.9–12.7	3.7–21.9	1.6–14.05	2.24–23.14
	Average	27.8	34.4	8.2	15.1	8.02	14.07

Table 10 MSE range comparison for all the sample input image

Input image		MSEDCTQ1	MSEDCTQ2	MSEDSTQ1	MSEDSTQ2	MSEDHTQ1	MSEDHTQ2
Lena.bmp	Range	9.7–112.3	15.9–120	47.2–157-10	66.2–160.1	3.0–31.0	5–35.9
Baboon.bmp	Range	35.3–116.3	50.7–122.8	58.5–158.3	74.2–159	3.9–63.8	6.6–70.3
Pappers_grey.bmp	Range	14.6–113.4	19.3–120.0	47.5–157.0	65.5–159.9	3.8–34.2	6.0–38.9

Table 11 CR range comparison for all the sample input image

Input image		CRDCTQ1	CRDCTQ2	CRDSTQ1	CRDSTQ2	CRDHTQ1	CRDHTQ2
Lena.bmp	Range	7.9–37.9	13.3–43.7	2.96–11.9	3.6–19.7	1.9–13.6	2.89–22.4
Baboon.bmp	Range	3.5–36.0	6.5–45.2	2.4–11.54	3.18–18.9	1.3–7.15	1.5–14.1
Pappers_grey.bmp	Range	7.7–39.5	13.8–44.64	2.9–12.7	3.7–21.9	1.6–14.05	2.24–23.14

Table 12 Input image comparison with respect to average MSE and average CR value

Input image	MSE Average	CR Average
Lena.bmp	71.2	17.4
Baboon.bmp	88.8	14
Pappers_grey.bmp	72.6	17.9

Table 13 Input image comparison with respect to average CC value

Input image	CC Average
Lena.bmp	0.899
Baboon.bmp	0.826
Pappers_grey.bmp	0.903

4.2 Future Scope

In today's era as the development and use of multimedia devices like smartphones, LED TVs are increasing, the quality of images and videos are also increasing. In current scenario, images and videos have billions of color as in full HD, 4 and 16 K resolutions. As the resolution and color depth of images and videos increases, the size of images and videos also increases. Hence, compression is required for storage purpose as well as for transfer in the network. In future, compression can be used in image compression as well it can be embedded into video compression like for online streaming, where compression level can be selected as per requirement. The present work can be extended for color images with some high computation complexity because present compression technique will be applied on each color component of color image. Compression and decompression are two separate phases in image compression, hence, the present work can be used for cryptography in future, where compression phase will be used for generating cypher image and decompression phase will be used for getting original image from cypher image.

References

1. Gonzalez RC, Woods RE (2002) Digital image processing, 2nd edn. Prentice Hall
2. Lata A, Singh P (2013) Review of image compression techniques. Int J Emerg Technol Adv Eng (IJETAE) 3(7):461–464
3. Kainmal AB, Manimurugan S, Devadas CSC (2013) Image compression technique: a survey. Int J Eng Invent 2:26–28
4. Jassim FA (2013) Image compression by embedding five modulus method into JPEG. Signal Image Process Int J (SIPIJ) 4(2):31–39
5. Chincholli P, Krishankumar E (2016) A comparative performance analysis of DCT, DWT-SVD, DCT-DWT-SVD digital water marking algorithms. Int J Technol Res Eng 3(8)

6. Malini S, Moni RS (2014) Use of discrete sine transform for a novel image denoising technique. Int J Image Process (IJIP) 8(4):204–213
7. Hu NC, Ersoy OK Generalized discrete Hartley transforms, China. In: 1991 international conference on circuits and systems, June 1991. IEEE, pp 130–132
8. Watson AB, Poirson A (1985) A separable two-dimensional discrete Hartley transform. Nasa Technical Memorandum 88203, December 1985
9. Kornblum JD (2008) Using JPEG quantization tables to identify imagery processed by software. Digit Investig (Elsevier) 5:S21–S25
10. Li ZN, Drew MS, Fundamental of multimedia. Pearson Educational International. ISBN 0–13–127256-X
11. Jassim FA (Member, IEEE) (2013) A novel steganography algorithm to hide a grayscale BMP image in two grayscale BMP image for dual secrecy. In: 2013 2nd national conference on information assurance, pp 73–77. IEEE
12. Jassim FA, Quassim HE (2012) Five modulus method for image compression. Signal Image Process Int J (SIPIJ) 3(5)
13. Salomon D (2007) Data compression: the complete reference, 4th edn. Springer, Berlin

Chapter 24
Practical Design Considerations of DC/DC Converter Used in MPPT for Solar PV Systems

Kusum Lata Agarwal and Dhanraj

1 Introduction

The solar PV system not only consists of PV panels but also uses a power electronic converter for connecting its output to the grid or to the local load. The power electronics converters normally used are, the DC–DC converter and the DC–AC inverter for AC conversion. To extract the maximum power from the SPV module, the Maximum Power Point Tracking (MPPT) algorithm is incorporated with the DC–DC converter to step up/down the level of the solar PV array output voltage. The power semiconductor devices are used as switches where the internal resistance switch and internal resistance of practical inductors introduce new challenging problems like power loss, electromagnetic interference (EMI). Due to these problems, efficiency and power quality go down. To mitigate these issues, it is essential to create efficient DC–DC converter and effective algorithms for the MPPT.

The boost converter is used to uplift the level of DC voltage as per requirement. In literature, many authors have been worked out on simulation and hardware realization of the boost converter with and without solar photovoltaic application.

Kollimalla et al. [1] have presented the hardware prototype. They found the dynamic performance and stability of DC–DC boost converter using variable perturbation size adaptive P&O MPPT algorithm for rapid changes in solar insolation. In this algorithm, they used current perturbation, adaptive control and variable perturbation. The converter efficiency was not available in terms of numerical number but proposed in terms of dynamic performance of DC–DC boost converter with rapid

K. L. Agarwal (✉) · Dhanraj
Department of Electrical Engineering JIET, Jodhpur, Rajasthan, India
e-mail: kusum.agarwal@jietjodhpur.ac.in

Dhanraj
e-mail: dhanrajchouhan013@gmail.com

M. Shorif Uddin et al. (eds.), *Intelligent Energy Management Technologies*, Algorithms for Intelligent Systems,
https://doi.org/10.1007/978-981-15-8820-4_24

changes in irradiation. Singh et al. [2] have been demonstrated the boost converter-based solar PV fed brushless DC motor driven the water pump. In this, they have used the DC–DC boost converter to achieve the MPPT of the PV array. The input PV parameters (Panel voltage and panel current) have sensed and fed to the INC MPPT algorithm to generate the desired PWM signal in MATLAB/Simulink. In this, the MATLAB-based simulation work has been carried out. They did not discuss the efficiency of the DC–DC boost converter.

To operate the AC appliances using Solar PV array, the voltage of DC link capacitor at the input side of the inverter should be greater than or equals to $2\sqrt{2}$ time of phase voltage, as determined by Popa et al. [3]. In this, they have investigated that DC–DC boost converter can be the best choice to reduce the number of SPV modules are being connected in series. Balamurugan et al. [4] have proposed the design of boost converter for PV application. In this, to track the MPP point of solar PV panel, the Perturb and Observe (P&O) method has been used. The ATmegha328 microcontroller has used to extract the maximum power from the solar PV panel.

Saxena et al. [5] have proposed an integration of solar photovoltaic with battery to single-phase grid in which the boost converter has used to step-up the DC voltage for the dc-link capacitor. Sathya et al. [6] have designed and implemented a 12 V/24 V boost converter for the solar-powered LED lighting system. The output voltage has controlled by the use of a closed-loop feedback system. In this, they have observed a 95% efficient boost converter in hardware realization.

Many authors have tried to address the DC–DC converters interfacing a load with either solar PV panel/module/array or rectified DC supply in order to extract the maximum power from the given source. But is found that no single literature is there to present the hardware realization of the proposed DC–DC boost converters. The objective of this paper is to propose step-by-step design method to develop the hardware prototype for the proposed DC–DC boost converter.

2 Basics of Boost Converter

As the name implies, a step-up (Boost) converter produces a higher average output voltage than the DC input voltage V_{PV}. Many times, this converter can be used to match the load impedance with impedance of solar photovoltaic panel ($R_{\text{opt}} = V_{mp}/I_{mp}$) to transfer the maximum power or to operate the SPV panel at MPP point. Figure 1 shows the SPV panel and this is interfaced to R_L load with the help of DC–DC boost converter. The boost converter consists of an inductor at input, semiconductor switch (it could be BJT, IGBT, and MOSFET), diode and a capacitor.

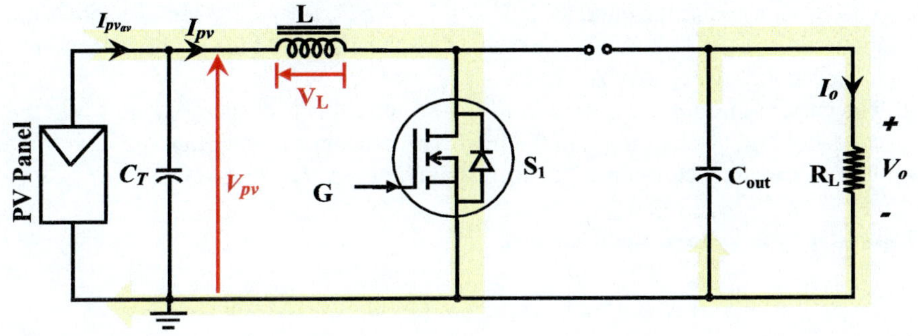

Fig. 1 Working of boost converter during DT_s time

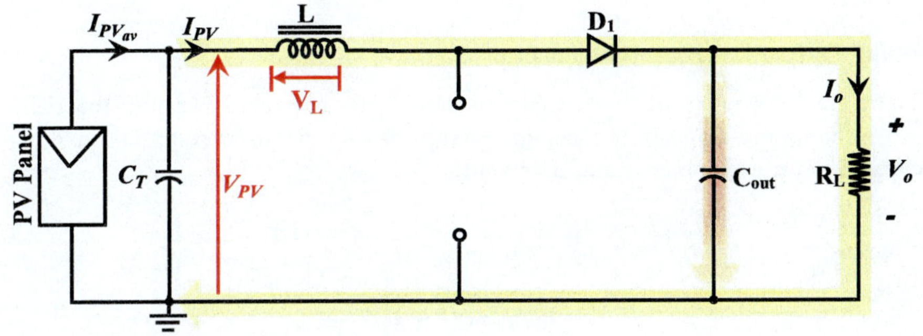

Fig. 2 Current paths during $(1 - D)T_s$ time in boost converter

2.1 Working of Boost Converter

During ON-Time

- During ON-time, the terminal current I_{PV} flows from PV module to an inductor and through the switch-Q as depicted in Fig. 1.

  The capacitor C_{out} continues to discharge through the load while the inductor L charges by extracting the energy from the SPV module.
  During ON-time, the voltage across the inductor is V_{PV} and current through it, rises linearly from a non-zero value to a peak value.

During OFF-Time

- During off-time, the inductor current flows continue and therefore the polarity of inductor voltage reverses and diode D conducts. And this current will charge the capacitor C_{out} and supply the load current I_o.

- The voltage across the inductor is $V_{PV} - V_o$. As $V_o > V_{PV}$, the voltage across the inductor is negative and this is expected because we need the positive portion to balanced out by negative portion.
- The current through inductor will decay because this is discharging now into the capacitor and discharging into the load. The capacitor is charging up.
- The average inductor current is the terminal current $\boldsymbol{I_{PV}}$ (Fig. 2).

Input–Output Voltage Relationship:

$$V_{pv} \times DT_s + \left(V_{pv} - V_o\right) \times (1 - D)T_s = 0$$

$$\boldsymbol{V_o = V_{PV}\left(\frac{1}{1 - D}\right)}$$

Input–Output Current Relationship:

The average current through the capacitor should have to be zero. Meaning that there is no charge building up in capacitor or capacitor is in steady state. Discharging capacitor current is taken as negative value.

$$(-I_o)DT_s + (I_{PV} - I_o) \times (1 - D)T_s$$

$$\mathbf{I_o = I_{PV}(1 - D)}$$

On dividing the equations,

$$R_L = \frac{V_o}{I_o} = \frac{V_{PV}\left(\frac{1}{1-D}\right)}{I_{PV}(1 - D)} = \frac{V_{PV}}{I_{PV}}\left(\frac{1}{(1 - D)^2}\right) = \frac{R_T}{(1 - D)^2}$$

$$\boldsymbol{R_L = \frac{R_T}{(1 - D)^2}}$$

2.2 *Effect of Boost Converter on MPPT*

The terminal resistance

$$\boldsymbol{R_T = R_L(1 - D)^2}$$

The terminal resistance or input resistance to the boost converter can be changed by varying the duty ratio. Let us draw the *I–V* characteristic and power curve of PV module as depicted in Fig. 3.

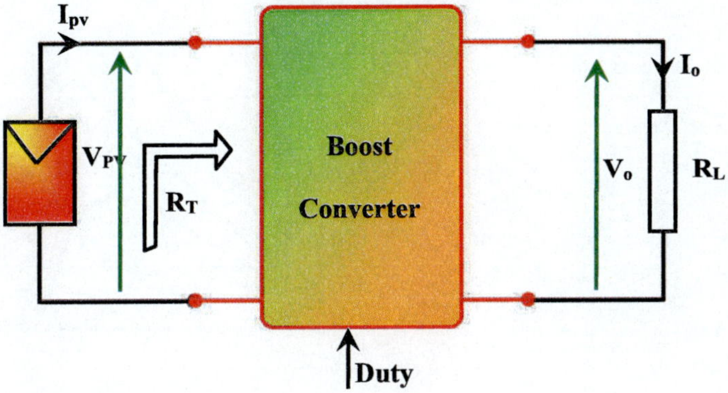

Fig. 3 PV module with MPP block for boost converter

Fig. 4 PV module I–V characteristics for boost converter

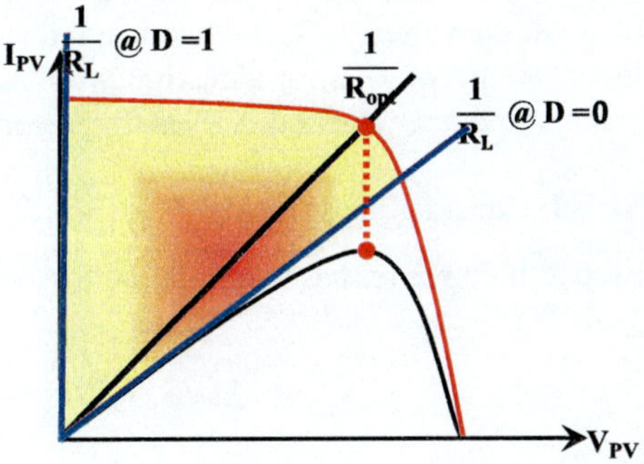

If duty $D = 1$, then.
$R_T = 0$, i.e., vertically short circuited and the load line is along with y axis.
If duty D = 0, then.
$R_T = R_L$.

With Boost converter the following point on PV curve can be traced. There is limitation with load line; consequently, there is limited load line (Fig. 4).

Hence, in boost converter the reachable points are lying between the load line corresponding to $D = 0$ and corresponding to $D = 1$ as shown in Fig. 5. From the above, it can be concluded that the maximum power point is not traced with Boost converter.

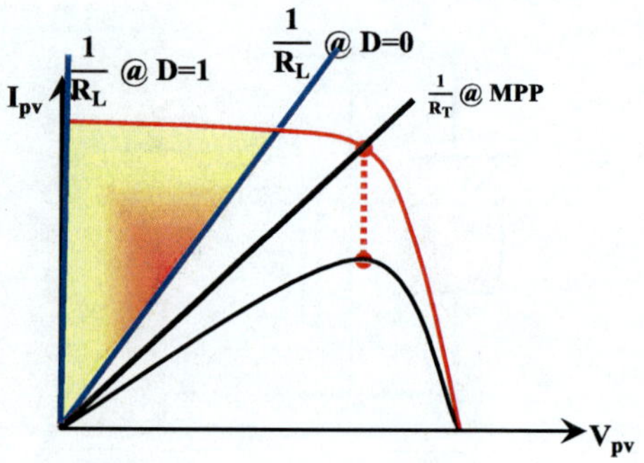

Fig. 5 PV module I–V characteristics with limited load line for boost converter

3 Design of Boost DC–DC Converter

The estimation of inductor L and capacitor C, where f_s is the switching frequency of boost converter; I_L is the average inductor current; ΔI_L is ripple content in the inductor current; $I_{L\,\text{Max}}$ is a maximum value of inductor current; $I_{L\,\text{Min}}$ is a minimum value of inductor current; ΔV_o is ripple content in the capacitor voltage.

Step 1: Consider the values of $P_{mpp}, V_{mpp}, I_{mpp}, I_{sc}$.

Step 2: If converter has zero internal losses then output voltage will be

$$V_{\text{in}} I_{\text{in}} = V_o I_o = \frac{V_o^2}{R_L} = P_{mpp}$$

$$\frac{V_o^2}{R_L} = P_{mpp} \Rightarrow V_o = \sqrt{P_{mpp} * R_L}$$

Step 3: For given output voltage, the duty of converter will be

$$V_o = \frac{V_{\text{in}}}{1 - D} \quad \Rightarrow D = 1 - \frac{V_{\text{in}}}{V_o}$$

Step 4: The inductor current is given by

$$I_L = \frac{V_{\text{in}}}{(1 - D)^2 R_L}$$

If the change in inductor current ΔI_L is 5% then

$$\Delta I_L = 0.05 * I_L$$

Since,

$$\Delta I_L = \frac{V_{\text{in}} D}{L f_s}$$

Determine L from the above expression.

Step 5: Maximum and Minimum Inductor Currents.

The maximum current through the inductor in Boost DC–DC converter is given by

$$I_{L\,\max} = I_L + \frac{\Delta I_L}{2}$$

The minimum current through the inductor in Boost DC–DC converter is given by

$$I_{L\,\min} = I_L - \frac{\Delta I_L}{2}$$

Step 6: Value of Output capacitance, if the ripple is not more than 1% then

$$C = \frac{D}{R_L f_s \frac{\Delta V_o}{V_o}}$$

3.1 Designing of Practical Inductors (L) and Selection of Capacitors (C) for Proposed Converter Topologies

The practical inductor consists of a wound conductor coil on a ferromagnetic material. This combination yields an inductance (L) that offers reluctance to a change in current, and therefore the current through an inductor cannot change instantaneously. The rate of change of current through an inductor (di_L/dt) is determined by the inductance and the voltage dropped across the inductor, given by the expression: $V = L * di_L/dt$. Furthermore, the use of ferromagnetic material as the inductor core allows energy to be stored in the inductor. When a positive voltage has appeared across the inductor, the current increases and meanwhile energy is dumping into the inductor. It is an essential characteristic that makes the inductor useful in the DC–DC converter since it acts as both a current-ripple filter and an energy-storage element.

When the switch is closed, the current flowing to the load increases and energy is also stored in the inductor. When the switch is opened and the output is disconnected from the input, the stable output current is maintained by drawing energy from the inductor. Since inductance determines the di_L/dt, its value is selected to achieve desired limits to the ripple current (I_{ripple}), providing a steady output current. The inductor can only hold a finite amount of energy before the ferromagnetic material will saturate, the inductance decreases, and ripple current increases. When making an inductor selection, it is important to check that the current at which the core saturates (I_{sat}) is greater than the application's peak inductor current, ($I_{pk} = I_{out} + I_{ripple}/2$). In this thesis, all topologies work in continuous conduction mode (CCM). For CCM, the practical value of inductor should be greater than the minimum value of required inductor values.

3.1.1 Design of Inductors

The inductor is the energy storing element and practical inductor has some value of its wire resistance. The wire resistance should be very small as possible therefore; the proper design of practical inductor plays an important role in MPP of SPV panels or modules. The following steps are used to design the inductors used in DC–DC converter.

Step 1 (Inductor Value): the value of the inductor has to be calculated. The calculation of inductors is done for DC–DC converter topologies.

Step 2 (Area Product): From the current waveform through the inductor, the energy-storage requirement for the inductor is estimated. The energy that needs to be stored in the inductor is given as

$$E_L = \frac{1}{2} L I_m^2$$

where E_L is the maximum energy in joules that needs to be stored in the inductor and I_m is the peak inductor current in amperes. Based on the energy to be handled by the inductor core, the area product is given as

$$A_p = A_c A_w = \frac{2E_L}{K_w K_c J B_m}$$

where:

$K_w = 0.6$ for single-winding inductor and 0.3–0.4 for multiple-winding inductors

$$K_c = 1$$

J = Current density for copper = 3×10^6 A/m²
B_m = Flux density in Tesla = 0.25 T for ferrites, 1 T for CRNGO and 1.2 T for CRGO.

The selected core should have an area product greater than that calculated by the above equation.

Step 3 (Permeance): The permeance of a core with air gap l_g is given by

$$\wedge = \frac{\mu_o \mu_r A_c}{l_m + \mu_r l_g}$$

where:

A_c = core cross-sectional area of the selected core (in m^2)

μ_r = relative permeability of the selected core material, for ferrite core it is generally 2000

l_g = introduced air gap length (in meter)

l_m = magnetic path length of the selected core (in meter)

Step 4 (Number of turns): The number of turns required for desired value of inductance is given by

$$N = \sqrt{\frac{L}{\wedge}}$$

Step 5 (Gauge of Wire): the cross-section area of the wire can be calculated as

$$a = \frac{I_{rms}}{J}$$

where, I_{rms} is the RMS value of the current flowing through the inductor which is equal to I_m/K_c and $K_c = I_m/I_o$. The gauge of wire (SWG) selected should have a cross-section area that is greater than that calculated by the above expression.

Step 6 (Available Window Area check): The inequality $A_W K_W > aN$ should be satisfied or else repeat the calculation for the number of turn and gauge of wire after choosing the next bigger core. Note that the value '*a*' used for checking the above inequality should be the actual cross-section area of the wire and not the calculated value in step-5.

3.2 Selection of MOSFET Switch and Driver Circuits

Most DC–DC converters are designed for continuous current operation. Note that as the switching frequency increases, the minimum size of the inductor to produce continuous current and the minimum size of the capacitor to limit output ripple both decrease. Therefore, high switching frequencies are desirable to reduce the size of both the inductor and the capacitor. The tradeoff for high switching frequencies is increased power loss in the switches. Increased power loss in the switches means that heat is produced. This decreases the converter's efficiency and may require a large heat sink, offsetting the reduction in the size of the inductor and capacitor. The selection of power MOSFET based on that the r_{dON} should be very low and voltage, current ratings of MOSFET.

Electrical isolation is provided with the use of integrated circuit (IC) package TLP250, TLP350 manufactured by TOSHIBA. This IC consists of a GaAlAs light emitting diode and an integrated photo detector. The schematic and pin configuration diagrams of TLP250 are depicted in Fig. 6.

Figure 6a clearly shows the input LED side and the receiving photodetector as well as the totem-pole driver stage. Pins 1 and 4 are not internally connected to anything and hence are labeled NC meaning no connection. Pin 8 is V_{CC}—the positive supply. Pin 5 is GND—the ground supply or the return path for the driving power supply. The supply voltage must be at least 10 V. The maximum voltage is dependent on the operating temperature. If the temperature is lower than 70 °C, up to 30 V can be used. For temperatures between 70 and 85 °C, up to 20 V can be used. However, there should not be a need to use higher than 20 V anyways (Fig. 7).

The Pins 2 and 3 are the inputs to the LED, anode and cathode, respectively. Like regular LEDs, it has an input forward voltage and a peak forward current. The

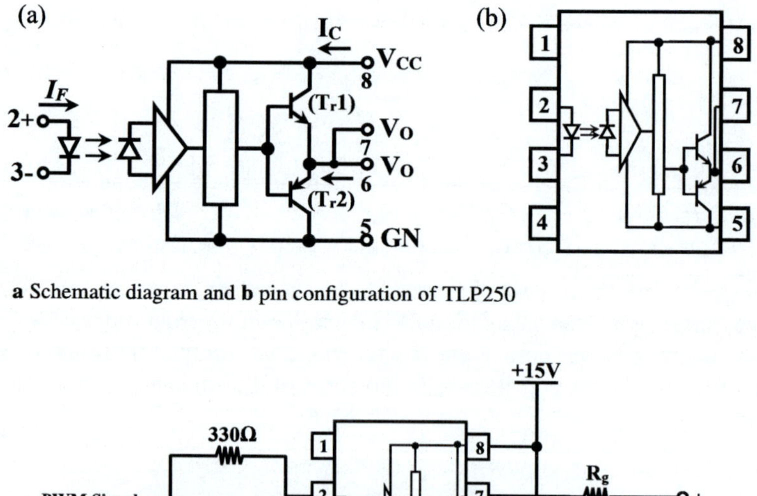

Fig. 6 **a** Schematic diagram and **b** pin configuration of TLP250

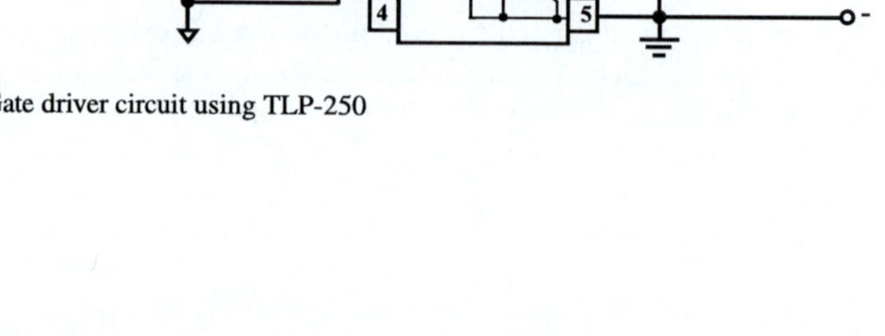

Fig. 7 Gate driver circuit using TLP-250

forward voltage will typically be between 1.6 and 1.8 V. The forward current should be less than 20 mA. The threshold input current for output transition from low to high is typically 1.2 mA but may be as high as 5 mA. Thus, 10 mA current should be good. Even though pins 6 and 7 are shown to be internally connected, the output should be taken from pin 6 as the image—datasheet—shows pin 6 labeled as V_o (Output). The output voltage will tend to rise to supply voltage when high (it will actually be slightly lower) and fall to ground level when low.

3.3 Design of Snubber Circuit for MOSFET

There is a danger of exceeding the voltage and the current rating of the MOSFET switch during turn-OFF and turn-ON instants, respectively. If there is an inductive load in the source side or if there is a significant amount of lead inductance associated with the source or drain terminal, then when MOSFET is being turned OFF, the current through the device could decrease rapidly to zero within the fall time of the device. As a consequence, a large voltage spike due to Ldi/dt will occur across the device and cause V_{DS} of the MOSFET to have a large spike during the fall time. This may damage the device. Therefore, it becomes necessary to limit the voltage spike during fall time.

On the other hand, during turn-ON of the device, due to the presence of any capacitive load or parasitic capacitance across the switch, there will be a huge surge current through the device which could damage the device. Therefore, it becomes mandatory to limit the current spike through the device during turn-ON.

The Snubber circuits modify the device switching characteristics and in doing so, reduce the device transients. The main task of the Snubber circuit is to absorb energy from the reactive elements. The parasitic capacitance and inductance cause large turn-on/off oscillation in the PWM input signal. The Snubber circuit improves the input signal shape.

In the MPPT operated DC–DC converter, the input current is fairly constant and it can be seen that the switch (MOSFET) is connected in the line or between the line and ground terminal. When the MOSFET is turned OFF, the input current or switch current i_T will fall to zero in a period of time corresponding to the fall time t_f mentioned in datasheet of MOSFET. Due to this, the voltage across MOSFET V_{ds} will shoot up to a large value because of the Ldi/dt phenomenon where L could be the lead inductances associated with the drain and source leads. Therefore, it is required to limit the voltage across the drain-source of the MOSFET and it should be made to rise gradually during the fall time t_f as depicted in Fig. 8b. To decrease the rate of rise of voltage V_{ds}, one can connect a capacitor C_s across MOSFET as indicated in Fig. 8a. Sometimes this Snubber circuit is also called as shunt Snubber because the capacitor C_s is connected in shunt with the device or MOSFET. Without loss of generality, one can assume that the switch current is falling linearly as shown in Fig. 8c during the t_f time and the drain-source voltage across the device is rising linearly as depicted in Fig. 8c during the fall time when the shunt Snubber is connected. From Fig. 8a,

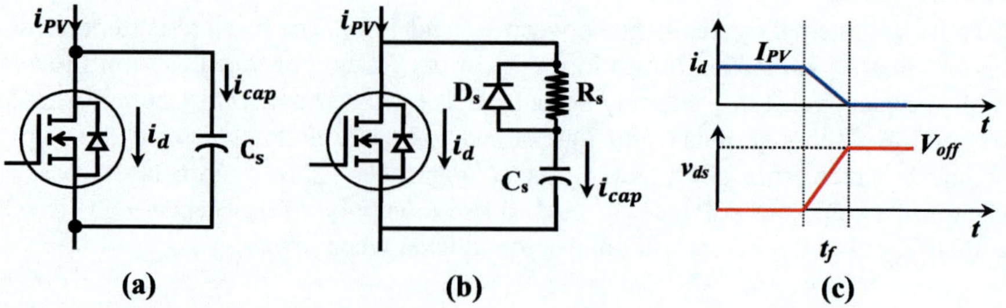

Fig. 8 Turn-off or shunt Snubber circuit for power MOSFET

$$i_{PV} = i_d + i_{\text{cap}}$$

where i_d is the current through the MOSFET and i_{cap} is the current through the shunt capacitor. During the fall time period of t_f, the current i_d through MOSFET is given by

$$i_d = i_{PV} - i_{PV}\frac{t}{t_f}$$

$$i_d = i_{PV}\left(1 - \frac{t}{t_f}\right)$$

The current through the capacitor is given by

$$i_{\text{cap}} = i_{PV} - i_{PV}\left[1 - \frac{t}{t_f}\right] = i_T \times \frac{t}{t_f}$$

The voltage V_{ds} across the device is the same as the voltage across the capacitor C_s. therefore

$$v_{\text{ds}} = \frac{1}{C_s} i_{\text{cap}} \cdot dt$$

$$v_{\text{ds}} = \frac{1}{C_s} i_{PV} \times \frac{t}{t_f} \cdot dt$$

The voltage across MOSFET or capacitor can be obtained during fall time by integrating the above equation within the fall time and then applying the boundary condition. At the end of fall time, the voltage across the MOSFET should be V_{PV} or V_{in} then the value of capacitor will be

$$C_s = \frac{i_{PV} t_f}{2V_{\text{in}}}$$

The problem of voltage spike during turn-off process of the MOSFET is resolved by the circuit of Fig. 8a but it will create a serious problem during the turn-ON of the MOSFET. When the MOSFET is turned ON again, the capacitor C_s will discharge through MOSFET and will result in a large current spike through the MOSFET which can damage the power MOSFET. Therefore, to control the current through MOSFET, a resistor R_s is introduced in series with capacitor C_s.

The resistor R_s should provide the function of current limiting only during turn-ON of the MOSFET. However, during turn-OFF of MOSFET, R_s is not needed. Therefore, to deduce the dissipation in R_s during turn-OFF, a diode is placed across R_s as shown in Fig. 8b so that R_s comes into effect only during turn-ON when C_s discharges through R_s and MOSFET.

3.4 *Choose a Snubber Capacitance Value Which Meets Two Requirements*

(1) It can provide a final energy storage greater than the energy in the circuit inductance
$\frac{1}{2}C_s V_{\text{off}}^2 > \frac{1}{2}L_i i_{PV}^2$, and therefore $C_s > \frac{L_i\, i_{PV}^2}{V_{\text{off}}^2}$.
Where, L_i is circuit inductance and V_{off} is the voltage across the device when it is turn-OFF.

(2) It produces a time constant with the Snubber resistor that is small compared with the shortest expected on-time for the MOSFET switch.

$$R_s C_s < \frac{t_{\text{ON}}}{5}$$

4 Selection of Snubber Resistor R_s

When the MOSFET is turned on, it should carry the load current plus the capacitive current discharge from C_s which is equal to V_{off}/R. therefore,

$$\frac{V_{\text{off}}}{R_s} + i_{PV} < I_{dm}$$

where, I_{dm} is the maximum drain-current rating of the MOSFET.

Rearranging the above inequality, the following inequality is obtained

$$R_s > \frac{V_{\text{off}}}{I_{dm} - i_{PV}}$$

It is also important to ensure that the capacitor discharges fully before the next charging when the MOSFET turn-OFF. Therefore, there is a minimum duration of time during which time the MOSFET should remain ON so that the capacitor can fully discharge. Therefore

$$t_{\text{ON}} > 5R_sC_s$$

where, t_{ON} is the minimum time for which MOSFET should remain in the ON-state.

Rearranging the above inequality, the following inequality is obtained:

$$R_s < \frac{t_{\text{ON}}}{5C_s}$$

From the above two inequalities, the range of choice of R_s is given by

$$\frac{V_{\text{off}}}{I_{dm} - i_L} < R_s < \frac{t_{ON}}{5C_s}$$

The power rating of Snubber resistor is given by

$$P_{R_s} = \frac{1}{2}C_s V_{\text{off}}^2 f_{sw}$$

where, f_{sw} is the switching frequency of the MOSFET.

5 Conclusions

The significant contributions of the presented paper are as follows:

1. The need of DC–DC converter in between solar PV module and resistive load has been investigated through mathematical calculation.
2. The investigation on need of DC–DC converter revealed that DC–DC converter with MPPT technique modifies the effective resistance being faced by Solar PV module.
3. The parameters (Value of L, C and Duty) of the proposed DC–DC converters have been determined keeping in mind that these converters must work in linear region and in continuous conduction mode (CCM).
4. The most important parameter of DC–DC converter interfacing SPV module with load is the internal resistance of inductor. The internal resistance of inductor may

restrict the maximum power transfer. Hence, the internal resistance of inductor can be lowered by using Litz wire or large cross-sectional copper wire.

5. The internal resistance of high speed switch (MOSFET/IGBT) must be low. The careful selection of power MOSFET plays a major role in energy extraction from SPV module.
6. To drive the Power MOSFET, a robust optocoupler (TLP-250) based driver circuit has been developed in the laboratory. The ringing effect has been minimized by connecting a snubber circuit across the power MOSFET.

References

1. Kollimalla SK, Mishra MK (2014) Variable perturbation size adaptive P&O MPPT algorithm for sudden changes in irradiance. IEEE Trans Sustain Energy 5(3):718–728
2. Singh B, Kumar R (2016) Solar PV array fed brushless DC motor driven water pump. In: IEEE 6th international conference on power systems (ICPS), March 2016
3. Popa D-L, Nicolae M-S, Nicolae P-M, Popescu M (2016) Design and simulation of a 10 MW photovoltaic power plant using MATLAB and Simulink. In: IEEE international power electronics and motion control conference (PEMC), September 2016, pp 378–383
4. Balamurugan M, Narendiran S, Sahoo SK, Shanmugam PK, Naidu RC (2016) Design of boost converter for PV applications. In: Proceedings of fifth international conference on soft computing for problem solving, April 2016, vol 2, pp 619–628
5. Saxena N, Singh B, Lal A, Vyas (2017) Single-phase solar PV system with battery and exchange of power in grid-connected and standalone modes. IET Renew Power Gener 11(2):325–333
6. Sathya P, Natarajan R (2013) Design and implementation of 12V/24V closed loop boost converter for solar powered LED lighting system. Int J Eng Technol (IJET) 5

Chapter 25
Machine Learning-Based Solar Energy Forecasting: Implications on Grid and Power Market

Chandershekhar Singh and Ajay Kushwaha

1 Introduction

In modern world, electrical power is a primary need and is counted in basic needs of society. With the increase of residential and industrial loads, generation (or supply) must increase at same rate to balance demand-supply. Conventional power plants (thermal) have been employed since various decades to satisfy energy requirements. It was a thinking that coal (fuel) is in abundance and it will never get exhausted. Apart from it, supply from conventional generation increases greenhouse gas (GHG) to atmosphere. However, there are methods to control/reduce GHG in conventional plants but it will be better option that fossil fuel plants may be replaced by renewable energy resources.

As of now, solar and wind energy generations are developed technology. They have been installed and maintained on a large scale for power production. Despite of intermittent nature of output power, lot of resource on their design and development are considered by various nations.

Solar energy is vast and spread over every part of land and water on earth, but it is not constant. It varies with respect to place and time due to geographical factors, atmospheric conditions, etc. Global solar power installation has reached above 500 GW, with annual increase of 100 GW only in year 2018. India has also increased its installation of 10.8 GW in year 2018, and adding total solar capacity to 33 GW [10].

Now a day's most of electrical grids are centralized in various nations. They transfer power from power plants (mainly thermal) to consumer end; however, decentralized dispatch scenario is increasing significantly due to extension of renewable generation (or distributed generation). There is a need to increase electricity generation amongst grid at different levels and this may be challenging to grid management.

C. Singh (✉) · A. Kushwaha
Department of Electrical Engineering, JIET, Jodhpur, India
e-mail: chandershekhar333@gmail.com

M. Shorif Uddin et al. (eds.), *Intelligent Energy Management Technologies*, Algorithms for Intelligent Systems,
https://doi.org/10.1007/978-981-15-8820-4_25

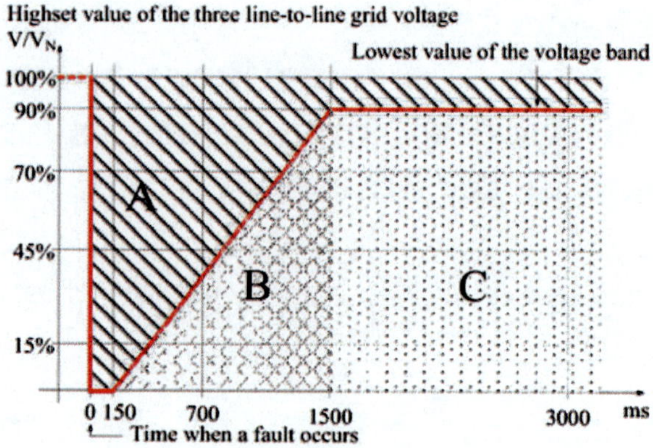

Fig. 1 Low voltage ride through requirement for grid connection

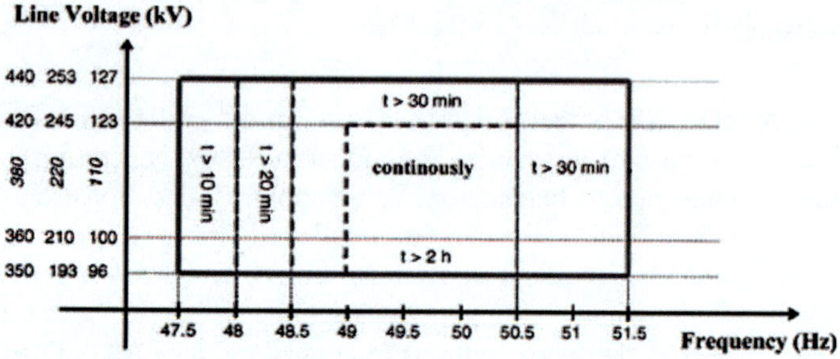

Fig. 2 Line voltage and frequency operating limits of generating unit

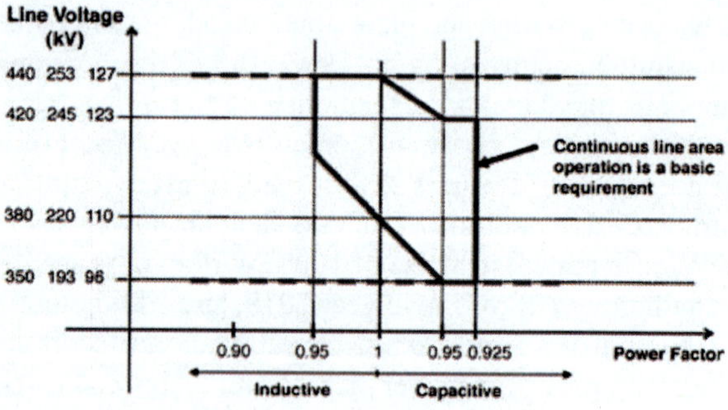

Fig. 3 Voltage and power factor operating limits for generation unit

The challenge for electrical grid operators will be to optimize generation location and dispatch to balance demand with energy supply and also maintain stability.

In solar technology, flat-panel PV, Concentrated Solar Power (CSP) and concentrated PV (CPV) system are mostly employed. They have ability to compete with

Table 1 Forecasting methods for the optimization of power output/Irradiance

Author	Forecasting time horizon	Methods	INPUT Parameters	OUTPUT
1zgi et al. [1]	0–60-min ahead	Artificial Neural Network—Levenberg Marquardt Back-Propagation Algorithm	750 Wp solar PV Panel—Ambient and cell temperature —Diffuse solar Irradiation	Solar power
Kardakos et al.[2]	Day and Intra-day ahead	Seasonal-Autoregressive Integrated Moving Average —Artificial Neural Network with multiple input	Solar radiation	PV power
Mellit et al. [3]	Hourly	Artificial Neural Network—Multi Layer Perceptron Architecture	Solar irradiance —Air temperature	Power of PV modules
Yona et al. [4]	24-h ahead	Based on insolation forecasting by utilizing weather reported data, fuzzy theory, and Radial Basis Neural Network	Weather reported data (Humidity, cloud amount, and temperature) —Insolation forecast data with fuzzy	PV power output
Almonacid et al. [5]	1-h ahead	Dynamic Artificial Neural Network: Non-linear Autoregressive Neural Network	Global solar irradiance —Air temperature —PV power output Module	PV power output generator

(continued)

fossil fuel-based production in near future. However, natural variability (intermittency) of the solar power and high cost of energy storage raises concerns regarding reliability and feasibility of this technology.

2 Technical Issues with Grid

Grid interconnection (loading) and disconnection (unloading) of generation units is one of the technical issues that needs to be considered for power dispatch.

The incoming generation unit and existing power-sharing (supplying) system have to adjust their power output such that dynamics at the point of interconnection would remain within limits. If new power generation unit is added it should synchronize

Table 1 (continued)

Author	Forecasting time horizon	Methods	INPUT Parameters	OUTPUT
Cornaro et al. [6],	24-h ahead	Statistical Model (ST) and Model Output Statistics (MOS) based on an ensemble Artificial Neural Networks by master optimization process	Statistical model: STNNS —Ordinal day number —Daily irradiation —Average daily temperature —Clearness index MOS model: ECMWF-MOSNNS —Ordinal day number —Daily irradiation —Average daily temperature —Clearness index —Daily irradiation prediction provided by	Global horizontal irradiance
Lima et al. [7]	24-h ahead	NWP-WRF model by using Artificial Neural Networks	Weather forecast outputs supplied by WRF model —Observational ground data	Surface solar irradiance
Hossain et al. [8]	Day ahead —1-h ahead	Extreme Learning Machine (ELM)	Daily average solar irradiance —Ambient temperature —Module temperature —Wind speed —PV output data	PV power
Mellit, A. MassiPava, V. Lughi [9]	Hourly and Daily	Adaptive feed-forward back-propagation network (AFFNN)	Solar Insolation time series	Hourly Insolation

its voltage, frequency, phase sequence as per grid codes. It also has to maintain his loading gradually and reach up to sharing limit of power allocated. It is so because ramp rate of loading has to maintain in limits, this ensures safety of generation equipments. The active and reactive power controlled by DC to DC converters and inverters at point of connection. These converters maintain parameter dynamics in range. In wind energy system, back to back converters are used to control dynamics.

Generation unit is also required to operate in normal and faulty conditions. In case of high transients/dynamics of grid or generation parameters, generation unit is made to trip. There are standard conditions when generation could remain connected to grid. This is fault ride through requirement.

It can be observed from the figure that generation unit may remain connected if fault is cleared within 150 ms. Also if voltage due to fault restore its operating magnitude in specified time (as straight line), then generation remains in operating mode, otherwise unit may be disconnected from grid [11].

Frequency and voltage operating range must be in limits for continuous operation of generation unit. In the figure, standard line voltages are 110, 220 and 380 volts and their range of continuous operation is 100–123, 210–245 and 360–420, respectively, if frequency is maintained 49–50 Hz. Deviation from standard magnitudes increases dynamics/disturbance in system, hence more deviation in parameters will lead to more disturbance of system. This gives less time for system to withstand disturbances. Hence system becomes prone to tripping of generation unit [12].

These disturbances in voltage and frequency create flicker. When load changes are more dynamically and demand–supply difference (changes) becomes fast then power fluctuation arises. This creates power quality problems.

Uneven demand pattern and less regulated generated power (especially in case of renewable) create grid instability in terms of frequency limit violation.

Reactive power is regulated to maintain system voltages within limits. In case of faults, voltage is restored by increasing reactive power to the system. The reactive power supply is provided by generator units, static capacitor banks, FACTS, etc.

In the figure, generation units supplying inductive VARs increases voltage magnitude and capacitive VARs decreases voltage [12].

Now all problems start with demand–supply difference, it is needful that load demand information must be specified. As the consumer is free to vary its load, therefore, there is only left provision to study its demand pattern/behavior. Hence power is regulated with respect to demand pattern of consumer.

It is a common practice to forecast demand behavior and manage supply resource accordingly.

If power forecasted becomes same to that needed by consumer, then generator, then load flow analysis can be carried out for stability analysis of Grid easily. Also there will be better stability condition as power injection is defined and maintained at a same level.

In thermal, energy/power output is controlled/regulated and fed to injection point of grid in order to obtain stable grid interconnection.

If solar power could be predicted, then interconnection will be more reliable. For controlling power, power electronics devices can be employed to obtain stable interconnection.

For solar DC to DC converters, grid inverter controls parameter of solar power with respect to Grid parameters. The power electronic devices selected must have burden as that of installed solar capacity.

3 Forecasting and Power Market

Solar energy is highly dependent on weather conditions especially cloud structure and day/night cycles. Clouds cause significant ramps in solar insolation and PV power output. Therefore, integration of electricity produced by solar power systems requires accurate energy evaluation and time horizon forecasts. Overall energy production pattern does not follow the energy forecasted.

Solar forecasts on multiple time horizons play a fundamental role in

1. Storage management of PV systems.
2. Control systems in buildings, such as solar thermal plants
3. Power scheduling and regulation.

It allows grid operators to study load pattern in order to optimize power wheeling and plan maintenance activities at the production sites and take necessary measures to protect production from extreme events.

Depending on the purpose, different information are needed, such as

1. Long-term historical data of expected energy yield (to assess sites where solar power is abundant)
2. Real-time data set of energy yield (optimizes real-time energy production management)
3. Forecasted site irradiance data.
4. Real-time datasets on weather conditions.

If generation accuracy increases, it can be compared with conventional generation resources that are scheduled and dispatched for power market.

Overall, accuracy in short-term forecasting yields:

(i) Exact Supplier (power) bidding in market and forecast returns from it.
(ii) Exact change in grid condition/grid parameter (such as frequency, reactive power, voltages) for which a load flow study program can be run and take an observation of grid parameter, which leads to grid stability.

Forecasting of supply is needed for ensuring power to be traded in power market. The Generation supply bids have power (in MW), time span (for dispatching) and price of power in bidding offer. Better forecasting ensures estimation of real-time magnitude of power. More the reliable prediction, more reliability to market players. Hence economic efficiency/incentives of transactions increase. There are various methods for estimation/predication of power generated with respect to state variables such as global irradiance, humidity, cloud, etc.

Apart from grid interconnection, grid stability is also essential; stability refers to frequency regulation, voltage regulation, energy imbalance, contingency analysis in case of outage, etc. Frequency regulation is maintained by demand supply balance. Hence reserves are needed to balance them. These reserve services are also traded apart from real power market.

Price bidding is normally for centralized market. In Indian system, discoms and few retailers along with some big industries are "buyer" and all generations units are "seller." Here power bid is placed by every generator (government/private) in market and buyers fulfill its power requirement and place his bid (prices and MW). Market settlement is made by government entity and then transmission lines are ensured/scheduled for bulk power transmission [13].

The generation price bid must be such that it becomes economical and reliable to buyer. Here cost of power is overall components of various costs, such as capital cost of installation, operation and maintenance, along with market returns (profit) whereas reliability depends on technology of generation [14]. Solar technology produces intermittent power that is not reliable in power market (in terms of fixed MW placed).

Hence solar power bid can be placed in market as minimum limit value of range in which generated solar power varies. But as difference between minimum and maximum is very large, therefore, this is another challenge to estimate about magnitude of power placed in power market.

If forecasted solar power is close to real power produced then power placed in market will be a curve of "power vs. time." This variation of power can be managed by system operator in its load flow studies. At last, payment for power transaction will be on the basis of total energy transaction at the end of schedule.

4 Machine Learning-Based Solar Generation Models

As power output of solar PV modules depends on atmospheric and geographical conditions, its power output can be predicted. There are various scenarios in which PV modules behave in different way. These behaviors are modeled by different methods, some of these methods are persistence method, statistical method, machine learning approach and hybrid approach [15].

In persistence model, solar insolation and climatic conditions of each day are considered to be same. Daily pattern of solar irradiance throughout the day of same geographical location is fixed. It does not depend on change of climatic conditions such as temperature, pressure, cloud variation, humidity, wind flow, suspension particles of air, etc. This method gives standard data that may be different from power output of module of next coming day. These data are used as benchmark for the analysis of power output when metrological data are considered in input.

In statistical model, more than one variable is considered for evaluation of power output. The solar insolation variable is combined with one of the above mentioned metrological variables and then power output of PV modules is determined. There are various models in which various combination of temperature, pressure, cloud variation, humidity, etc., are made with solar irradiance data (throughout the day) of a particular geographical location, and used as input variables for power output function.

The accuracy of this method depends on the accuracy of metrological data and method of regression used. The simple and multiple regressions give better results than single input (persistence model) method [16].

As the input data of solar irradiance and metrological have nonlinear nature, nonlinear approaches of determination of relation or functions are used further for better results. Artificial neural network is one of the approaches, which consider wide range of nonlinear inputs and determined by different models. These models are trained by real-input and -output data. Input data consist of more than one input and activation function is selected on behalf of nature of nonlinearity of input and output data. There are no assumptions made for training of models, therefore, they produce better output than statistical methods.

In this method, there is input layer that accepts input data, hidden layers (which may be more than one) determine the effect of input on output defined by weights and bias. Output layer produces output data in the neural network model.

Initially, these weights are randomly selected for training of model and activation function squeezes sum of all signal and sent to output layer. For training, reference input–output data are used. Output of model is compared with reference output, further weights and bias are modified and again output is determined. The weights are modified so as to reduce error between new output and reference output than previous error. In this process, weights are modified each time to obtain better output and it continues till the obtained output matches with reference output. On completion of training, weights are obtained that validate reference input–output data. After training of model, it can be used to obtain output from new set of input data. There are various types of neural network layer interconnection or architecture, some of them are multilayer perceptron, multilayer feed forward neural network, radial basis function neural network, recurrent neural network, adaptive neuro-fuzzy inference systems (ANFIS), etc. These architectures have been used for solar power output analysis. Auto-regressive integrated moving average (ARMA), Support vector machine (SVM), ANFIS are frequently used for forecasting [15]. Some of the details of forecasting methods have been tabulated for the optimization of power output/irradiance.

Above methods have been used for various inputs at different geographical locations to forecast hourly power output of PV modules. Machine learning refers to methods to analyze real-time response to real-time changes in defined inputs, the methods are normally combination of regression, neural, fuzzy, genetic, recognition of weather input, etc., also reinforcement learning method. Reinforcement learning normalize inputs and information is processed. Its decision-making system performs action/evaluation and output is obtained, the environment (inputs and state variables) respond on output produced. This information policy reward makes tuning in decision strategy/action to tune new output [17]. New output is a function of change in input and state variables. This method may include above different machine learning techniques to produce outputs constraint to environment limits.

5 Conclusion

As solar power is essential to be utilized for generation expansion, its intermittent nature should be controlled to regulated output. Grid integration is controlled by power converters, relevant algorithms are employed for gate signal of converters to control voltage, frequency, etc. Accuracy in forecasting infers to reliability in power market trade and less congestion conditions in transmission. As solar power output is forecasted correctly, overall power market economics improves. This may be achieved by machine learning algorithms while incorporating solar insolation, weather data for short-term forecasting. There are various methods that have reduced error in forecasted output to great extent. When neural, fuzzy and GA are combined with one another and implemented to analyze the effect of weather on power output, they have determined almost same output as real power produced in some cases. SVM and reinforced learning along with previous algorithms are recent areas for forecasting power output.

References

1. ErcanIzgi et.al (2012) Short–mid-term solar power prediction by using artificial neural networks Solar Energy 86 (2012) 725–733
2. Kardakos EG et al (2013) Application of Time Series and Artificial Neural Network Models inShort-term Forecasting of PV Power Generation Power Systems Laboratory, Department of Electrical and Computer Engineering Aristotle UniversityofThessaloniki54124. Thessaloniki, GREECE
3. A. Mellit et.al (2013) Artificial neural network-based model for estimating the produced power of a photovoltaic module Renewable Energy 60 (2013)
4. Atsushi Yona,et.al(2013) Determination Method of Insolation Prediction WithFuzzy and Applying Neural Network for Long-TermAhead PV Power Output Correction IEEE transactions on sustainable energy, vol. 4, no. 2, April 2013
5. Almonacid F et al (2014) A methodology based on dynamic artificial neural network for short-term forecasting of the power output of a PV generator. Energy Convers Manag 85(2014):389–398
6. C. Cornaro et.al (2015) Master optimization process based on neural networks ensemble for 24-h solar irradiance forecast Solar Energy 111 (2015) 297–312
7. Francisco J.L. Lima et.al (2016) Forecast for surface solar irradiance at the Brazilian North-easternregion using NWP model and artificial neural networks Renewable Energy 87 (2016) 807e818
8. MonowarHossain et.al (2017) Application of extreme learning machine for short term output power forecasting of three grid-connected PV systems Journal of CleanerProduction 167 (2017) 395e405
9. A. Mellit, A. MassiPavan, V. Lughi(2019) Short-term forecasting of power productionin a large-scale photovoltaic plant. Solar Energy, pages 402–413
10. http://www.ieapvps.org/fileadmin/dam/public/report/statistics/IEAPVPS_T1_35_Snapshot2019-Report.pdf
11. B. Wu, Y. Lang, N. Zargari, S. Kouro, —Power Conversion and Control of Wind Energy System||, John Wiley 2011,
12. G. Abad,.J. Lo´pez, M. A. Rodrı´guez, L. Marroy, —Doubly Fed Induction Machine, Modeling and Control for Wind Energy Generation.||, John Wiley, 2011, pp 304, 481-483

13. Daniel S. Kirschen, Goran Strbac, 'Fundamentals of Power System Economics', John Wiley & Sons 2004
14. Bhattacharya Kankar, Bollen Math, Daalder Jaap E (2001) Operation of Restructured Power Systems. Springer, US
15. Utpal Kumar Das, Kok Soon Teya, Mehdi Seyedmahmoudiana, SaadMekhilef, Moh Yamani Idna Idris, Willem Van Deventerc, Bend Horanc, Alex Stojcevski, 'Forecasting of photovoltaic power generation and model optimization: A review'
16. Oudjana SH, Hellal A, Mahamed IH. Short term photovoltaic power generation forecasting using neural network. In: Proceedings of the 11th international conference on environment and electrical engineering (EEEIC) 2012. IEEE; 2012.pp. 706–11
17. Mohssen Mohammed, Muhammad Badruddin Khan, EihabBashier Mohammed Bashier, 'Machine Learning Algorithms and Applications' CRC Press, Taylor & Francis Group, 2017

Chapter 26
Performance Comparison of Minimum Shift Keying and Gaussian Minimum Shift Keying in Additive White Gaussian Noise Environment

Bablu Kumar Singh and Supriya Vyas

1 Introduction

Minimum Shift Keying (MSK) is a continuous phase modulation scheme. The modulated carrier does not contain phase discontinuities and frequency changes at carrier zero crossings. It is typical for MSK that the difference between the frequency of logical 0's (f0) and 1's (f1) is equal to half the data rate. MSK modulation makes the phase change linear and limited to $\pm$ (π/2) over the symbol interval. Due to the linear phase change effect, better spectral efficiency is achieved. That means that MSK is ordinary FSK with the modulation index set to 0.5 [1]. In an MSK, signal properties can be observed are the envelope, is constant, phase is continuous between the bit transitions and form of FSK signal, which involves two frequencies.

1.1 Gaussian Filter

The one-dimensional Gaussian filter has an impulse response expressed with standard deviation as parameter:

$$g(x) = \frac{1}{\sqrt{2\pi} \cdot \sigma} \cdot e^{\frac{-x^2}{2\sigma^2}} \tag{1}$$

Here, σ is the standard deviation (Figs. 1 and 2).

B. K. Singh (✉) · S. Vyas
M.B.M Engineering College, JNV University, Jodhpur, Rajasthan, India
e-mail: bablukumar.singh@gmail.com

S. Vyas
e-mail: supriyavyas77@gmail.com

M. Shorif Uddin et al. (eds.), *Intelligent Energy Management Technologies*, Algorithms for Intelligent Systems,
https://doi.org/10.1007/978-981-15-8820-4_26

Table 1 Parameters of MSK

Parameters for MSK	Values set
MSK symbols	1000
Samples per symbol	16
Symbol rate (Hz)	400 k
Filter length (symbols)	8

Gaussian filtering is used to remove noise and details of a signal. Gaussian filters have the properties of having no overshoot to a step function input while minimizing the rise and fall time [2]. Graphically, we can see below a familiar bell-shaped Gaussian distribution.

The Gaussian function defines the probability distribution for noise or data and it is used for smoothing waveform by avoiding abrupt changes in the varying amplitude of the signal.

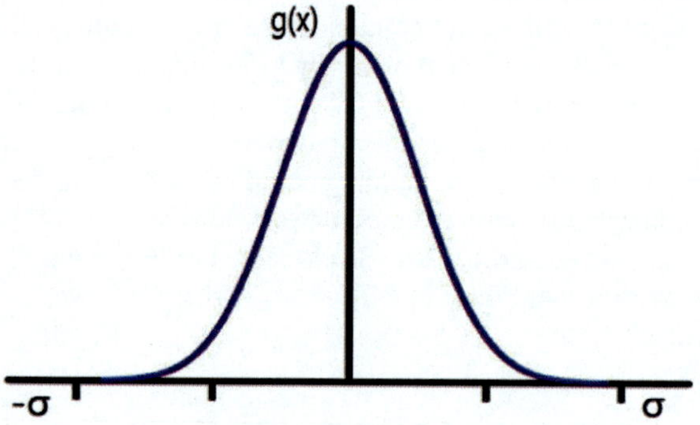

Fig. 1 Gaussian filter response

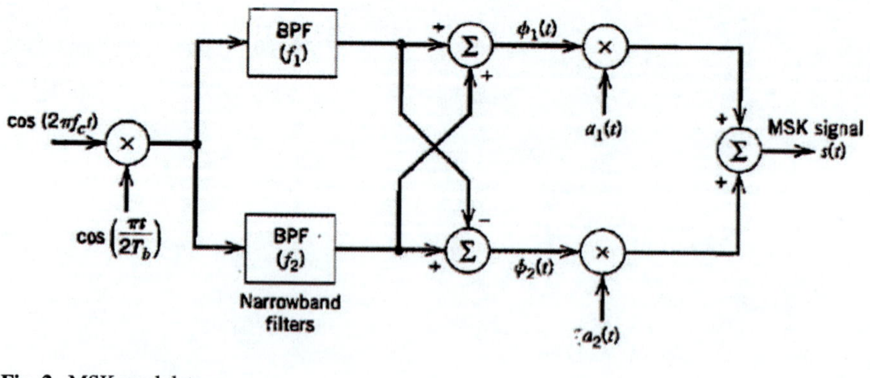

Fig. 2 MSK modulator

1.2 Raised Cosine Filter

This type of filter often used for pulse shaping in digital modulation due to its skill to minimize inter-symbol interference (ISI). The frequency response |H(f)| of a perfect/ideal raised cosine filter is symmetrical about the center of 0 Hz [3]. In our application in MSK modulation and demodulation, the overall magnitude response of the Nyquist filter is split between the transmitter and receiver. The phase response compensates for the transmitting filter phase so to make it linear phase with following equations below.

$$\mathrm{G(f)} = \mathrm{G_T(f)G_R(f)} \tag{2}$$

$$|\mathrm{G}_T(\mathrm{f})| = |\mathrm{G}_R(\mathrm{f})| = \sqrt{|\mathrm{G(f)}|} \tag{3}$$

$$\angle|\mathrm{G_R(f)}| = -2\pi f \tau_o \text{-} \angle \mathrm{G_T(f)} \tag{4}$$

Here, GT (f) and GR (f) are the transfer functions of the transmitting and receiving filters accordingly. The receiving filter in this case is matched to the transmitting filter to maximize the signal-to-noise ratio (SNR) at the sampling time instances at the receiver. The transmitting and receiving filters are typically considered to be linear phase with a nominal delay of that is required to make the filters physically realizable [4] (Figs. 3 and 4).

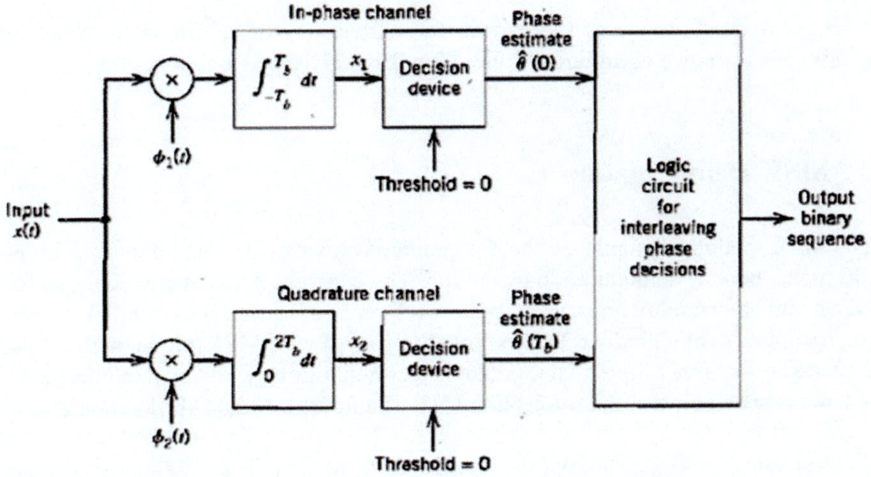

Fig. 3 MSK demodulator

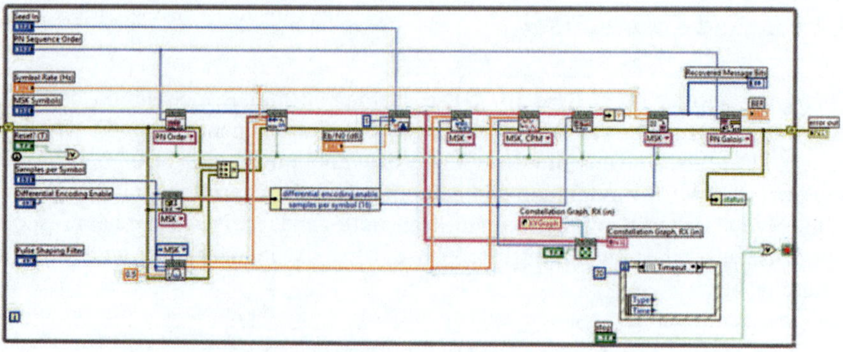

Fig. 4 LabView code for MSK modulation and demodulation

2 MSK Modulator

The MSK modulation is a constant envelope signal with continuous phase that results from modulating the instantaneous frequency with rectangular pulses. MSK is considered to be a special case of Offset QPSK (OQPSK) with half sinusoidal pulse weighting rather than rectangular [5]. The block schematic diagram of an MSK modulator is shown in the figure.

The expression of MSK signal is

$$S_{MSK}(t) = A\cos[(2\pi f_c t + \phi_n(t)], nT_b \leq t \leq (n+1)T_b \tag{5}$$

Here, A is amplitude of the carrier, fc is frequency of carrier, $\phi_n(t)$ which is continuous is known as additional phase function. *Tb* is the symbol period.

3 MSK Demodulator

The MSK modulated signal can be demodulated either by a coherent or a differential (noncoherent) demodulation technique. The coherent demodulator scheme, for QPSK and other modulations, needs precise reference of carrier, phase, and frequency for optimum demodulation. This is not that easy for an MSK modulated signal. Hence a suboptimal differential modulated technique along with threshold detection is a popular choice in cellular telephony [1]. The figure of the MSK demodulator is shown below.

Depending upon the value of the phase state θ(0), the inphase component equals +g(t) or – g(t), where

$$g(t) = \begin{cases} \sqrt{\frac{2E_b}{T_b}} \cos\left(\frac{\pi t}{2T_b}\right), & -T_b \le t \le T_b \\ 0, otherwise \end{cases} \tag{6}$$

And depending upon the value of the phase state θ(Tb), the quadrature component equals +g(t) or – g(t), where

$$g(t) = \begin{cases} \sqrt{\frac{2E_b}{T_b}} \cos\left(\frac{\pi t}{2T_b}\right), & -T_b \le t \le T_b \\ 0, otherwise \end{cases} \tag{7}$$

Probability of error/Bit error rate for coherent MSK is given as:

$$\frac{1}{2} erfc\left(\sqrt{\frac{E_b}{N_o}}\right) \tag{8}$$

And the bit error rate for coherent GMSK [5] is:

$$\frac{1}{2} erfc\left(\sqrt{\frac{\alpha E_b}{2N_o}}\right) \tag{9}$$

4 Simulation of MSK in LabVIEW

MSK Modulator–Demodulator is designed in LabView and various parameter values such as Samples per second is set to 16, symbol rate is kept 400 kHz and MSK symbol is kept as 1000. The Eb/No value is then varied continuously and the corresponding bit error rate is noted down first without using any filter and then using both Gaussian and Raised cosine filters, respectively.

The theoretical results can be calculated using BER expression shown in Eqs. 8 and 9 to verify with the LabView results.

5 Results

A comparative analysis of simulated curves for BER vs. Eb/No (signal to noise ratio) of MSK with and without filters is as shown below.

6 Conclusion

Performance comparison of Minimum Shift Keying's demodulated signal using Gaussian filter and raised cosine filter has been performed in this paper and the simulated result as shown in Fig. 5, shows that with increase in Eb/No, BER decreases and at lower Eb/No the performance of Gaussian filter is much better as compared to raised cosine filter it can also be verified with constellation graph shown in Figs. 6 and 7, respectively.

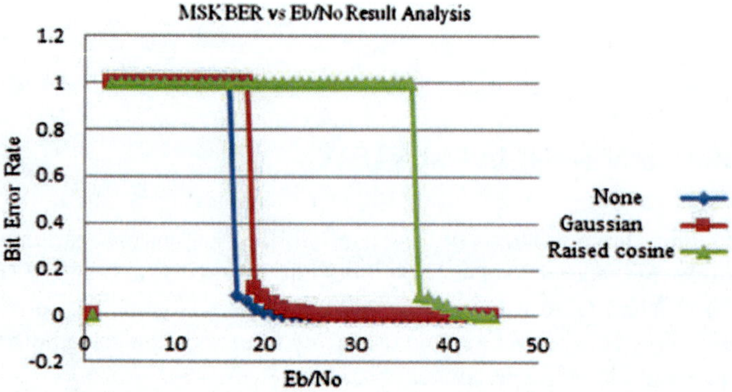

Fig. 5 Performance analysis of BER vs Eb/No result analysis using LabView

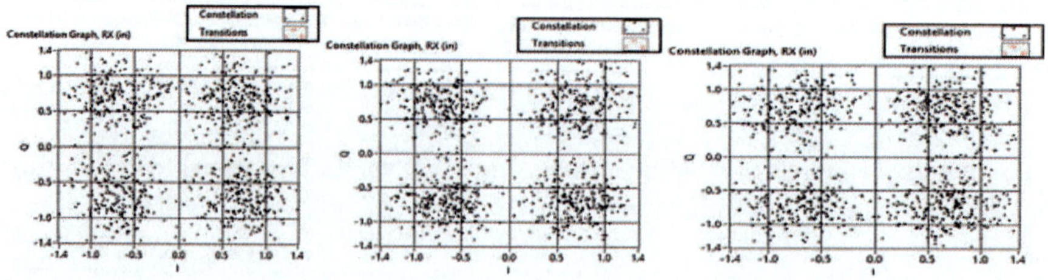

Fig. 6 Constellation of MSK signal at 20 Eb/No without filter, Gaussian Filter and raised cosine filter, respectively

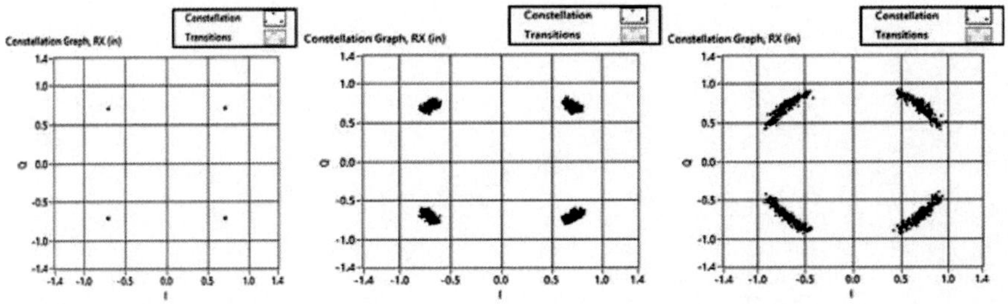

Fig. 7 Constellation of MSK signal at 40 Eb/No without filter, Gaussian Filter and raised cosine filter, respectively

References

1. Couch II LW (2001) Digital and analog communication systems, 6th Edn. Prentice-Hall, Inc
2. Cheng S, Iyer Sehshadri R, Valenti MC, Torrieri (2007) The capacity of noncoherent continuous-phase frequency shift keying, in Proc. Conf. on Info. Sci. and Sys (CISS)
3. Tapkir PR, Singh SB, Thune NN (2016) Analysis and implementation of minimum shift keying (MSK) modulation on FPGA platform. In: International Conference on Automatic Control and Dynamic
4. Chen Y, Li Z, Liu P, Liu S (2017) The BER performance comparison of MSK and GMSK schemes for short-range visible light communication. In: IEEE 9th International Conference on Communication Software and Networks (ICCSN)", Guangzhou, pp. 611–614
5. Shamee B, Christen L, Nuccio S, Yang J, Willner A (2010) Gausian Minimum Shift Keying for spectrally efficient and dispersion tolerant optical communications," CLEO/QELS: Laser Science to Photonic Applications. San Jose, CA, pp 1–2
6. Molnar G, Babic M, Sipus Z (2018) Sharpened raised-cosine FIR filters, in 41st International Convention on Information and Communication Technology, Electronics and Microelectronics (MIPRO), Opatija, pp. 0111–0116
7. Symon H (2013) Digital communication Systems. Wiley 4th Ed

Chapter 27
MATLAB-Based Comparative Analysis of Alternative PV Models

Khamma Kanwar and Jayashri Vajpai

1 Introduction

In the recent decades, Solar Photovoltaic (PV) systems have become the fastest-growing alternative source of energy. These systems are environment friendly and require less maintenance. Photovoltaic effect converts the solar radiation into the electrical energy in solar cells. PV arrays have number of cells in series and parallel or a group of panels. PV systems require little maintenance and are pollution free. PV systems have standalone and grid connected applications such as, light sources, battery charging, solar water pumping stations, hybrid vehicles, military and satellite systems [1].

The nonlinear characteristic of PV cell can be modeled on the basis of Shockley equation. Both single diode and double diode models are commonly used to simulate PV characteristics. The single diode model simulates the PV cell characteristics accurately. The efficiency of the solar energy conversion is directly related with the maximum power extraction (MPP) from the PV system [2].

MPP is based on realizing the dynamic or static method of optimization. In the dynamic method, the MPP is achieved by tracking of position and movement of the sun. This approach may not be suitable for energy conversion at a small to medium power range due to its high cost and energy consumption. In the static method (for small power range), the MPP is tracked with the help of DC–DC converter by adjusting continuously the operating point. In the literature, many methods have been reported to track the MPP of PV generator. The Perturb and Observe (P&O) method is widely used approach for tracking MPP [3].

K. Kanwar · J. Vajpai (✉)
M. B. M. Engineering College, Jai Narain Vyas University, Jodhpur, India
e-mail: jvajpai@gmail.com

K. Kanwar
e-mail: kkjoya.jnvu@gmail.com

M. Shorif Uddin et al. (eds.), *Intelligent Energy Management Technologies*, Algorithms for Intelligent Systems,
https://doi.org/10.1007/978-981-15-8820-4_27

Table 1 Comparison of datasheet parameter of three different PV systems

	Mathematical model	Inbuilt PV array	User-defined PV array
Model	KC200GT	KC200GT	KC200GT
Maximum Power [W]	200.143	200.143	200.143
Open circuit voltage V_{oc} [V]	32.9	32.9	32.9
Short circuit current I_{sc} [A]	8.21	8.21	8.21
Voltage at MPP V_{mp} [V]	26.3	26.3	26.3
Current at MPP I_{mp} [A]	7.61	7.61	7.61
Diode saturation Current Io [A]	$9.825*10^{-8}$	$2.3246e^{-10}$	$8.9892e^{-16}$
Open-circuit voltage/temperature coefficient K_V [V/K]	−0.1230	−0.35502	−0.1230
Short-circuit current/temperature coefficient K_I [A/K]	0.0032	0.06	0.0032
Ideality constant a	1.3	0.97736	0.64611
Shunt resistance R_p [Ω]	415.405	150.6921	65.4647
Series resistance R_s [Ω]	0.221	0.34483	0.47804
Number of cell in series N_s	54	54	54

Over the last decades, many researchers have studied the characteristics of PV array by using simulation approach. Nguyen and Nguyen [4] have explained the step by step procedure of mathematical modeling of PV array in MATLAB/SIMULINK under different environment condition. Villalva et al. [5] have developed mathematical model to find out the best I–V equation for PV array by adjusting the I–V curve at three points, viz., open circuit voltage (Voc), MPP and short circuit current (Isc). Bouraiou et al. [6] have presented mathematical modeling of one diode and two diode models of PV array and carried out simulation under different temperature and irradiance conditions. The two algorithms have been compared experimentally by Houssamo et al. [7] to maximize the output power from a PV system for the given set of conditions using inbuilt PV array block in MATLAB. AbdelHady [8] has used the inbuilt PV array and developed a MATLAB Simulink model that evaluates the system performance when it is tied either with local grid or National grid. Some of the researchers have used inbuilt PV array block and others have developed the mathematical model of PV array. However, none of them has discussed which model gives the best response.

In this paper, the mathematical model of PV array has been simulated in MATLAB/SIMULINK on the basis of model equation. Then comparison of manufacture datasheet parameters and output response is made between the developed mathematical PV array model and inbuilt PV array block available in MATLAB/SIMULINK in order to verify the accuracy of the inbuilt model.

2 Methodology

The proposed comparative analysis of alternative PV models is carried out in the following subsections.

2.1 Mathematical Modeling of PV Array

An overview of PV system is shown in Fig. 1 [3]. It consists of PV array along with DC–DC Power converter that is connected to a PWM generator driven by MPPT algorithm.

The four parameters model equivalent circuit diagram of PV array comprises of a current source, a diode, a series resistance and a parallel resistance. The antiparallel diode with the photo current source represents the theoretical model of ideal PV cell. I–V characteristics may be expressed as follows [6]:

$$I = I_{pv} - I_0\left[\exp\left(\frac{V + IR_s}{V_t a}\right) - 1\right] - \frac{(V + IR_s)}{R_p} \quad (1)$$

where

I_{pv}	Photocurrent (A)
I_0	Diode saturation current (A)
$V_t = N_s kT/q$	Thermal voltage of the array
N_s	Number of cells connected in series
q	Electron charge ($1.60217646 * 10^{-19}$C)
k	Boltzmann constant ($1.3806503 * 10^{-23}$J/K)
T	Module operating temperature (K)
a	Diode ideality constant
Rs	Series resistance(ohm) and
Rp	Parallel resistance (ohm)

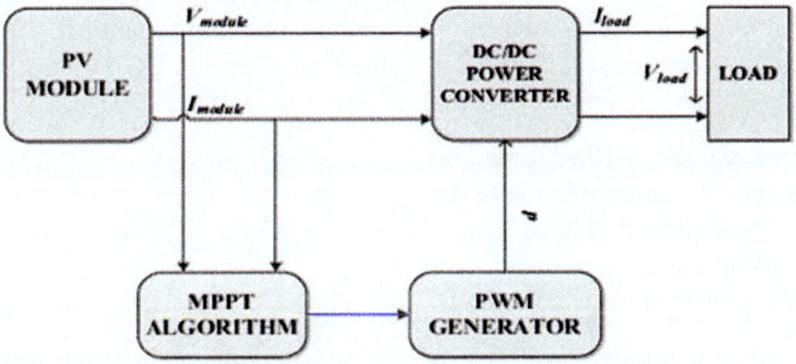

Fig. 1 Overview of PV system [3]

The photocurrent of the PV cell depends linearly on the solar irradiation and temperature according to the following equation:

$$I_{pv} = (I_{pv,n} + K_I \Delta_T) \frac{G}{G_n} \tag{2}$$

where $I_{pv,n}$ photo current at the nominal condition generally 25 °C and 1000 W/m^2, $\Delta_T = T—T_n$

T_n Nominal temperatures (K)

G Irradiation on the device surface (W/m^2), and

G_n Nominal irradiation

The diode saturation current is described by:

$$I_0 = \frac{I_{sc,n} + K_I \Delta_T}{\exp\left(\frac{V_{oc,n} + K_V \Delta_T}{a V_t}\right) - 1} \tag{3}$$

$I_{sc,n}$ Short circuit current at nominal condition (A)
$V_{oc,n}$ Open circuit voltage at nominal condition (V)
K_V Open circuit voltage/temperature coefficient K_V [V/K]
K_I Short circuit current/temperature coefficient K_I [A/K]

Using these equations, the mathematical model of PV array has been developed in MATLAB in many papers. The mathematical equation-based model of PV array is simulated in this paper for the purpose of comparison.

2.2 *Maximum Power Point Tracking*

The I–V characteristics of PV array are nonlinear and show unique MPP under a particular set of environment conditions, and this peak point keeps changing with solar irradiation and ambient temperature. The efficiency of PV systems is generally very low. For improvement of efficiency, the PV source impedance should match with the load impedance for every operating condition, thereby requiring Maximum power point tracking (MPPT) which optimizes the operating point [6].

The maximum power is set by adjusting the Duty Cycle of DC-DC converter. The MPP can be tracked by various algorithms like:

a. Perturb and observe (P&O) method
b. Incremental Conductance method
c. Short circuit current method
d. Open circuit voltage method
e. Ripple correlation control method

P&O method is the most widely used MPPT technique on account of simplicity, easy implementation and effectiveness.

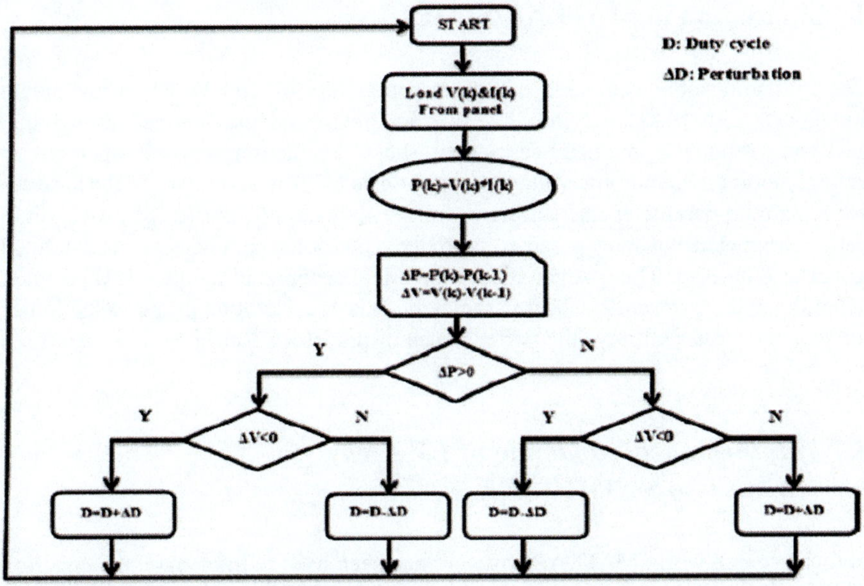

Fig. 2 The flowchart of the P&O algorithm[8]

Perturb and Observe—In this approach, the voltage from the array is adjusted in small bidirectional steps by the controller to observe the changes in the output power. Further adjustments are done in the direction in which the power increases. However, the P&O method can result in oscillations of power output due to the rise and fall of the curve of power against voltage below the MPP [7]. The flowchart of P&O algorithm is shown in Fig. 2 [8]

2.3 Modeling of DC/DC Converter

DC/DC converter is employed to convert a fixed voltage DC source into a variable-voltage DC output. It provides high efficiency, good acceleration control and fast dynamic response. There are mainly four basic topologies such as boost converter, buck converter, buck-boost converter and Cuk converter [9]. However, since the objective of this research work is comparison of models, the boost converter topology has been employed for simplicity. The input and output voltage relationship can be represented by following equation:

$$V_{out} = \frac{1}{1-D} V_{in}$$

Here D is Duty Cycle, V_{in} is input voltage and V_{out} is output voltage.

3 Simulation in MATLAB/SIMULINK

The following table compares datasheet parameters of 200 W PV array model KC200GT with MATLAB inbuilt PV array block and user-defined model that have same reference parameters as per datasheet like maximum power, open circuit voltage, short circuit current, voltage and current at MPP and number of cells in series but remaining parameters are different. In user-defined model, there is provision to define parameters but some parameters are fixed and some parameters can be defined as per as datasheet. The mathematical model is developed in the preceding section using the manufacturer datasheet parameters. It is expected that proper selection of model may further improve the performance of model KC200GT.

3.1 *Mathematical Modeling of PV Array in MATLAB/SIMULINK*

In this research work, MATLAB software has been used to implement the equation of PV cell for developing a mathematical model of PV array KC200GT as shown in Fig. 3. This model was designed for supplying maximum power to resistive load with MPPT under particular environment conditions using Simulink blocks (Fig. 4).

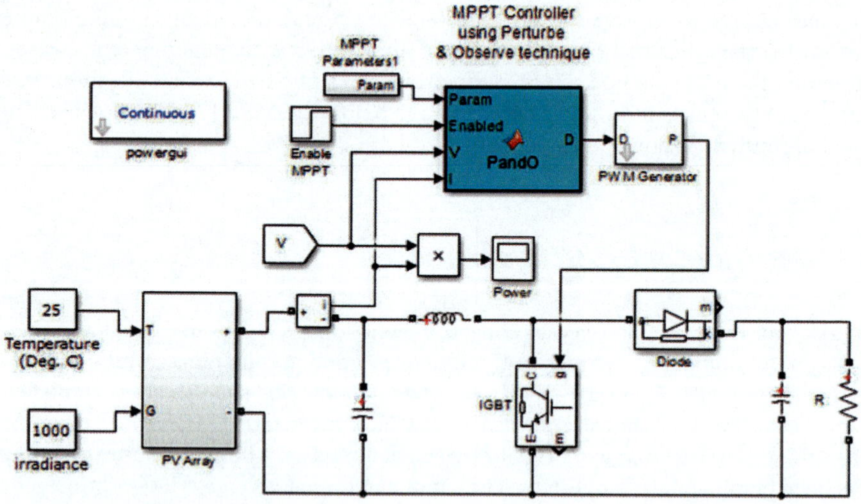

Fig. 3 MATLAB/Simulink model developed by employing mathematical equations of PV system

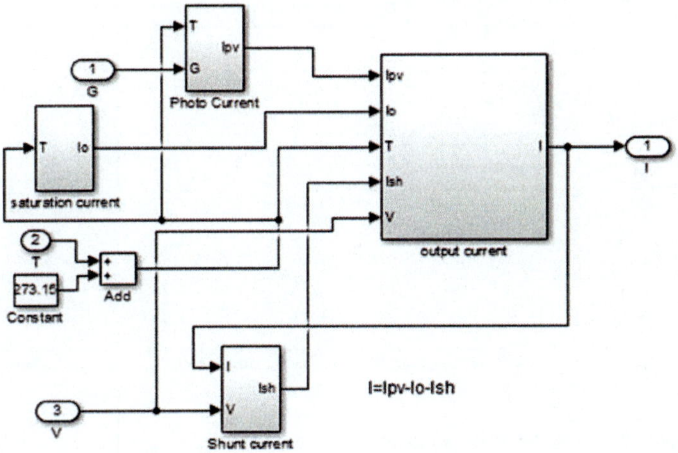

Fig. 4 Subsystem of PV array

3.2 *Inbuilt PV Array Block Available in MATLAB/SIMULINK*

The simulation was also carried out using the inbuilt PV array block model KC200GT already available in MATLAB as shown in Fig. 5. The operating conditions were designed for giving maximum power to the resistive load with MPPT. In this block, some parameters are different from datasheet. Hence it is implemented by employing

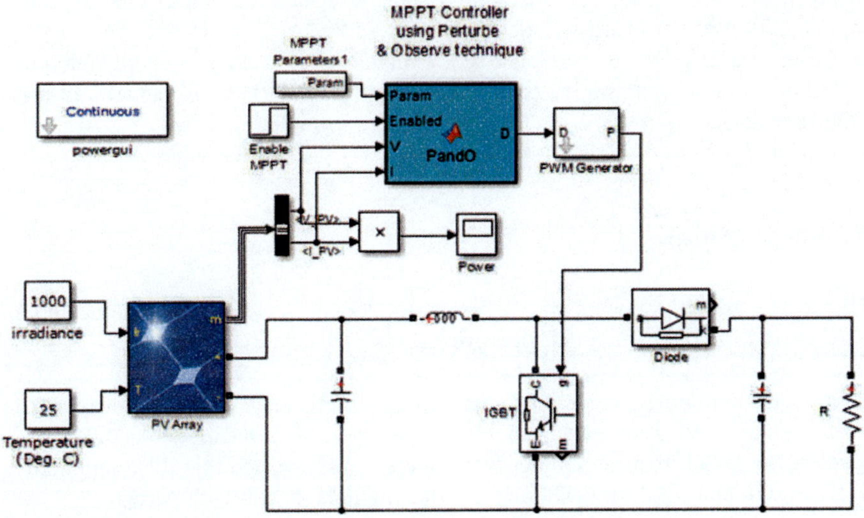

Fig. 5 PV system using Inbuilt PV array block

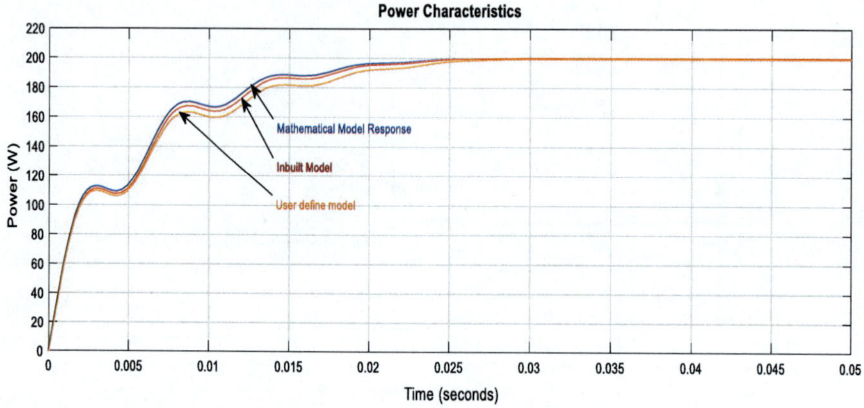

Fig. 6 Output power characteristics of three different systems

the user-defined model. The performance of the developed mathematical model is compared with the inbuilt PV array block model and user-defined PV models.

3.3 Simulation Results

The simulation is done for the three different systems as follows: In the first system, the PV model is implemented using mathematical equations in Simulink. In the second system, the PV model is implemented using inbuilt PV array block. In the third system, PV model is implemented using user-defined block. The simulated system output power characteristics are shown in Fig. 6.

It was found that the performance of PV model developed by using mathematical equations is better than the system designed by inbuilt PV array model and user-defined model.

4 Conclusions

The three different models of PV array KC200GT have been implemented in MATLAB Simulink. Then comparative analysis of datasheet parameters of these three different models of PV array has been carried out. The P&O algorithm is used for giving the maximum power to the constant load with the help of boost converter. The power characteristics obtained by mathematical equation-based modeling of PV array gives better result than remaining two models. It is expected that this comparative analysis will be useful for designing accurate models for research purpose.

References

1. Rahim KNA, Ping HW, Selvaraj J (2012) Photovoltaic module modeling using simulink/matlab. In: The 3rd International Conference on Sustainable Future for Human Security SUSTAIN, vol 17, pp 537–546. Elsevier (2013). https://doi.org/10.1016/j.proenv.2013.02.069
2. Surya Kumari J, Sai Babu Ch (2012) Mathematical Modeling and Simulation of Photovoltaic Cell using Matlab-Simulink Environment. Int J Electr Computer Engineering (IJECE) 2(1): 26–34, ISSN 2088-8708
3. Jazayeri M, Uysal S, Jazayeri K (2014) Evaluation of Maximum Power Point Tracking Techniques in PV Systems using MATLAB/Simulink, Sixth Annual IEEE Green Technologies Conference, pp 54–60. IEEE computer society. https://doi.org/10.1109/GREENTECH.2014.21
4. Nguyen XH, Nguyen MP (2015) Mathematical modeling of photovoltaic cell/module/ arrays with tags in Matlab/Simulink. https://doi.org/10.1186/s40068-015-0047-9
5. Villalva MG, Gazoli JR, Filho ER (2009) Comprehensive Approach to Modeling and Simulation of Photovoltaic Arrays, IEEE TRANSACTIONS ON POWER ELECTRONICS, 24, NO. 5, pp. 1198-1208, 2009. https://doi.org/10.1109/tpel.2009.2013862
6. A. Bouraiou, M. Hamouda, A. Chaker, M. Sadok, M. Mostefaoui, S. Lachtar: Modeling and Simulation of Photovoltaic Module and Array based on One and Two Diode Model Using Matlab/Simulink
7. I. Houssamo, F. Locment, M. Sechilariu: Experimental analysis of impact of MPPT methods on energy efficiency for photovoltaic power systems, Elsevier Electrical Power and Energy Systems, VOL. 46 pp. 98–107, http://dx.doi.org/10.1016/j.ijepes.2012.10.048
8. R. AbdelHady: Modeling and simulation of a micro grid-connected solar PV system, science direct water science, VOL. 31, pp. 1–10, 2017, http://dx.doi.org/10.1016/j.wsj.2017.04.001
9. Esram T, Chapman PL (2007) Comparison of Photovoltaic Array Maximum Power Point Tracking Techniques. IEEE Trans Energy Convers 22(2):439–449. https://doi.org/10.1109/TEC.2006.874230
10. A. Kumars, K. L. Ratnakar, B. S. Yogananda and B. R. Kamath.: Simulation of Photovoltaic Cell and MPPT Controllers and their Analysis, International Journal of Engineering Science and Innovative Technology (IJESIT) Vol. 4, pp. 253–263, 2015
11. Badii G, Ali H, Othman H, Rachid D (2014) Experimental Study of MPPT Algorithms for PV Solar Water Pumping Applications. https://doi.org/10.5339/qfarc.2014.ITSP0107
12. Saoud MS, AbbassI HA, Kermiche S, Ouada M (2013) Improved incremental conductance method for maximum power point tracking using cuk converter, WSEAS Trans Power Syst 8(3)

Chapter 28
Nuclear Fusion: Energy of Future

Jayesh Nehiwal, Harish Kumar Khyani, Shrawan Ram Patel, and Chandershekhar Singh

1 Introduction

Fusion of light nuclei is by far the foremost process of energy release of energy in the universe. The sun, like all of the stars, is nerved by the fusion of hydrogen nuclei into one helium nucleus. The process of fusion releases the huge energy of 26.72 MeV or 6.7 MeV per nucleon, which means around 180 MWh energy per gram. Hence, the energy released in fusion reaction per unit mass of fuel is nearly seven times the burning of coal. This efficaciously exemplifies the ultimate economic tackling of matter which is chosen by nature for its prime energy release process. Further, these and other fusion mechanisms are also liable for the generation of elements in the stars. Nuclear fusion is a crucial and highly tempting option for a future terrestrial supply of energy, since this source of energy would be environmentally agreeable and the fuel comprising in it is practically limitless. Out of the nearly 80 plausible fusion reactions, this reaction (Figs. 1, 2 and 3)

$D + T \rightarrow {}^{4}He\ (3.517\ MeV) + n\ (14.069\ MeV)$

between the isotopes of hydrogen, i.e., deuterium, D, and tritium, T, is of utmost use. Around 20% of energy is carried away by alpha particle ^{4}He, whereas the remaining 80% of energy is released as the kinetic energy of neutron n. The concentration of Deuterium on Earth is practically limitless and is available in the form of heavy water in the oceans which is 0.015% in natural hydrogen. Tritium does not exist due to its radioactivity characteristics with a half-life of 12.35 years, so for a reaction of D–T we have to generate Tritium through the breeding reaction of Lithium

$^{6}Li + n \rightarrow T + {}^{4}He + 4.8\ MeV$

$^{7}Li + n \rightarrow T + {}^{4}He + n' - 2.47\ MeV$

J. Nehiwal (✉) · H. K. Khyani · S. Ram Patel · C. Singh
Jodhpur Institute of Engineering & Technology, Jodhpur, India
e-mail: jayeshnehiwal@gmail.com

M. Shorif Uddin et al. (eds.), *Intelligent Energy Management Technologies*, Algorithms for Intelligent Systems,
https://doi.org/10.1007/978-981-15-8820-4_28

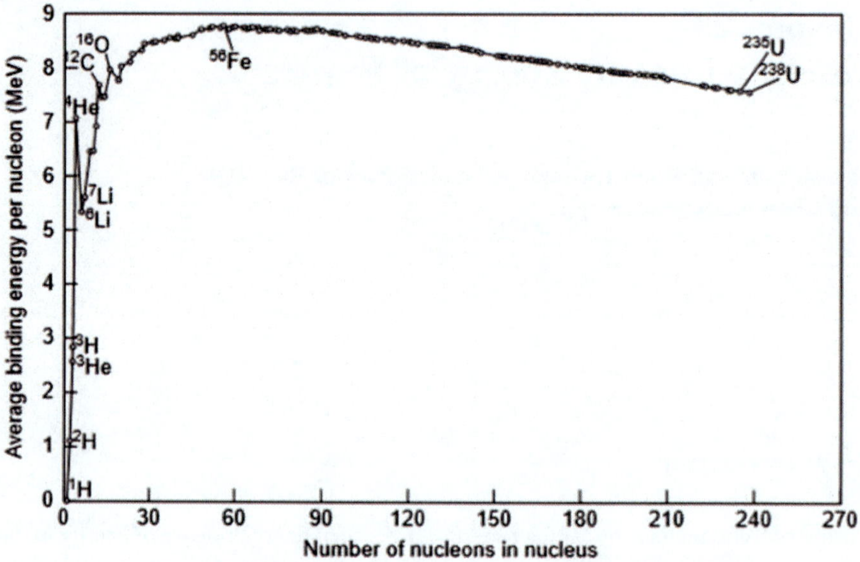

Fig. 1 Average nuclear energy binding curve (Courtesy of Wikimedia Commons)

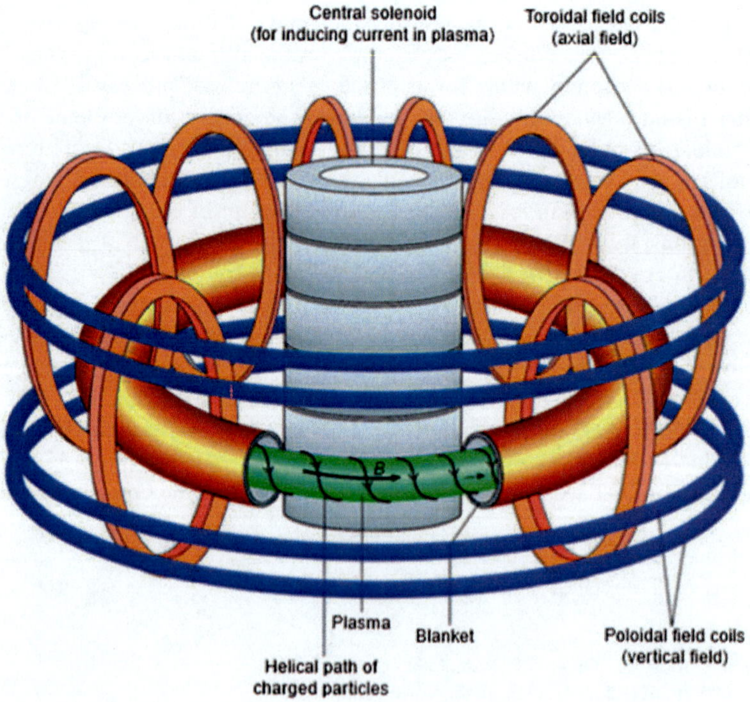

Fig. 2 Schematic diagram of Tokamak (Springer series in Material Science, vol. 214, pg. 314)

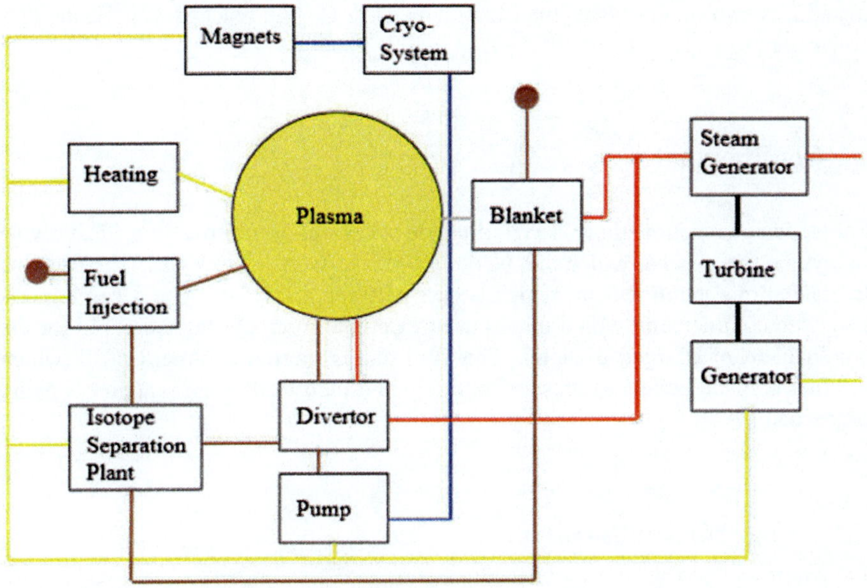

Fig. 3 Flow chart of fusion reactor (reference [3])

where n′ is a slow neutron, which has a contribution for the generation of tritium. Natural lithium comprises 7.42% of ^{6}Li and 92.58% of ^{7}Li. So we use deuterium and lithium as fuel for fusion as they are cheap. According to the present scenario, the resources of deuterium and lithium would be available for more than 30 million years. The radioactive tritium is processed internally only in [1]. For the fusion of two atoms, there is a requirement of binding energy. This binding energy is the converse energy of mass defect according to Einstein's theory of $E = (\Delta m)\,c^2$. This is the energy required for disassembling the nucleus from an atom into separated nucleons and the energy spent on separating each nucleon is average binding energy [2]. For all such elements, the binding energy is shown in Fig. 1. From which it can be interpreted that average binding energy for hydrogen isotopes and helium atom are relatively low and hence fusion can be quite easily possible to meet the conditions [2].

Under normal conditions of temperature and pressure, the fusion cannot take place. For fusion to take place atoms are to be heated so that the motion of electrons and nuclei will increase and electrons will get separated. The state of matter where electrons and nuclei are not together is called a plasma state. The plasmas are very hot and move with a very high velocity which makes the reaction very tough to control as we have to overcome the high repulsive forces. So it is worth to gravitate that fusion depends on plasma temperature and plasma density [3]. For the fusion ignition, there is a requirement of the crossing energy value of the triple product of temperature, density and energy confinement time. The triple product of D–T reaction is two orders lesser than that of D–D & D–^{4}He reaction [3]. Hence it is convenient to go for a fusion of D–T.

2 Confinement

For the fusion reaction, there is a requirement to control the plasma state, i.e., proviso to confine the plasma, which can be done by mainly three processes. The first one is gravitational confinement, which occurs between stars only. The second one is magnetic confinement, which makes use of Lorentz force of magnetic field for the confinement of charged particles. The third one is inertial confinement, in which the fuel is made pellets to pressurize highly and then further the reaction is being proceeded [3].

2.1 Magnetic Confinement

It is evident that no solid material can sustain heavy temperatures of 100 million Kelvin so this dilemma is solved by stating the theory that all charged particles can be confined under magnetic fields. It inspired to go for toroidal closed arrangement.

Only toroidal confinement also has limits to let escape some massive plasma so a better technique is to provide superposition of strong toroidal field with poloidal field [1]. The high temperature plasma experience Lorentz force, which is responsible for confinement. Tokamak is a device or, more precisely, it is reactor which is of donut shape surrounded by magnetic coils [4, 8, 9] as shown in Fig. 2.

2.2 Inertial Confinement

It is entirely different school of thought for fusion process. It compresses the fuel into small solid pellets and due to extreme denseness when these pellets are heated then the possibility exists for microexplosion. Altogether here the confinement is not actually reached by external fields but the inertia binds the atoms together [3, 10]. Basically fuel pellet is enclosed by the layer of another material and extreme intense beams of laser are subjected on the charged particle of high energy. The layer that is the outer one gets heated up and evaporates. The products formed in evaporation get mobile outward, and the pellet is compressed inward, due to momentum conservation (inertial binding) [3]. The initiation of such fusion is done with the help of laser beam and this can even also be done directly by laser driving the pellet itself and indirectly by making use outer another layer usually gold layer and then laser driving.

3 Alternative Fusion: Muonic Fusion

This is an idea which is being investigated in few laboratories, to produce muons, which are so called as the heavy sisters of the electron. The muon is used to inject into a mixture of deuterium–tritium gas and there exist little probability that muon will get captured by the tritium or the deuterium atom and there will be formation of the deuterium–tritium molecule. Due to the heavy mass of muon, the dimensions will be smaller as compared with normal molecule with bounded electrons [3]. Hence the fusion reaction may undergo as the nuclei are closer to each other. But there is problem to this as well, as the cost involved in production of muons is too much and can catalyze fusion reactions about 200 times only [5].

4 Fusion Energy Conversion and Its Transfer

With the help of light isotopes, it is possible to have a fusion reaction possible, i.e., D–T fusion and this identification was responsible reason which has inspired to work on various other factors to utilize this energy as a source of electrification. And hence the need of design of such power plant is there. The features like steam generator, turbine and current generator will be same as that in conventional fossil-fuelled or nuclear power plants. A simple flowchart telling about the material and energy flow can be depicted from Fig. 3. The fuel is injected and the conditions required for fusion with system confinement are done which further leave the plasma and they are stopped in the blankets which are the plasma surrounding modules [3].

5 Blanket System

This is the key system that converts fusion energy into electrical energy. Its main functions are energy extraction, tritium breeding and radiation shielding, which means the transferring of energy from fusion to thermal is being done, it ensures self-sufficiency and it protects external subsystem from neutron and gamma radiation which may cause an explosion. The heat which is produced in the blanket is transported via water or helium to the steam generator and then it is fed to grid. A small part of the same is utilized to power other components of plant. Cryo-system requires the electrical power which produces low-temperature helium for the super-conducting magnets, the current drive, the current in the magnets and the heating systems of plasma [2].

6 India's Role in Fusion Energy

R. Srinivasan and the Indian DEMO Team reviewed that the contribution to energy from the fusion reactor is expected to rise beyond 2060. It will be beneficial if the research program gets accelerated so that fusion reactors can be realized well before 2050 [6]. The multibillion project of ITER which was launched in 2007 to build a Tokamak reactor and to check its feasibility, around 35 countries of the world took their participation for the research and India is also part of that. Around 58% of the ITER project has been completed and is been expected to generate power by 2035 [11].

7 Future Directions

Fusion energy is itself very unique in energy sector and challenging for R&D. ITER is the biggest fusion reactor experiment that expects around 500 MW of power generation in 400 s of pulse operation. The technology involving breeding capacity of blanket is important and is still ongoing R&D but other factors of self-sufficiency of tritium, economic, environmental friendliness, safety are also important [3, 7].

8 Conclusion

Fusion energy is safe and clean energy but needs efforts in implementing for the applications. There is availability of various materials to fuse them like D, T, ^{3}He, but need extreme conditions to initiate the reaction. The reactor for such fusion is still under research and the issues are expected to be solved by ITER and DEMO by 2025. Taming the fusion energy is a promising challenge but its rewards are incalculable for the future energy needs of mankind.

References

1. Schumacher (2001) Status and problems of fusion reactor development", Springer-Verlag, ISSN 0028–1042, Volume 88, Issue 3, pp 102–112, March 2001
2. Yican W, Sumer S (2018) Fusion Energy Production. Elsevier- Comprehensive Energy Systems, vol 3, pp 539–589
3. Hamacher T, Bradshaw AM (2001) Fusion as a Future Power Source: Recent Achievements and Prospectus. World Energy Council, 18th Congress, Buenos Aires
4. Kikuchi M., Lackner K. Tran MQ Fusion Physics Vienna. IAEA 2012
5. Alvarez LW et al (1957) Phys Rev 105:1127–1128
6. Srinivasan R and the Indian DEMO Team (2010) Role of Fusion Energy in India J. Plasma Fusion Res. SERIES, Vol. 9 (2010)

7. Mehlhorn TA (2014) National Security Research in Plasma Physics and Pulsed Power: Past, Present, and Future. IEEE Transactions on Plasma Science, 42(5):1088–1117
8. Meade DM (1996) Recent Progress on the Tokamak Fusion Test Reactor. Journal of Fusion Energy,, 15(3–4):163–167 ISSN 0164-0313
9. Cohn DR, Schwartz J, Bromberg L, Williams JEC (1988) Tokamak Reactor Concepts Using High Temperature, High-Field Superconductors. J Fusion Energy 7(1):91–94
10. Ido S (1983) The Laser Fusion Reactor. J Fusion Energy, 3(5–6):459–463 ISSN 0164-0313
11. thehindubusinessline.com

Chapter 29
Medical Image Processing Using Soft Computing Techniques and Mathematical Morphology

Pratik Bhansali and Sandip Mehta

1 Introduction

Today, the mass information produced by medical imaging has considerably increased. All the medical examinations carried out like for the preparation of neurosurgical operations or for the study of cerebral pathologies (cerebrovascular accident, tumor, sclerosis in plate, etc.) can represent several hundreds of megabytes (100 Mo = more than 0.8 billion bits 0 or 1). Physicians and cognitive scientists need reliable tools to assist them in their decision-making and in interpreting the mass of information created.

The developments in magnetic resonance imaging (MRI) have allowed us to know and evaluate the different pathologies of human beings with a degree of detail never seen before. Many post-processing techniques have now been developed to efficiently visualize large volumes of data and study the functionality of certain organs, such as the brain, using the images generated by CT and MRI. Almost all postprocessing tools require a spatial coherence of the data. If the acquisitions are multiple in times, studying the same volume several times, it is necessary to ensure the spatial agreement of the voxels by means of image processing techniques.

To interpret image is to produce symbolic description of it, that is, to recognize and describe the different entities that make it up. Among the interpretation tools, segmentation (or tagging) is a vital link in many applications and quantitative analyses [1–4]. Segmentation is defined as the partitioning of an image into constituent regions (also called classes or subsets), which are homogeneous with respect to one or more

P. Bhansali (✉) · S. Mehta
Department of Electrical Engineering, Department of Electronics and Communication Engineering, Jodhpur Institute of Engineering & Technology, Jodhpur (Rajasthan), India
e-mail: pratik.bhansali01@gmail.com

S. Mehta
e-mail: sandip.mehta@jietjodhpur.ac.in

M. Shorif Uddin et al. (eds.), *Intelligent Energy Management Technologies*, Algorithms for Intelligent Systems,
https://doi.org/10.1007/978-981-15-8820-4_29

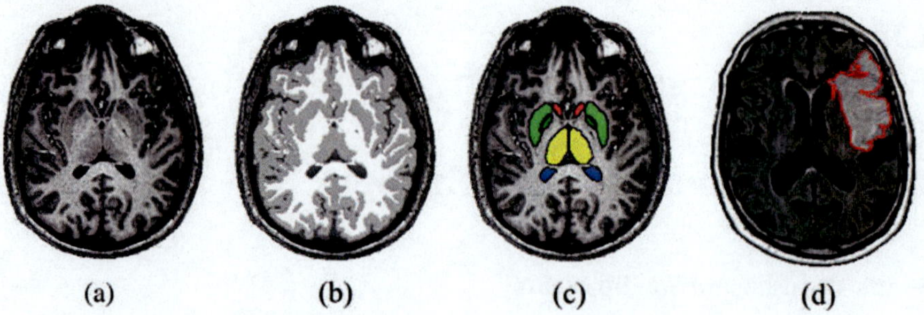

Fig. 1 MRI acquisition (image **a**), tissue segmentation (image **b**), segmentation into structures (image c) and segmentation of lesion (image **d**)

characteristics such as intensity, texture or functionality [5]. One of the typical use cases of segmentation is the distinction of neuroanatomical zones, in which it is used in order to separate areas of cerebral interest such as: hippocampus, frontal cortex, areas of language, among others [6, 7].

The goal is, for one or more images, to give each voxel a label to give a description. In cerebral MRI segmentation, we generally consider different level of descriptions (see Fig. 1).

Tissue segmentation, which aims to describe the composition of each voxel among the three main topics of the brain [8], the segmentation in structures, which describes the membership of each voxel with a known brain structure: thalamus, putamen, ventricular system, etc. and the segmentation of the cerebral lesions (tumors, multiple sclerosis, cerebral vascular accidents) in the clinical setting, which aims to study their location, volume, etc. Segmentation is, therefore, a central tool in both the clinical and the neuroscience fields.

2 Segmentation in Medical Imaging

In medical imaging, it is necessary to take measurements on acquired images. Whether anatomical or functional, the image provides accurate information about patient such as: level of tracer uptake, volumes of different organs or regions, depth of injury, etc. In many of these cases, the interest of the measure lies in a particular area of the image associated with an anatomical region. In order to be able to make these measurements, we use what we call segmentation techniques [9, 10].

Automatic image analysis and classification, in general, is based on the characteristics of the image. These are basically two: the signal strength of the pixels and their geometric arrangement in the image. Out of these characteristics, the intensity is probably the most used, along with the information of contours that gives us knowledge about the borders between the different objects of the image.

The techniques based on intensity base operation in the assumption that a type of fabric has a characteristic and own brightness. This is generally not true, so it is not possible to segment only by these techniques, requiring a postprocess to eliminate regions with an intensity similar to the object of interest that do not belong to it [11].

In the literature on MRI segmentation, the methods are usually divided into two categories: monospectral methods, if there is a single acquisition or modality of image, and multispectral, if they use one or several images with different characteristics but with geometric correspondence for the same anatomical volume [12, 13].

The difficulty of segmenting an image is due to the structural complexity of the MRI images and the contrast often insufficient to extract the structure of interest without any knowledge of priori neither on its form nor on its location. Thus a priori knowledge provided by the atlases probabilistic models will allow to better differentiate close anatomical structures contrast and spatial location [14].

Many approaches have been proposed to segment cerebral MRIs. They can be classified into two groups: methods supervised, requiring interaction with the user and unsupervised methods, where the parameter estimation is performed automatically.

3 Fuzzy C-Means

Fuzzy C-Means (FCM) is a fuzzy unsupervised classification algorithm, it introduces the basic concept of fuzzy set into class definition: every point in the dataset belongs to each cluster with a certain degree, and all clusters are characterized by their center of gravity. It uses a criterion of minimizing intraclass distances and maximizing interclass distances, but giving a certain degree of membership to each class for each pixel [15, 16].

This algorithm requires a prior knowledge of number of clusters and then generates the classes through an iterative process by minimizing an objective function. Thus, it makes possible to obtain a fuzzy partition of image by giving each pixel a degree of belonging to a given class [17]. The cluster with which a pixel is associated is the one with the highest degree of membership. The main steps of the Fuzzy C-means algorithm are:

1. The arbitrary fixation of a membership matrix.
2. The computation of class centroids.
3. The readjustment of the membership matrix according to the position of the centroids.
4. Calculation of the minimization criterion and return to Step 2 if there is no convergence of criteria.

4 Fuzzy C-Means-Based Segmentation

Let $X = \{x_1, x_2, x_3, \ldots x_N\}$ a set of N observations to classify in K classes where each observation $x_i (i = 1, 2, 3, \ldots ..N)$ can be represented by a set of D attributes: $x_j = (x_{j1}, x_{j2}, x_{j3}, \ldots x_{jD})^T$.

K Classes are represented by a vector of class centers

$V = (v_1, v_2, v_3, \ldots v_k)$, where $v_k = (v_{k1}, v_{k2}, \ldots, x_{kD})^T$, is the center of class k.

Each observation x_i is defined by its degree of belonging μ_{ik} to the class k, that $\mu_{ik} \in [0, 1]$. We can then define a partition matrix $U = [\mu_{ik}]$ of dimension $(N * K)$.

The FCM algorithm consists in partitioning the N observations into K classes in order to minimize the similarity of observations within each class. It results in the minimization of the following objective function:

$$J_{FCM}(U, V) = \sum_{k=1}^{K}\sum_{i=1}^{N} \mu_{ik}^{m} d^2(x_i, v_k) \tag{1}$$

Under constraints following:

$$0 < \sum_{i=1}^{N} \mu_{ik} < N, \sum_{k=1}^{K} \mu_{ik} = 1 \tag{2}$$

m: is the fuzzification factor or fuzzy factor such that $1 < m < \infty$.

$d(x_i, v_k) = x_i - v_k$ is the distance between the observation i and the center of the class k.

Generally the distance used is Euclidean:

$$d(x_i, v_k) = \sqrt{\sum_{j=1}^{D} \left(x_{ij} - v_{kj}\right)^2} \tag{3}$$

The first constraint ensures that no class should be empty and the second is a normalization constraint which ensures that the sum of the degrees of membership of each observation to all classes equal to 1.

Blurred partitioning is achieved by an iterative optimization of the objective function given by Eq. (1), with an update of membership levels and centers of classes as in the case of the K-means algorithm.

Update of the following degrees of membership:

$$= \frac{1}{\sum_{j=1}^{K} \left(\frac{d(x_i, v_k)}{d(x_i, v_k)}\right)^{\frac{2}{m-1}}} \tag{4}$$

Hence the formula for updating class centers:

$$v_k = \frac{\sum_{i=1}^{N} \mu_{ik}^{m} * x_i}{\sum_{i=1}^{N} \mu_{ik}^{m}} \tag{5}$$

In what follows, we present the algorithm of the FCM method:

(1) Set the parameters: K: number of classes; ε: threshold representing the convergence error (for example: = 0.001);m: degree of blur, usually taken as 2; Tmax: max iteration; t = 1

(2) Initialize the matrix membership degrees $_0$ by random values in the interval [0, 1]:

Calculate the matrix $_0$ by the relation (5)

Calculate the objective function $_0$ by the relation (1);

(3) Update the degree of membership matrix by the relation (4)

(4) Update the matrix (the centers) by the relation (5);

(5) Calculate objective function by the relation (1);

(6) Increment the value of t;

$$t = t + 1 \tag{6}$$

(7) Repeat steps 3–6 until you meet the stopping criterion which is:

$$(J_t - 1 - J_t) < \varepsilon_{\text{out}} \geq J\text{max} \tag{7}$$

Finally, a final step is necessary when the desired result is a non-blurred. This is called "defuzzification." When we do not want to highlight the pixels where the degrees of membership are approximately the same for each class, a natural way to proceed is to consider that the final class of a pixel is that for which degree of membership is maximal.

However, this algorithm that requires knowledge of the number of classes is not robust and its effectiveness depends strongly of the initialization stage of the class centers because the iterative process can easily provide a locally optimal solution. Thus to improve the results of the classification, several methods are adopted for optimizations of the result.

5 Particle Swarm Optimization

The clustering of images is a main task in the analysis and recognition of which intend to divide pixels of an image into different classes such that the pixels of same class are similar as possible and the pixels of diverse classes are also dissimilar as possible. In spite of the efforts devoted to this problem leading to abundant literature, it remains a very difficult task because of the lack of information, the distribution of the pixels as well as a single and precise definition of a cluster.

PSO algorithms have shown their effectiveness in solving optimization problems whose research space is large. Swarm particle optimization is a popular metaheuristic approach based on swarm intelligence. The particle swarm optimization algorithm launches the search with a population of solutions, where each is called "particle." The latter is characterized by a speed of movement and a position in the search space. During the search process, each particle moves to change its position in the search space based on its current speed, current position, and best position found in past iterations, and best position found by the searcher "Swarm." Its displacement allows it to update its position and speed of movement at each iteration.

6 PSO-Based Segmentation

First, the performance function (fitness) is defined to start with our PSO-based approach, which reflects the ability of individuals to adapt to the problem under consideration. This function, which the algorithm seeks to minimize, is closely dependent on the objective function.

Further, a particle is from the best overall solution and its best solution, the greater the variation in its velocity, in order to move the particle towards the best solutions.

The random variables p1 and p2 can be defined as follows:

$p1 = c1 * r1, p2 = c2 * r1.$

With r1 and r2 following a uniform law on [0…1]. c1 and c2 are constants that represent a positive acceleration, with $c1 + c2 \leq 4$.

The algorithm runs as long as a convergence criterion has not been reached. This may be:

- A fixed number of iterations.
- Depending on the fitness.

When the speed variation is close to 0.

The movement of the agents is controlled in terms of *pbest* and *gbest*. The speed of agent *i* is updated first taking into account the current position (px_i, py_i) with respect to its *pbest*:

$$v_{xi} = \begin{cases} vx_i - rand() * c_1 if px_i > pbestx_i \\ vx_i + rand() * c_2 if px_i < pbestx_i \end{cases} \tag{8}$$

Then it is updated with respect to *gbest* according to:

$$v_{xi} = \begin{cases} vx_i - rand() * c_1 if px_i > gbestx \\ vx_i + rand() * c_2 if px_i < gbestx \end{cases} \tag{9}$$

The parameters c_1 and c_2 will be replaced by constant 2 which produced better results although they left as a future topic the investigation of their optimal value. In the simplified version of these simulations (Current Simplified Version), the agents updated their speeds by means of the equation:

$$v_{i,j} = v_{i,j} + 2 * rand() * \left(pbest_{i,j} - p_{ij}\right) + 2 * rand() * \left(gbest_j - p_{ij}\right) \quad (3)$$

where the subscript i, j denoted agent i and dimension j.

In this work, the fuzzy segmentation is considered as common optimization problem then PSO algorithm is used:

The objective function of each of the same function algorithm of the PSO method:

$$J_{FCM}(U, V) = \sum_{k=1}^{K} \sum_{i=1}^{N} \mu_{ik}^{m} d^2(x_i, v_k) \quad (10)$$

7 Morphological Operation for Postprocessing

Morphological techniques probe the given image with a small template or shape called structuring element (S.E.) [19]. The S.E. is sited at all possible locations in the image and it is then compared with corresponding pixels of neighborhood.

After applying both FCM- and PSO-based algorithms for the purpose of segmenting human brain MRI image morphological operations are applied to extract the region of interest, i.e., to the removal of nonbrain region form the segmented image.

Morphology process images with the help of previously chosen special shapes, which are generally smaller than the image and are called the structural element, which acts as an operator on an image to produce a result. The shape, size and orientation of the structural element are chosen based on prior knowledge taking into account the relevant geometric structures that are present in the image and the objective pursued with the morphological operation implemented.

Each structure element requires the definition of a point of origin (or reference) for its application as a morphological operator. This allows the structure element to be related in a particular way to the pixels of the image.

After applying FCM- and PSO-based clustering segmentation technique, the clustered information of the image achieved is further processed through morphological operation to achieve the desired region of interest, i.e., the brain image from the dataset of MRI chosen for segmentation.

First, an initial mask is created to filter the background of images with a threshold and then the small regions of binary images are filtered. The attributes of the image obtained are compared by the reference image to calculate the precision.

8 Results

The database is obtained from "Department of Telecommunications Brno University of Technology" form signal processing laboratory [20]. This database contains a set of actual medical images of MRI examination of several brain exams. There are total 12 sets of training data available, i.e., MRI image sets of 12 different patients. Each image set of a patient consists of 257 slices of image with resolution of 400 × 400 px and 16 bit grayscale bit depth. Each examination is composed with the set of sequences of modalities, where each sequence constitutes all the images. The number of sequences differs according to the examination.

The image sets contain not only original MRI images but also segmented labeled image slices of each patient done by the human expert with the accuracy of 99.540% + 0.07756 which acts as are ground truth, comparing to which we validate the accuracy of our segmentation methods. The simulation results on the random database images of patient number 9 after applying both the proposed methodologies are presented and then a comparative analysis on both visual and statistical ground is done to validate the work.

To find the accuracy of our simulated result, we use the contingency matrix that evaluates the performance of system by measuring the accuracy [21]:

$$Accuracy = \frac{TP + TN}{TP + TN + FP + FN} \tag{11}$$

A comparative visual analysis of both the methods, i.e., FCM- and PSO-based segmentation is presented on Image slice number 70, taken from image dataset of patient number 9 to showcase our methodology which can be implemented on all the image slices of the dataset. Here the initially defined clusters are varied to judge the preeminence of the proposed PSO-based approach for segmenting the brain image MRI.

Image slice no.: 'Image_70':

Method- "Fuzzy C Means"

(a) Cluster = 3
The above shown images in Fig. 2 from image (a) to image (h) revels the sequence of step by step appearance of results obtained as while performing Fuzzy C means.

Here the cluster no. is initialized to be three (3) and the process for segmenting the brain image is followed.

(b) Cluster = 4
The above shown images in Fig. 3, i.e., from image (a) to image (h) revel the sequence of step by step appearance of the result when the cluster number is initialized to be 4 and the process for segmenting the brain image is followed.

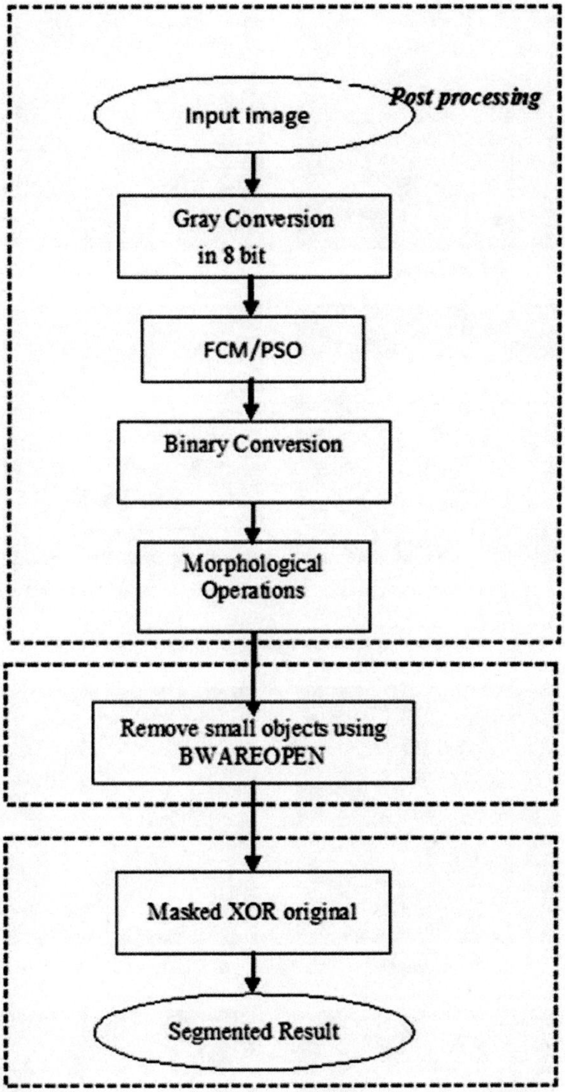

Fig. 2 Flow diagram for the proposed work

(c) Cluster = 5

The process is repeated by initializing the cluster number to be 5. The images in Fig. 4 from image (a) to image (h) revel the sequence of step by step appearance of the results obtained.

Method—"Particle Swarm Optimization (PSO)"

(a) Cluster = 3

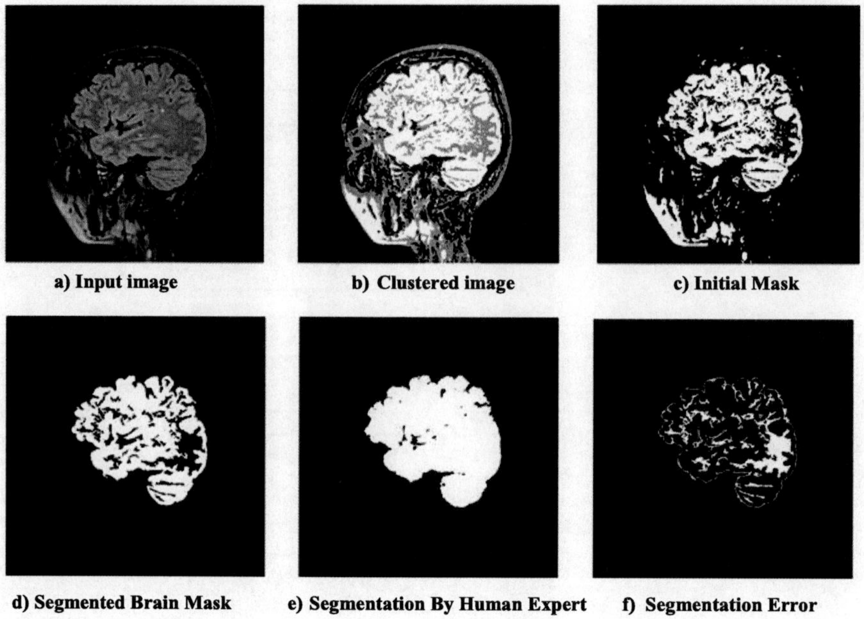

Fig. 3 Segmented brain image processed by FCM with cluster = 3

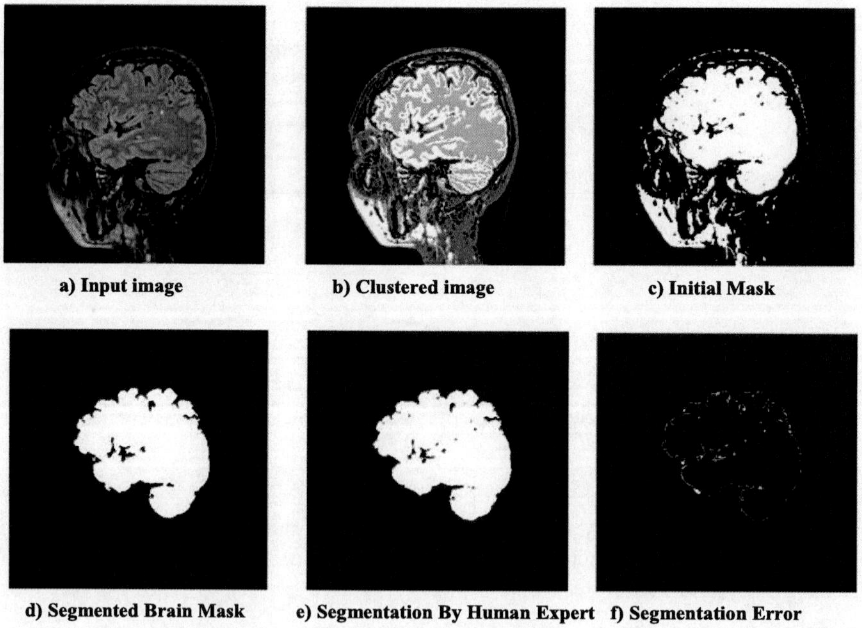

Fig. 4 Segmented brain image processed by FCM with cluster = 4

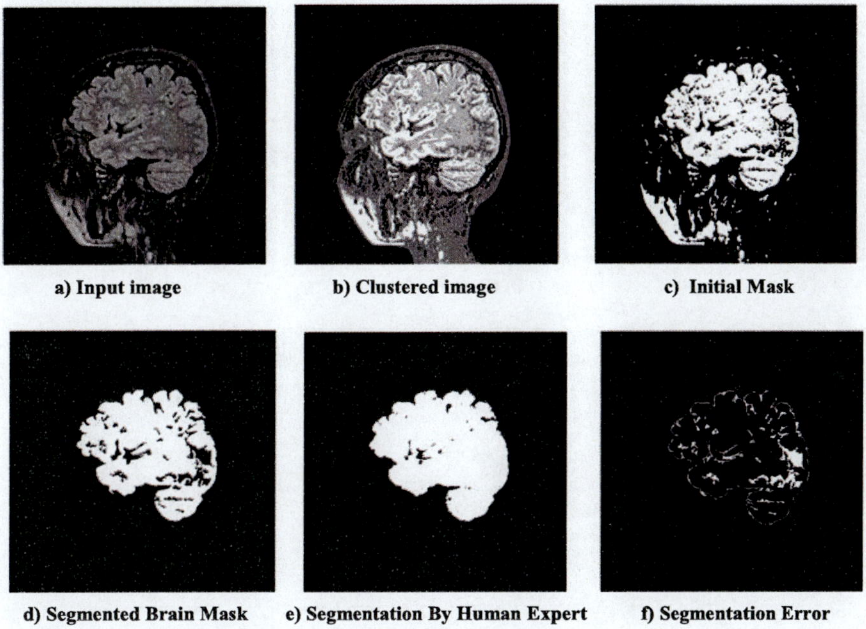

Fig. 5 Segmented brain image processed by FCM with cluster = 5

The above shown images in Fig. 5, i.e., from image (a) to image (h) revel the sequence of step by step appearance of the results obtained while performing PSO.

Here the cluster number is initialized to be 3 and the process of segmenting the brain image is followed.

(b) <u>Cluster = 4</u>

The process is repeated by initializing the cluster number to be The images in Fig. 6, i.e., from image (a) to image (h) revel the sequence of step by step appearance of results obtained.

(c) <u>Cluster = 5</u>

The process is repeated by initializing the cluster number to be 5 and the process for segmenting the brain image is followed.

Table 1 shows the comparative analysis of the accuracy obtained after applying both Fuzzy C Means and proposed PSO segmentation methods on different random image slices of patient number 9 and its variation with respect to the change in the initial cluster defined.

From the statistical evaluation shown in Table 1, a graph is plotted between cluster number and accuracy which shows the comparative behavior of FCM- and PSO-based segmentation approach on image no._70, i.e., “70” number slice of patient 9, which was also visually evaluated in the previous segment. The same can be followed for all the other image slices present in the dataset.

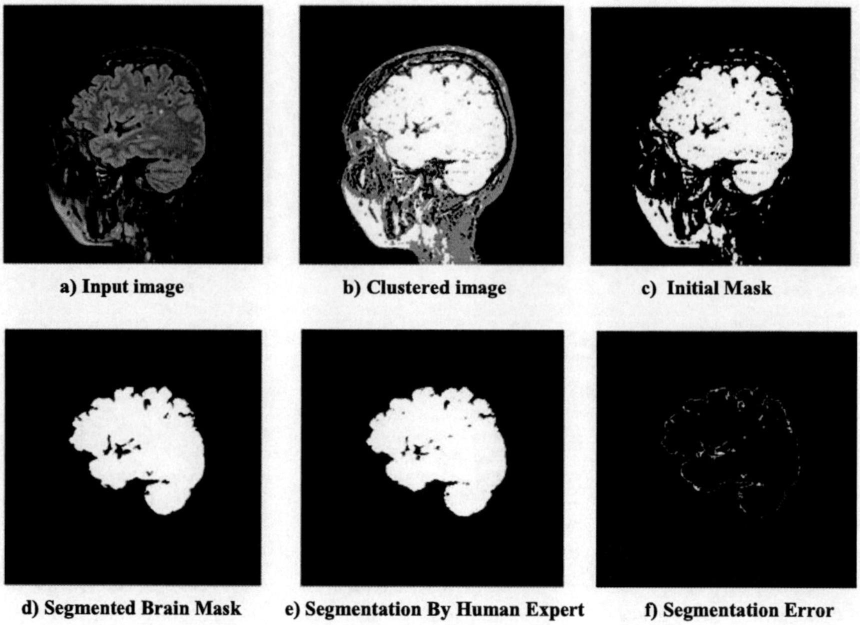

Fig. 6 Segmented brain image processed by PSO with cluster = 3

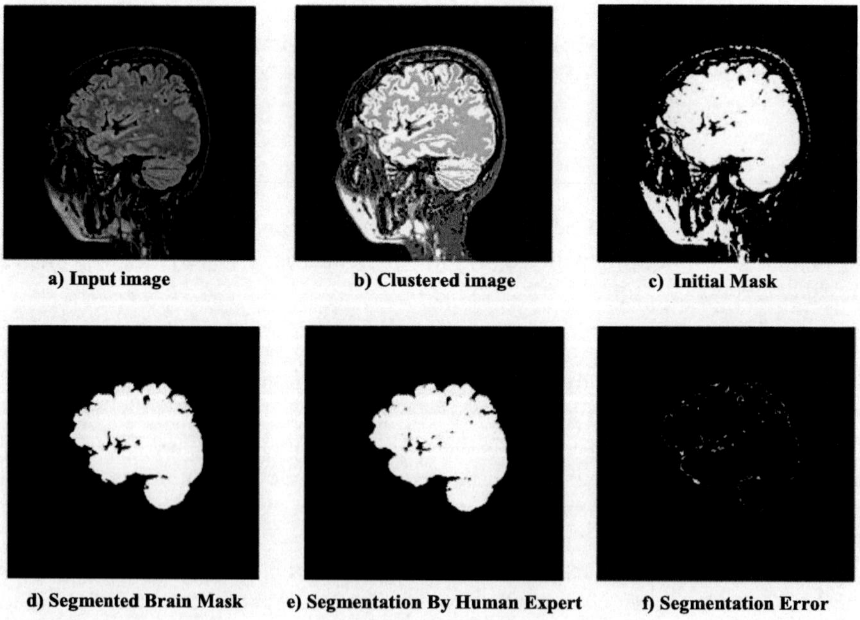

Fig. 7 Segmented brain image processed by PSO with cluster = 4

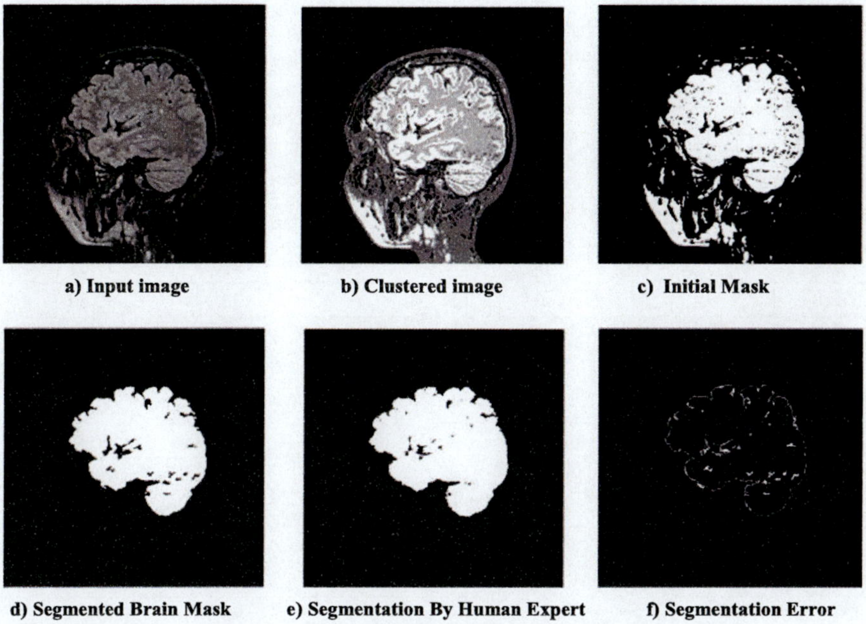

Fig. 8 Segmented brain image processed by PSO with cluster = 5

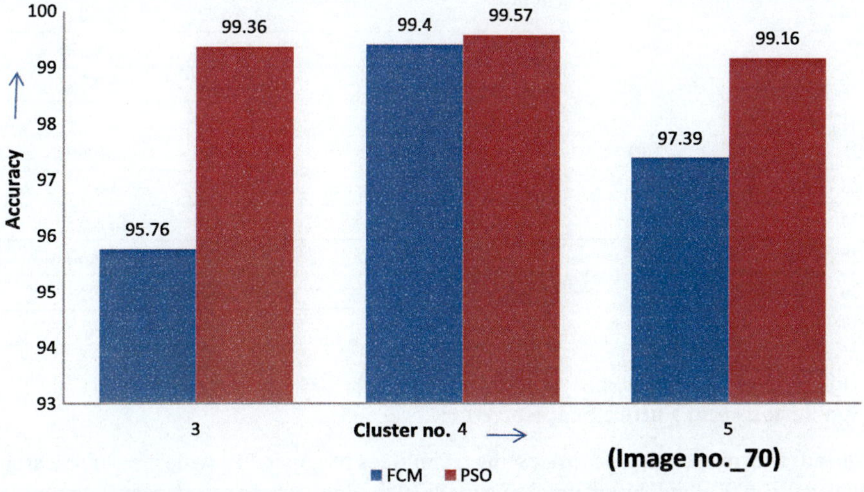

Fig. 9 Graphical Representation between cluster number and accuracy

Table 1 Result for patient number 9

S. no.	Image number	Cluster number	Accuracy	
			FCM	PSO
1	50	3	98.02	99.3
2		4	99.40	99.44
3		5	99.22	99.48
4	60	3	97.99	99.53
5		4	99.65	99.67
6		5	98.94	99.59
7	70	3	95.76	99.36
8		4	99.50	99.57
9		5	97.39	99.16
10	80	3	96.98	99.48
11		4	99.62	99.67
12		5	95.71	99.43
13	100	3	96.12	99.08
14		4	99.11	99.13
15		5	98.78	99.13
16	120	3	87.26	94.3
17		4	94.3	94.42
18		5	94.26	94.22
19	150	3	95.32	98.32
20		4	98.72	98.74
21		5	98.14	98.18
22	160	3	95.45	98.69
23		4	98.89	98.91
24		5	98.38	98.58
25	200	3	98.50	99.52
26		4	99.51	99.6
27		5	99.11	99.57

Conclusions and Future Scope of Work

The advancement in Image Processing techniques has opened a wide area of research especially in the application involving medical imaging where accuracy in diagnosing images plays a very important role. This paper has used two soft computing techniques, which involve FCM and PSO method followed by mathematical morphology to extract the region of interest from the brain MRI images.

The analysis on a patient's human brain MRI image having 257 different image slices has been done using MATLAB software. The preliminary results obtained from the proposed method may be used to achieve better results in terms of accuracy or segmented error of the MRI image. Further, it is evident from the study carried out that the use of the proposed PSO algorithm allows increasing the performance of the segmentation method under random variation of initial number of cluster. It is found that more stable outcome was obtained in terms of accuracy as compared to FCM method which gives unstable outcome with random variation of the initial number of cluster.

The proposed method applied on MRI database images which may be explored on different types of medical records like Ultrasonic images, X-Ray, CT scan, etc. However, further research and development is needed to improve the existing technique and introduce new possible solutions that may preserve the sharp edges in the template image.

References

1. Gonzalez RC, Woods RE (2002) Digital Image Processing. 3nd edition, Prentice Hall
2. Alasdair MC (2004) Introduction to Digital Image Processing with MATLAB. Cenage Learning
3. S Jayaraman, S Esakkirajan and T Veerakumar, "Digital Image Processing", Tata McGraw Hill, 2009
4. K. R. Castleman, "Digitial Image Processing", Pearson, 1996
5. Rogowska, J., 2000. Overview and Fundamentals of Medical Image Segmentation-5
6. Worth AJ, Makris N, CavinessJr VS, Kennedy DN (1997) Neuroanatomical segmentation in MRI: technological objectives. Int J Pattern Recognit Artif Intell 11(08):1161–1187
7. Despotović, I., Goossens, B. and Philips, W., "MRI segmentation of the human brain: challenges, methods, and applications". *Computational and mathematical methods in medicine, 2015*
8. Yazdani, Sepideh, RubiyahYusof, AmirhoseinRiazi, and AlirezaKarimian. "Magnetic resonance image tissue classification using an automatic method." *Diagnostic pathology* 9, no. 1, pp. 207, 2014
9. Pohle, R. and Toennies, K.D., "Segmentation of medical images using adaptive region growing". In *Medical Imaging* (pp. 1337–1346). International Society for Optics and Photonics, July 2001
10. Balafar MA, Ramli AR, Saripan MI, Mashohor S (2010) Review of brain MRI image segmentation methods. Artif Intell Rev 33(3):261–274
11. Lladó, X., Oliver, A., Cabezas, M., Freixenet, J., Vilanova, J.C., Quiles, A., Valls, L., Ramió-Torrentà, L. and Rovira, À.," Segmentation of multiple sclerosis lesions in brain MRI: a review of automated approaches". *Information Sciences, 186*(1), pp. 164–185, 2012
12. Cabezas M, Oliver A, Lladó X, Freixenet J, Cuadra MB (2011) A review of atlas-based segmentation for magnetic resonance brain images. Comput Methods Programs Biomed 104(3):e158–e177
13. Balafar, M.A, " Fuzzy C-mean based brain MRI segmentation algorithms". *Artificial Intelligence Review*, pp. 1–9, 2014
14. Fedorov, Johnson, EswarDamaraju, Alexei Ozerin, Vince Calhoun and Sergey Plis. "End-to-end learning of brain tissue segmentation from imperfect labeling" International Joint Conference on Neural Networks (IJCNN), 2017
15. Bai Xiangzhi, Sun Chuxiong, Sun Changming (2019) Cell Segmentation Based On FOPSO Combined With Shape Information Improved Intuitionistic FCM. IEEE Journal of Biomedical and Health Informatics 23(1):449–459
16. Tao Lei, XiaohongJia, Yanning Zhang, Senior Member, IEEE, Lifeng He, Senior Member, IEEE, Hongying. "Significantly Fast and Robust Fuzzy C-Means Clustering Algorithm Based

on Morphological Reconstruction and Membership Filtering" IEEE TRANSACTIONS ON FUZZY SYSTEMS,vol- 26, no.- 5, pp: 3027 – 3041, 2018
17. Hiba Amin Mohammed Ali, Mohamed A.A. Ahmed, Eltahir Mohamed Hussein "MRI Brain Tumor Segmentation Based on Multimodal Clustering and Level Set Method" International Conference on Computer, Control, Electrical, and Electronics Engineering (ICCCEEE), 2018
18. Kennedy J (2011) Particle swarm optimization. In: Encyclopedia of machine learning, pp 760–766. Springer US
19. Reinoso O, Sebastián JM, Aracil R, Torres F (2001) Morphological operations with subpixel resolution on digital images. Machine Graphics and Vision 10(1):89–102
20. Signal Processing Laboratory, Department of Telecommunications Brno University of Technology, 2011. http://splab.cz/en/research/konference-a-workshopy/challenge-2013/challenge-3-brain-tissue-analysis
21. Robert S, Mustofa AA, Christy Atika Sari, De Rosal Ignatius Moses Setiadi, Eko Hari Rachmawanto (2018) MRI Image Segmentation using Morphological Enhancement and Noise Removal based on Fuzzy C-means.5th International Conference on Information Technology, Computer, and Electrical Engineering (ICITACEE)

Chapter 30
Analysis of Hotspot Development in Power Transformer and Its Life Estimation

Vinit Mehta and Jayashri Vajpai

1 Introduction

The operational efficiency and the economic viability of power systems are governed largely by the functioning and the cost of its constituent power transformers. The reliability of power systems is adversely affected by the failure and maloperation of power transformers. These occur due to the failure of insulation caused by high stress under abnormal or critical operating conditions or in cases when heat generated in a power transformer is not dissipated efficiently by the surrounding medium. Generally, hotspots are developed in the power transformers when the heat dissipation is not uniform or effective, leading to thermal stress.

This paper presents a technique for estimating the loss of life of transformer by modeling the thermal stresses that are responsible for the deterioration of their quality and performance and employing them to calculate accelerated aging. The most important factor among these is the hotspot temperature (HST), which is a major reason for the loss of life of transformer. The HST of a transformer primarily depends on the ambient temperature, the rise in the top oil temperature (TOT) over the ambient temperature and the rise in the winding HST over the top oil temperature. HST values for different load conditions can be estimated by considering appropriate computational model on the basis of the thermal characteristics of the transformer and the cooling system.

This paper proposes a computational model that has been simulated on MATLAB/Simulink and that has provision for evaluation of the HST for every hour in a given load cycle. This is employed to estimate the aging acceleration factor. The percentage loss of life is predicted on the basis of these values. Further, by providing Oil Natural Air Forced (ONAF) cooling arrangement during peak load period, the

V. Mehta (✉) · J. Vajpai
Electrical Engineering Department, J.N.V. University, Jodhpur, Rajasthan, India
e-mail: vinit741@gmail.com

M. Shorif Uddin et al. (eds.), *Intelligent Energy Management Technologies*,
Algorithms for Intelligent Systems,
https://doi.org/10.1007/978-981-15-8820-4_30

saving in percentage loss of life is determined. The proposed model has been used to predict the loss of life of a 315MVA power transformer in operation at 400 kV GSS, Surpura, Jodhpur (Rajasthan, India). After Introduction, the paper includes five more sections, that present the state of art, proposed methodology, algorithm, MATLAB/Simulink model, results and discussion.

2 State of Art

In order to overcome the abnormal operating conditions and to increase the transformer loading capacity, different calculation procedures for estimating the winding hotspot temperature with reference to load changes have been proposed by many authors.

The base of their work is primarily IEEE or IEC guidelines. IEEE Guide for Loading Mineral-Oil-Immersed Transformers [1] is applicable to loading mineral oil-immersed distribution and power transformers, with different types of construction, along with special considerations for the degree of conservatism involved in the loading. This has paved the way for understanding and developing models for the simulation of thermal characteristics of power transformers as attempted in this paper. Hashmi et al. [2] have performed the steady-state calculations using IEC guidelines to determine the hotspot temperatures of distribution and power transformers in the worst environment due to long summer periods. Amoda et al. [3] have presented an investigation into the adequacy of the IEEE HST model, when the model parameters are to be determined from measured field data.

Shiyou et al. [4] have analyzed the mechanism of thermogenesis and thermolysis of transformer along with the position of the hotspot temperature. Further, the calculation of the loss of insulation life of dry-type transformer has been carried out on the basis of HST. Silva and Bastos [5] have addressed the influence of simplifications to be made on the geometries of power transformers for the performance of thermodynamic simulations to diminish the computational time and to obtain the magnetic fields, temperatures and heat flow in the interior of the transformer. Humayun et al. [6] have proposed the demand response and dynamic thermal rating-based optimization model for efficient capacity utilization and life management of transformers during contingencies. Kweon et al. [7] have estimated the hotspot temperature in the power transformer by the optical fiber sensors and validated it by the conventional heat run test. Gouda et al. [8] have introduced the HST- and TOT-based thermal model under linear and nonlinear loads. Vanegas and Mahajan [9] have determined the thermal characteristics, load profiles and acceleration factor equation of an oil-immersed current transformer and compared the estimated and expected values of the aging acceleration factor.

DejanSuja et al. [10] have presented an accurate temperature calculation method based on the thermal-electrical analogy that considers nonlinear thermal resistances at different locations within a power transformer. Yong Liang [11] has developed a graphical tool for predicting TOT using a semiphysical model to assess the effect of solar radiation and wind velocity on the prediction of TOT.

Muhamad et al. [12] have investigated the effect of HST on the overall transformer temperature distribution and its effect on the heat dissipation to the surface of transformer tank, for condition monitoring purposes. Further, simulation of mineral oil-filled distribution transformer (ONAN type) has been done by using Finite Element Method Magnetism software. Longnv et al. [13] have presented an accurate computational alternative for hot-spot temperature-rise estimations in a single-phase autotransformer to compute the stray losses using finite-element method (FEM), along with the average surface convection heat transfer coefficients of the structural parts of the transformer. Radakovic et al. [14] have developed a detailed thermal-hydraulic network model for determining the value of the hotspot factor and the HST using FEM. Srinivasan [15] has proposed a semiphysical model comprising of variable environmental conditions for the estimation of HST in transformer and along with a MATLAB/Simulink-based valid model. Takami et al. [16] have done an online monitoring of the transformer using FEMLAB and MATLAB software to estimate the HST of oil-immersed power transformers.

In general, many simplifying assumptions have been made in the various proposed methods for calculating the HST of power transformers, as reported in the standards documentation and published literature. The aim of this paper is to develop a comprehensive computational methodology for thermal model of power transformer, for estimating accelerated aging factor and loss of life, along with ease of application offered by the MATLAB software.

3 Proposed Methodology

This section presents the salient features of the proposed thermal modeling of a three-phase power transformer by using MATLAB software and uses it to estimate the loss of life of the power transformer. The theoretical aspects of the estimation of loss of life are first described, followed by the algorithm.

3.1 Thermal Modeling of Top Oil Temperature Rise

The rise of top oil temperature over ambient temperature is an indication of continuous loading of transformer. An increase in the load increases the losses thus increasing the overall temperature. The rate of change of temperature depends upon the overall thermal time constant of the transformer, which in turn depends upon the heat capacity of the transformer, i.e., the mass of the core, coils, and oil, and the rate of heat exchange from the transformer. The change of top oil temperature is modeled as a first-order differential equation as follows [10].

$$T_{TO}\frac{d\Delta\theta_{TO}(t)}{dt} = \Delta\theta_{TO}(u) - \Delta\theta_{TO}(t) \quad (1)$$

where, T_{TO} is the top oil time constant in minutes, $\Delta\theta_{TO}(t)$ is the top oil temperature rise over ambient temperature in °C, $\Delta\theta_{TO}(u)$ is the final top oil temperature rise in °C, and t is the time referenced to the time of the loading change.

Equation (1) is solved to obtain the following exponential response from the initial temperature state to the final temperature state [10],

$$\Delta\theta_{TO}(t) = [\Delta\theta_{TO}(u) - \Delta\theta_{TO}(i)]\left[1 - e^{-t/T_{TO}}\right] + \Delta\theta_{TO}(i) \quad (2)$$

where, $\Delta\theta_{TO}(i)$ is the initial top oil temperature rise in °C.

The final rise in the top oil temperature depends upon the load factor and can be approximated by the following equation:

$$\Delta\theta_{TO}(u) = \Delta\theta_{TO}(r)\left[\frac{K^2R + 1}{R + 1}\right]^n \quad (3)$$

where, $\Delta\theta_{TO}(r)$ is the full load top oil temperature rise over ambient temperature in °C, R is the ratio of load loss at rated load to no-load loss, K is the ratio of the specified load to rated load, n is an empirically derived exponent that depends upon the cooling method. The IEEE loading guide [1] recommends the use of $n = 0.8$ for natural convection and $n = 0.9$ to 1.0 for forced cooling. The top oil time constant at the considered load is given by the following:

$$T_{TO} = 60 * \frac{C_{th-oil} * \Delta\theta_{TO}(r)}{q_{tot}} \quad (4)$$

where, q_{tot} is the total supplied losses in W, and C_{th-oil} is the equivalent thermal capacitance of the transformer oil in W-h/°C.

The equivalent thermal capacitance of the transformer oil is given by the following equation:

$$C_{th-oil} = 0.48 * M_{Oil} \quad (5)$$

where, M_{Oil} is the weight of the oil in kg.

3.2 *Thermal Modeling of Hotspot Temperature Rise*

The increase in the transformer current due to losses increases the oil and winding temperature. The change of hotspot temperature is modeled as a first-order differential equation shown in Eq. (6) [10]:

$$T_{HS}\frac{d\Delta\theta_{HS}(t)}{dt} = \Delta\theta_{HS}(u) - \Delta\theta_{HS}(t) \quad (6)$$

where, T_{HS} is the hotspot time constant in minutes, $\Delta\theta_{HS}(t)$ is the hotspot temperature rise over top oil temperature rise in °C, $\Delta\theta_{HS}(u)$ is the final hotspot temperature rise in °C and t is the time referenced to the time of the loading change.

This can be solved to obtain

$$\Delta\theta_{HS}(t) = [\Delta\theta_{HS}(u) - \Delta\theta_{HS}(i)]\left[1 - e^{-t/T_{HS}}\right] + \Delta\theta_{HS}(i) \quad (7)$$

where, $\Delta\theta_{HS}(i)$ is the initial hotspot temperature rise in °C.

Based on the IEEE model, the final rise in the hotspot temperature considering the load factor can be obtained by the following equation:

$$\Delta\theta_{HS}(u) = \Delta\theta_{HS}(r)[K]^{2m} \quad (8)$$

where, $\Delta\theta_{HS}(r)$ is the rated hotspot temperature rise over top oil temperature and m is an empirically derived exponent that depends on the cooling method.

The winding hotspot time constant can be calculated as follows:

$$T_{HS} = 2.75 * \frac{\Delta\theta_{HS}(r)}{(1 + P_e) * S^2} \quad (9)$$

where, T_{HS} is the winding hotspot time constant in minutes at the rated load, P_e is the relative eddy current losses (W), S is the current density in A/mm^2 at rated load.

Finally, the hotspot temperature is calculated by adding the ambient temperature, the top oil temperature rise over ambient, and the hotspot temperature rise over top oil. This can be expressed by the following equation [1].

$$\theta_H = \theta_A + \Delta\theta_{HS}(t) + \Delta\theta_{TO}(t) \quad (10)$$

where, θ_A is the ambient temperature in °C and θ_H is the ultimate hotspot temperature in °C.

3.3 *Estimation of Equivalent Aging Factor*

In oil-immersed transformers, paper or cellulose material along with oil forms the major insulation. Therefore, the insulation must maintain adequate dielectric strength against voltage surges and adequate mechanical strength against short-circuit forces.

As cellulose ages thermally in an operating transformer; three mechanisms contribute to its degradation, namely; hydrolysis, oxidation, and pyrolysis. The agents responsible for the respective mechanisms are water, oxygen, and heat. Each of these agents will have an effect on degradation rate so they must be individually controlled.

Water and oxygen content of the insulation can be controlled by the transformer oil preservation system but control of heat is left to transformer operating personnel.

Transformer insulation life is defined as the total time period between the initial state for which the transformer insulation is considered new and the final state for which dielectric stress or short circuit stress could occur in normal service and cause an electrical failure.

Experimental evidence indicates that the relation of insulation deterioration to time and temperature follows an adaptation of the Arrhenius reaction rate theory that has the following form [1]:

$$\text{Per unit life} = \text{A*exp}\left[\frac{\text{B}}{\theta_{\text{H}} + 273}\right] \tag{11}$$

where, A is a modified per unit constant and B is the aging rate. The temperature of 110 °C is selected for one per unit life.

The Aging Acceleration Factor (F_{AA}) per unit transformer insulation life is given by the following equation [1].

$$F_{AA} = \exp\left[\frac{15000}{383} + \frac{15000}{\theta_{\text{H}} + 273}\right] \tag{12}$$

The equivalent loss of life (in hours or days) at the reference temperature in a given time period for the given temperature cycle is given as follows.

$$F_{EQA} = \frac{\sum_{n=1}^{N} F_{AA_n} * \Delta t_n}{\sum_{n=1}^{N} \Delta t_n} \tag{13}$$

where, F_{EQA} is the equivalent aging factor for the total time period, n is the index of the time interval, N is the total number of time intervals and $F_{AAn}$is the aging acceleration factor for the temperature which exists during the time interval Δt_n.

3.4 *Estimation of Percentage Loss of Life*

The insulation per unit life curve is used to calculate the percent loss of total life of a transformer. The normal insulation life at the reference temperature is defined in hours or years. The values of normal insulation life for a well-dried, oxygen-free system are given in Table 1.

The percentage loss of life is given as follows:

$$\text{Percentage Loss of Life} = \frac{F_{EQA} * t*100}{\text{Normal Insulation Life}} \tag{14}$$

Table 1 Normal Insulation Life of a Well-Dried, Oxygen-Free 65 °C Average Winding Temperature Rise Insulation System at the Reference Temperature of 110 °C [1]

Basis	Normal insulation Life hours	years
50% retained tensile strength of insulation	65 000	7.42
25% retained tensile strength of insulation	135 000	15.41
200 retained degree of polymerization in insulation	150 000	17.12
Interpretation of distribution Transformer functional life test data	180 000	20.55

Further, by providing ONAF cooling arrangement during peak load period, the saving in percentage loss of life is determined. This can then be employed for the evaluation of residual life.

The algorithm for thermal modeling of the power transformer is now presented.

4 Algorithm

Step 1 Initialize the input variables of transformer on hourly basis. This includes load factor K, total losses q (W), ambient temperature $\Delta\theta_A$ (°C), OTI reading $\Delta\theta_{TO}(i)$(°C) and WTI reading $\Delta\theta_{HS}(i)$ (°C) every hour in a given load cycle.

Step 2 The rated values of rated top oil rise over ambient temperature $\Delta\theta_{TO}(r)$(°C), rated hotspot rise over ambient temperature $\Delta\theta_{HS}(r)$ (°C), exponent n & m, weight of oil M_{oil} (kg), current density S (A/mm^2), relative eddy current loss P_e (W), ratio of load loss at rated load to no-load loss R.

Step 3 The different values of final top oil temperature rise $\Delta\theta_{TO}(u)$(°C) are determined by placing the values of K, R and n on hourly basis, using Eq. (3).

Step 4 The different values of final hotspot temperature rise $\Delta\theta_{HS}(u)$(°C) are determined by placing the values of K and m on hourly basis, using Eq. (8).

Step 5 The top oil time constant T_{TO} (mins) is determined with the help of Eq. (4).

Step 6 The hotspot time constant T_{HS}(mins) is determined with the help of Eq. (9).

Step 7 The top oil temperature rise $\Delta\theta_{TO}$ (°C) is obtained from the application of Eq. (2).

Step 8 The hotspot temperature rise $\Delta\theta_{HS}$ (°C) is obtained from the application of Eq. (7).

Step 9 The hotspot temperature θ_H during a day on hourly basis is calculated using Eq. (10).

Step 10 The aging acceleration factor is obtained using Eq. (12).

Step 11 The equivalent aging of the power transformer is calculated by using Eq. (13).

Step 12 The percentage loss of life is obtained from the application of Eq. (14).

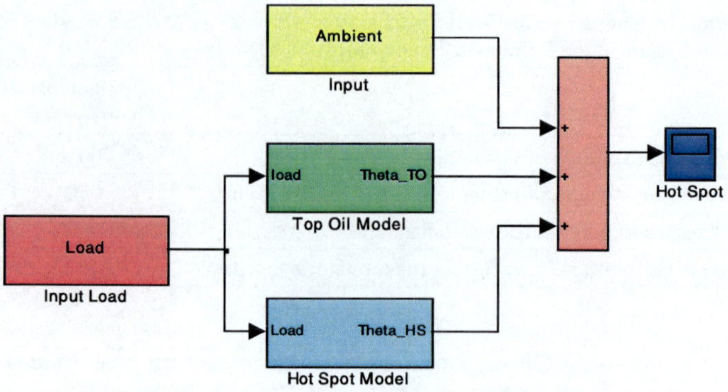

Fig. 1 Block diagram showing the thermal dynamic model of a power transformer

All the steps illustrated above, may be repeated for a load cycle that contains overload conditions during an hour for a day. The percentage loss of life thus calculated shows the amount of loss of life of transformer that can be used to estimate the reduced loss of life of the power transformer.

5 Matlab/Simulink Model

Figure 1 shows a simplified block diagram of the MATLAB/Simulink thermal dynamic model of a power transformer.

Equations (2) and (7) are solved using MATLAB program for determination of the top oil temperature rise $\Delta\theta_{TO}(t)$ and hotspot temperature rise $\Delta\theta_{HS}(t)$, respectively. At each discrete time interval of 60 min, the top oil temperature rise and the hotspot winding temperature rise are calculated.

The hotspot temperature θ_H is the sum of ambient temperature, top oil temperature rise and hotspot temperature rise. The measured hotspot temperature θ_H results for a 315 MVA, 400/33 kV transformer during the given load cycle are then used to determine the residual life of the transformer.

6 Results and Discussion

In order to validate the proposed model, data gathered under various load conditions from a real power transformer (315MVA) which are recorded in the month of January, have been used. In this study, work has been carried out in a power transformer situated at 400 kV GSS, Surpura, Jodhpur substation. The specifications, cooling arrangements and temperature measuring equipments of the proposed power transformer are as shown in Tables 2, 3 and 4, respectively.

Table 2 Specification of a 315 mva 400 kV/33 kV power transformer

Rated voltage (HV)	400 kV
Rated voltage (LV)	33 kV
Rated current (HV)	454.70 A
Rated current (LV)	1837.00 A
Current Density	0.28A/mm^2
No. of phase	3
Frequency	50 Hz
Connection Symbol	Yd11
Weight of core and coil	129400 kg
Weight of tank and fittings	32850 kg
Weight of Oil	64090 kg
Rated top oil rise over ambient temperature	45 °C
Rated hotspot rise over top oil temperature	55 °C
Ratio of load loss at rated load to no-load loss	2
Exponent 'n'	0.8
Exponent 'm'	0.9
No load loss	17500 W
Relative winding eddy current losses	152 W

Table 3 Cooling Equipment used in 315MVA 400 kV/33 kV power transformer

Oil pumps and fans capacity	Pump (600 gpm)	Fan (467 cum per min)
No. of oil pumps and fans (Running + Standby)	4 (2 + 2)	10 (8 + 2)

Table 4 OTI and WTI auxiliary contacts settings

OTI	Alarm	95 °C
	Trip	100 °C
WTI	Fan Start	85 °C
	Pump Start	95 °C
	Alarm	115 °C
	Trip	125 °C

The thermal behavior of the power transformer has been evaluated and verified using the top oil and hotspot temperature models. Results of these thermal models for a power transformer are discussed in the following section. The typical load factor, total losses, ambient temperature, OTI and WTI reading for a 315MVA, 400/33 kV power transformer located at 400 kV GSS, Jodhpur are shown in Table 5.

Table 5 Input data to a 315 MVA transformer

Clock time	Load factor (K)	Total losses (Watts)	Ambient temperature (°C)	OTI reading (°C)	WTI reading (°C)
6:00 am	0.57	685263	12	22	23
7:00 am	0.59	685905	13	25	24
8:00 am	0.61	696532	14	26	25
9:00 am	0.71	745623	14	24	26
10:00 am	0.78	795623	15	25	26
11:00 am	0.79	797010	15	29	29
12:00 pm	0.87	800100	17	28	29
01:00 pm	0.89	812356	18	27	30
02:00 pm	0.96	889654	18	26	31
03:00 pm	1.08	895698	18	25	32
04:00 pm	1.01	891258	18	30	32
05:00 pm	0.99	889932	17	31	31
06:00 pm	0.98	890008	15	32	30
07:00 pm	1.04	895498	14	35	30
08:00 pm	0.87	801124	14	36	28
09:00 pm	0.85	790050	14	25	27
10:00 pm	0.65	724613	13	26	24
11:00 pm	0.64	734212	13	29	23
12:00 am	0.55	686541	13	30	23
01:00 am	0.56	686689	12	32	22
02:00 am	0.57	689365	12	31	24
03:00 am	0.59	690635	11	35	25
04:00 am	0.61	696432	11	36	26
05:00 am	0.54	685252	12	35	24

The normal load profile of the 315MVA, 400/33 kV power transformer for 24 h in winter season (January) is shown in Fig. 2. It is clear from Fig. 2, that the maximum value of the load factor is 1.08 occurring at 03:00 pm.

The HST for every hour is evaluated by using a MATLAB program. The effect of heat dissipated due to various losses, i.e., constant and variable losses in a transformer on the useful life of a cellulose insulation material has first been estimated on a per unit basis. Cumulative loss of life has been calculated for varying load conditions with the understanding that one real day of operation will produce less or more aging than one day. Figure 3 shows the graphical representation of the hotspot temperature for a day including the variation of the ambient temperature, the top oil temperature rise over ambient, and the hotspot temperature rise over top oil and the results are shown

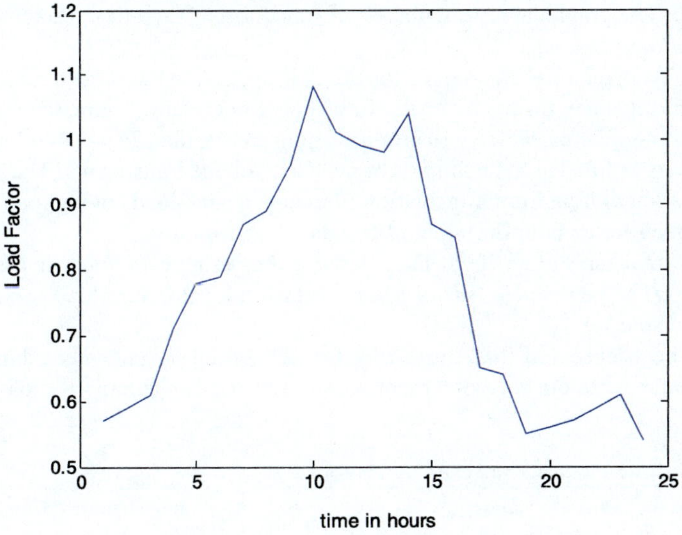

Fig. 2 Normal load cycle profile

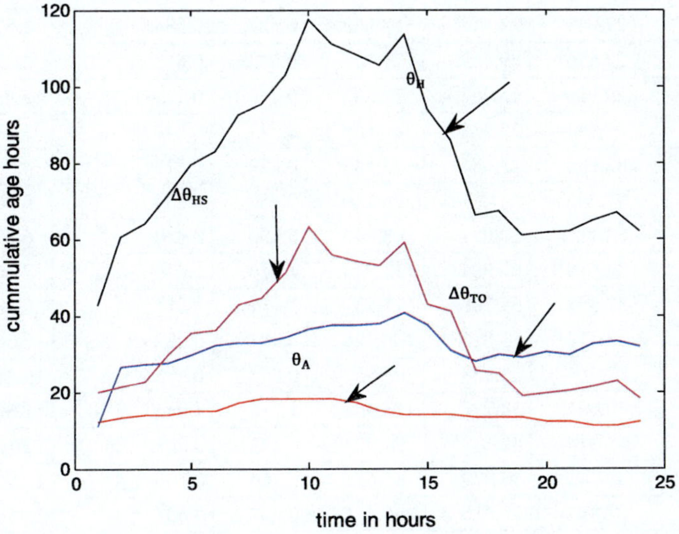

Fig. 3 Graphical representation of hotspot temperature, hotspot temperature rise, top oil temperature rise and ambient temperature

in Table 6. The graphical representation of transformer insulation life throughout a day is shown in Fig. 4.

Now, a peak load having load factor of 1.5 is applied to the same transformer for an hour during a day results in a rise of hotspot temperature. Again, the cumulative loss of life is predicted, which will produce aging greater than 24 h. Table 7 shows the result of loss of life due to the impact of peak load on the transformer. The graphical representation of transformer insulation life when a peak load having load factor of 1.5 is applied for an hour during the same day is shown in Fig. 5.

The normal insulation life at the reference temperature in hours is assumed to be 180000. The percentage loss of life is evaluated for different load conditions as shown in Table 7.

It has been observed that employing ONAF cooling arrangement during peak load period reduces the hotspot temperature to 135 °C. The cumulative loss of life is

Table 6 Analytical results for a 315 MVA transformer

Clock time	$\Delta\theta_{TO}$	$\Delta\theta_{HS}$	θ_H	F_{AA}	Aging Hours	Cumulative age hours
6:00 am	10.889	19.824	42.713	0.434	0.434	0.434
7:00 am	26.374	21.299	60.673	0.560	0.560	0.994
8:00 am	27.249	22.612	63.862	0.584	0.584	1.579
9:00 am	27.598	29.659	71.257	0.643	0.643	2.222
10:00 am	29.514	35.088	79.602	0.713	0.713	2.936
11:00 am	31.944	35.922	82.866	0.741	0.741	3.678
12:00 pm	32.856	42.686	92.543	0.829	0.829	4.507
01:00 pm	32.742	44.467	95.210	0.854	0.854	5.362
02:00 pm	34.111	50.931	103.042	0.930	0.930	6.292
03:00 pm	36.410	62.904	117.314	1.076	1.076	7.368
04:00 pm	37.349	55.788	111.137	1.011	1.011	8.380
05:00 pm	37.412	53.816	108.228	0.982	0.982	9.362
06:00 pm	37.713	52.838	105.551	0.955	0.955	10.317
07:00 pm	40.691	58.774	113.465	1.035	1.035	11.352
08:00 pm	37.346	42.678	94.024	0.843	0.843	12.196
09:00 pm	30.743	40.929	85.673	0.766	0.766	12.962
10:00 pm	27.834	25.316	66.151	0.602	0.602	13.565
11:00 pm	29.477	24.617	67.094	0.610	0.610	14.175
12:00 am	28.937	18.787	60.724	0.560	0.560	14.736
01:00 am	30.273	19.391	61.664	0.568	0.568	15.304
02:00 am	29.780	20.030	61.810	0.569	0.569	15.873
03:00 am	32.453	21.308	64.761	0.591	0.591	16.465
04:00 am	33.296	22.621	66.917	0.608	0.608	17.074
05:00 am	31.871	18.191	62.062	0.571	0.571	17.645

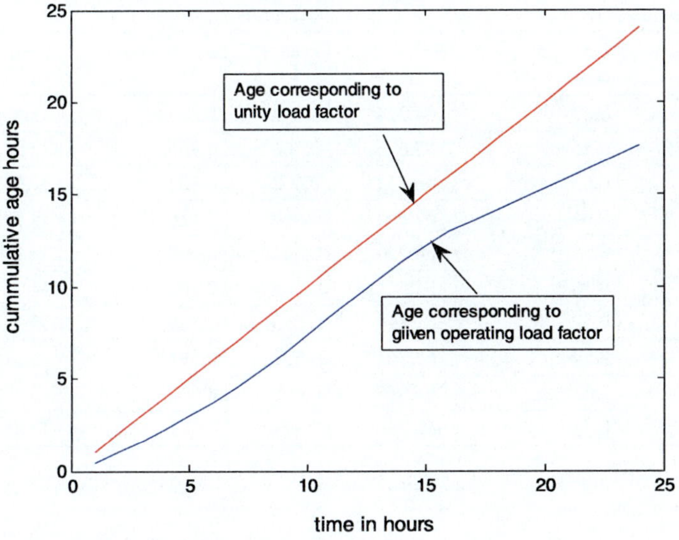

Fig. 4 Transformer insulation life

Table 7 Peak Load Applied to 315MVA Transformer

Clock Time	Load Factor (K)	θ_H	Faa	Aging hours	Cumulative Age Hours
6:00 am	0.57	42.713	0.434	0.434	0.434
7:00 am	0.59	60.673	0.560	0.560	0.994
8:00 am	0.61	63.862	0.584	0.584	1.579
9:00 am	0.71	71.257	0.643	0.643	2.222
10:00 am	0.78	79.602	0.713	0.713	2.936
11:00 am	0.79	82.866	0.741	0.741	3.678
12:00 pm	0.87	92.543	0.829	0.829	4.507
01:00 pm	0.89	95.210	0.854	0.854	5.362
02:00 pm	0.96	103.042	0.930	0.930	6.292
04:00 pm	1.01	111.137	1.011	1.011	27.793
05:00 pm	0.99	108.228	0.98	0.982	28.775
06:00 pm	0.98	105.551	0.95	0.955	29.730
07:00 pm	1.04	113.465	1.035	1.035	30.765

(continued)

Table 7 (continued)

Clock Time	Load Factor (K)	θ_H	Faa	Aging hours	Cumulative Age Hours
08:00 pm	0.87	94.024	0.843	0.843	31.609
09:00 pm	0.85	85.673	0.766	0.766	32.375
10:00 pm	0.65	66.151	0.602	0.602	32.978
11:00 pm	0.64	67.094	0.610	0.610	33.588
12:00 am	0.55	60.724	0.560	0.560	34.149
01:00 am	0.56	61.664	0.568	0.568	34.717
02:00 am	0.57	61.810	0.569	0.569	35.286
03:00 am	0.59	64.761	0.591	0.591	35.878
04:00 am	0.61	66.917	0.608	0.608	36.487
05:00 am	0.54	62.062	0.571	0.571	37.058

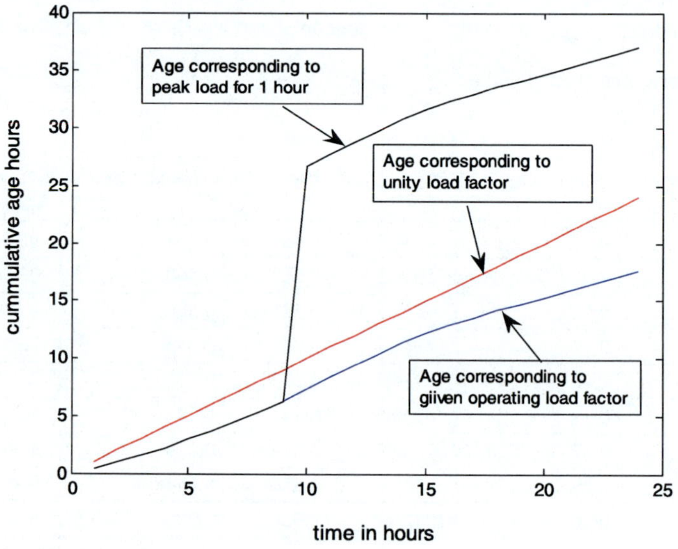

Fig. 5 Transformer insulation life corresponding to peak load for 1 h

now calculated again. This results in the saving in percentage loss of life as shown in Table 8. The results obtained from thermal model may be used to estimate the residual life of the power transformer.

The proposed model gives the approximate HST values that are in close agreement with the measured field data. It can be concluded that with the increase in the transformer temperature beyond thermal limits would reduce its life below the specified normal life.

Table 8 Cooling Equipment Used In 315mva 400 kV/33 kV Power Transformer

S. No.	Condition of Operation	% loss of life
1	Operating Load Factor every hour	0.0098
2	Peak load factor for an hour	0.0206
3	ONAF cooling during peak hour period	0.0153

Further, the study of insulation ageing is important, as it reduces both the mechanical and dielectric-withstand strength of the transformer. An ageing transformer is subjected to faults that result in high radial and compressive forces. Also, the conductor insulation gets deteriorated and becomes unable to sustain the mechanical stresses caused by a fault. Hence, it is the dominant factor in limiting the lifetime of the transformer. Providing proper cooling such as ONAF has great potential to save the percentage loss of life.

The thermal model proposed in this paper is dependent on the accuracy of estimated steady-state temperature rises. Therefore, it will be important to develop a variable time interval calculation method with estimation at smaller time intervals when there are dynamic changes in temperature and larger intervals when the steady state is achieved. However, further research and development are needed to improve the existing monitoring systems and introduce designs and applications that include better thermal modeling. These thermal models will allow the transformer manufacturers to provide better specifications and users to operate the transformers on appropriate loading by considering the ambient temperature conditions.

References

1. Guide for Loading Mineral-Oil- Immersed Transformers, IEEE Standard C57. 1991–1995
2. Murtaza H, Matti L, Seppo H (2013) Effect of Climate Change on Transformers Loading Conditions in the Future Smart Grid Environment. Open Journal of Applied Sciences 3:24–29. https://doi.org/10.4236/ojapps.2013.32b005. Available:http://www.scirp.org/journal/ojapps
3. Oluwaseun AA, Tylavsky DJ, McCulla GA, Knuth WA (2008) A New Model for Predicting Hottest-Spot Temperature in Transformers. IEEE Trans. Power Symposium, NAPS '08. 40th North American, pp. 1–8. https://doi.org/10.1109/naps.2008.5307407
4. ShiyouWANG,Youyuan WANG and Xuetong ZHAO, "Calculating Model of Insulation Life Loss of Dry-Type Transformer Based on the Hot-Spot Temperature", *IEEE* 11th International Conference on the Properties and Applications of Dielectric Materials, pp. 720-723, July 2015. https://doi.org/10.1109/icpadm.2015.7295373
5. Juliano R. da Silva and Joao P. A. Bastos, "Analysis of Power Transformer Geometry Simplifications on Electromagnetic and Thermodynamic Simulations", *IEEE Trans. on magnetics*, vol. 51, no. 3, article 8400404, March, 2015. https://doi.org/10.1109/tmag.2014.2358993
6. Muhammad Humayun, Mubbashir Ali, Amir Safdarian, Merkebu Z. Degefa, and MattiLehtonen, "Optimal Use of Demand Response for Lifesaving and Efficient Capacity Utilization of Power Transformers during Contingencies", *IEEE Trans.* Power & Energy Society General Meeting, pp. 1 - 5, 2015. DOI: https://doi.org/10.1109/pesgm.2015.7285627

7. Dong-JinKweon, Kyo-Sun Koo, Jung-Wook Woo and Joo-SikKwak, "A Study on the Hot Spot Temperature in 154 kV Power Transformers", Journal of Electrical Engineering & Technology, vol. 7, no. 3, pp. 312-319, 2012. Available:http://dx.doi.org/10.5370/JEET.2012.7.3.312
8. O.E. Gouda, G.M. Amer, W.A.A. Salem, "Predicting transformer temperature rise and loss of life in the presence of harmonic load currents", Ain Shams Engineering Journal, vol. 3, pp. 113–121, 2012. Available: www.elsevier.com/locate/asej
9. Diego M. RobalinoVanegas and Satish M. Mahajan, "Correlation between Hot-Spot Temperature and Aging Factor of Oil-Immersed Current Transformers", *IEEE Trans.* Power and Energy Society General Meeting - Conversion and Delivery of Electrical Energy in the 21st Century, pp. 1-5, 2008. DOI: https://doi.org/10.1109/pes.2008.4595981
10. Dejan Susa, MattiLehtonen, and HasseNordman, "Dynamic Thermal Modelling of Power Transformers", IEEE Trans. on power delivery, vol. 20, no. 1, pp. 197-204, January 2005
11. Yong Liang, "Simulation of Top-Oil Temperature for Transformers", Thesis and Project Report, Power Systems Engineering Research Center Cornell University, New York, February 2001. Available:www.pserc.wisc.edu
12. N. A. Muhamad, H. Kamarden and N. A. Othman, "Heat Distribution Pattern of Oil-filled Transformer at Different Hottest Spot Temperature Locations", *IEEE* 11th International Conference on the Properties and Applications of Dielectric Materials, pp. 979–982, July 2015. https://doi.org/10.1109/icpadm.2015.7295438
13. Longnv Li, ShuangxiaNiu, S. L. Ho, W. N. Fu and Yan Li, "A Novel Approach to Investigate the Hot-Spot Temperature Rise in Power Transformers", *IEEE Trans. on magnetics*, vol. 51, no. 3, article 8400204, March, 2015. https://doi.org/10.1109/tmag.2014.2359956
14. Zoran Radakovic, UrosRadoman, and PredragKostic, "Decomposition of the Hot-Spot Factor", IEEE Trans. on power delivery, vol. 30, no. 1, pp. 403 – 411, Feb, 2015. https://doi.org/10.1109/tpwrd.2014.2352039
15. M.Srinivasan and A. Krishnan, "Effects of Environmental Factors in Transformer's Insulation Life", WSEAS transactions on power systems, vol 8, Issue 1, Jan 2013
16. Kourosh Mousavi Takami, Hasan Gholnejadand JafarMahmoudi, "Thermal and hot spot evaluations on oil immersed power Transformers by FEMLAB and MATLAB software's", *IEEE Trans.* Thermal, Mechanical and Multi-Physics Simulation Experiments in Microelectronics and Micro-Systems, pp. 1–6, 2007. https://doi.org/10.1109/esime.2007.359924

Chapter 31
Photovoltaic Module Cleaning Prediction Using Deep Neural Networks

Kapil Panwar, Aditya Jindal, and Kusum Lata Agarwal

1 Introduction

With increase in population and per capita electrical consumption, world is quickly moving toward renewable sources for energy generation. Solar and wind are the most used renewable sources for generating electrical power. Solar photovoltaic power plants are increasing in numbers and most of them are in plain grounds or near to desert areas so as to obtain high irradiation profile for high yields. Due to plain ground areas, soil accumulation on PV module is a common problem. This soil accumulation on PV modules affects the generation. This creates a thin layer of hindrance for irradiation and resulting generates lower electrical power [1].

To mitigate generation loss due to soil accumulation over PV modules, PV modules need to be cleaned/washed. For cleaning of PV modules, required water should be of drinking water type, with normal pH (pH 6–7) so as to avoid any harm to its glass and its antireflection coating [2]. Hence treated water is used to clean PV modules.

As most of the solar PV plants are in remote area as well as near to desert area, there is already water scarcity. Due to water scarcity, modules cannot be washed on daily basis using water. On the other hand, if borewell water is being used along with water treatment plant then waste management is another challenge.

K. Panwar (✉) · A. Jindal
Mahindra Teqo Pvt. Ltd, Bengaluru, India
e-mail: panwar.kapil2@mahindra.com

A. Jindal
e-mail: jindal.aditya2@mahindra.com

K. L. Agarwal
Electrical Department, JIET Jodhpur, Mogra Khurd, India
e-mail: kusum.agarwal@jietjodhpur.ac.in

M. Shorif Uddin et al. (eds.), *Intelligent Energy Management Technologies*, Algorithms for Intelligent Systems,
https://doi.org/10.1007/978-981-15-8820-4_31

Cleaning of solar PV module should be done in such a way that it minimizes water consumption and keep plant in condition to generate optimum electrical power.

In this paper, artificial intelligence (deep neural network) technique has been used to develop a model that predicts when solar PV modules should be cleaned, and also which modules should be cleaned. As soil accumulation is different for near pathways, clusters and switch yard due to vehicle movement and totally different in middle of power plant [3]. By this water can be used wisely in cleaning only those modules which are contributing a lot in generation minimization, instead of cleaning whole plant.

2 Deep Neural Network

Artificial intelligence is a branch of science in which machines are made intelligent like humans, to take their own decision. AI is a broadly used term, It has three popular subbranches as which are shown in Fig. 1.

To make machines think like human, artificial brain is developed using artificial neurons. These artificial neurons work in same ways as human neuron works.

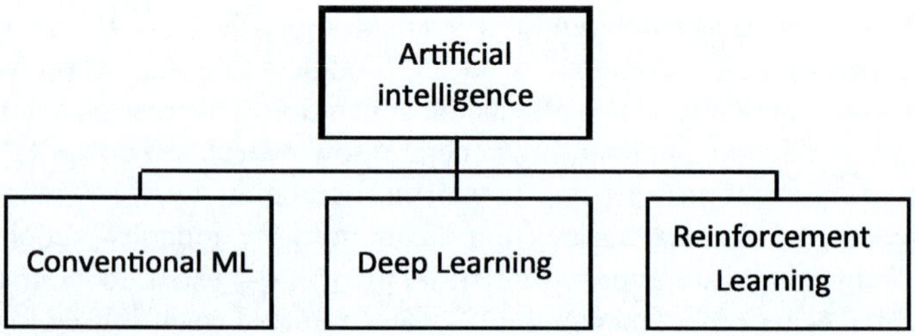

Fig. 1 Branches of artificial intelligence

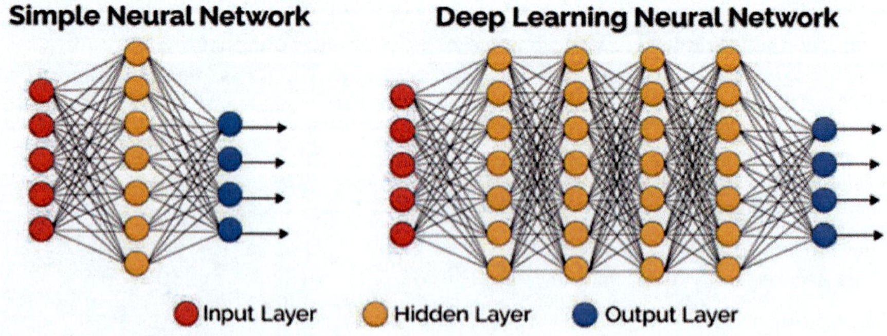

Fig. 2 Deep Neural Network [DNN]

Deep learning is a subbranch of AI in which a number of neurons are being used with multiple hidden layers to make successive learning more accurate. These algorithms require high-processing units like GPUs or TPUs to process millions of data and train the model. Deep Neural Networks are used for big data problems, computer vision and in solving complex problems.

3 Methodology

For this model creation, we selected one Solar PV plant located in the western region, India. In that solar PV, plant SMA make central inverter is installed along with Trina make PV modules.

3.1 Dataset Used for Model

In plant, there are lots of sensors available that measures weather parameters and DC current, voltage and power of Inverter. This dataset is of 1-min resolution and stored in local DB using SCADA. Hence 5 years of historical data are being used as input to this model. Parameters used are as per the below list:

- Global Horizontal Irradiation (w/m^2)
- Ambient Temperature (°C)
- Rain Fall (in mm)
- Humidity (%)
- Inclined Irradiation (w/m^2)
- PV Module surface temperature (°C)
- DC Current (A)
- DC Voltage (V)
- DC Power (kW).

3.2 Impact of Change in Irradiation on DC Current and DC Voltage

Using feature engineering, it is found that, change in irradiation directly impact on DC Current and DC Power, whereas change in temperature has an impact on DC voltage [4]. Sample IV curve has been shown in Fig. 3 and PV curve in Fig. 4.

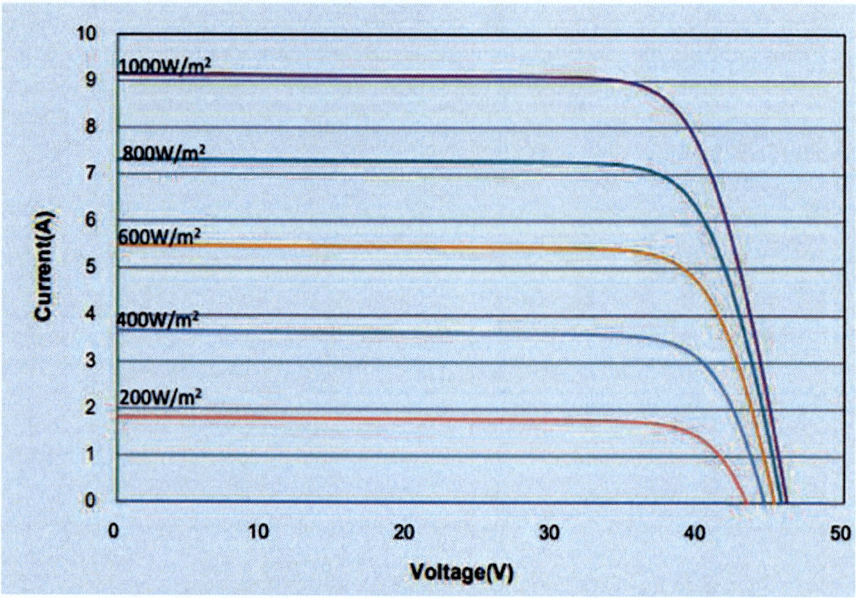

Fig. 3 IV curve of solar PV module [Generic]

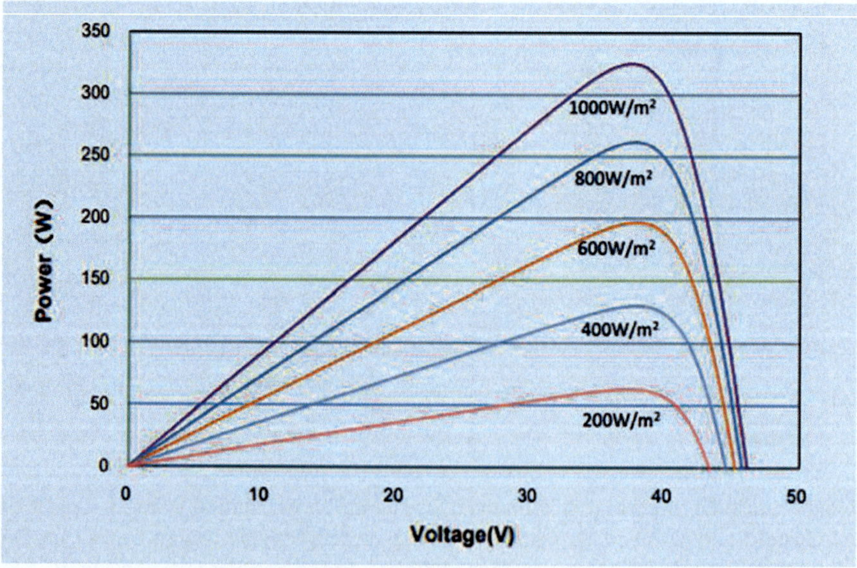

Fig. 4 PV curve of solar PV module [Generic]

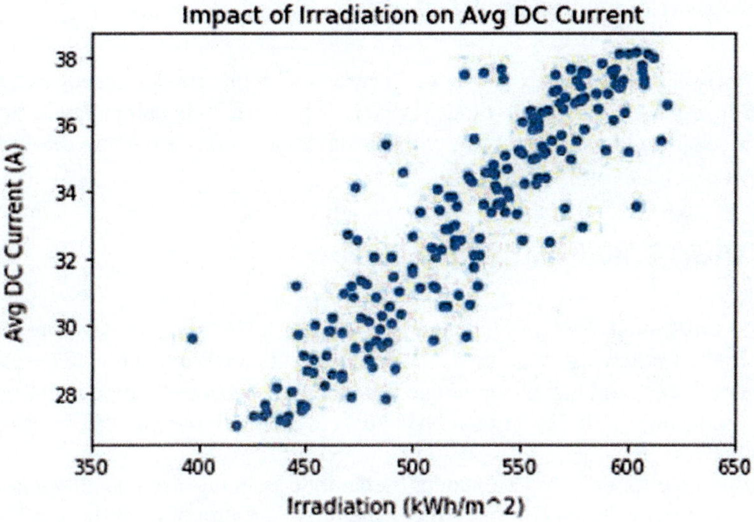

Fig. 5 Impact of irradiation on avg DC current is linear

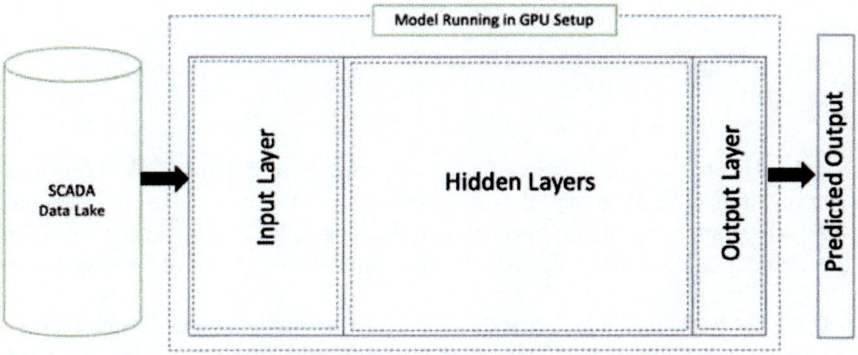

Fig. 6 Architecture of DNN model

3.3 Creating Model and Training

For solving this kind of problems, various kinds of models can be developed using machine learning techniques such as SVM, logistic regression and anomaly detection, etc. In this paper, deep neural networks are being used which is an advance technique of artificial intelligence branch. Inspired from human neurons, algorithmic neurons are developed and same as human brain, number of neurons are interconnected in layers to develop a deep neural network (Figs. 5 and 6).

3.4 Neural Network Architecture

Neural networks are made of layers of neurons. For this model, three sections of neuron layers are developed: Input layer, hidden layer and output layer. For the same, nine neuron input layers and two neuron output layers are being developed.

3.5 Model Training and Optimization

To build this model, Python language is being used. Neurons are developed using Ten-sorFlow framework. Training is the most critical part of any model. Due to huge amount of data, "mini batch" technique is being used to train this model. Developed model used SoftMax in last hidden layer so as to predict binary output for given set of parameters.

To optimize model "Auto Encoder" technique is being used as this technique optimizes time cycle as well as reduces memory consumption while training this model [5]. This technique reduces dimension of weights which are being updated in each learning cycle. This method is logically same as PCA which is used to reduce dimensionality for any machine learning model.

4 Result

To solve this problem, the generated model is being trained using 80% of historical data. To validate the accuracy of model, rest of the 20% data are fed to developed model and compared predicted versus actual cleaning daily data for a specific inverter. These validation data were selected arbitrary which were not part of training data so as to avoid biasness.

These predicted data and actual data of validation set give more clarity. Figure 7 represents the predicted versus actual module cleaning results.

This result represents that cleaning date predicted by model was 88% correct with actual cleaning dates whereas 12% times cleaning suggested dates were not matched with validation cleaning set. Similarly, noncleaning dates were predicted 90% accurately.

5 Conclusion

This model can predict when a set of modules need to be cleaned with 88% accuracy. This accuracy can be increased by tuning hyperparameter and including feedback in every new training cycle.

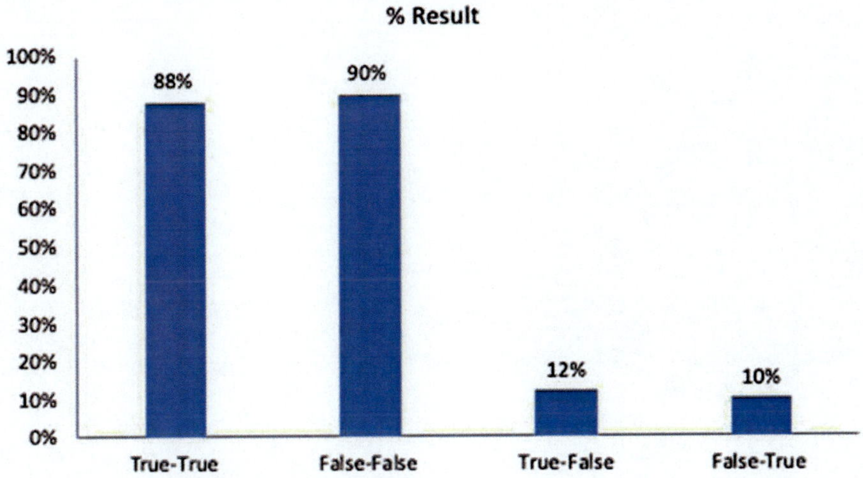

Fig. 7 Result of model with validation set

There is a future scope for training this model with multiple site datasets so as to make it generalized. Same model can be developed using the anomaly technique and can be compared with the existing model performance.

Acknowledgements We provide our special thanks to Mahindra Teqo Pvt. Ltd. for providing real data and opportunity to develop this model.

References

1. Wable SS (2017) Design and manufacturing of solar panels cleaning system. Int J Res Appl Sci Eng Technol (IJRASET) 5(VII)
2. Khadka, N (2018) Solar panel cleaner technology: a review. In: Conference: 5th international conference on developments in renewable energy technology (ICDRET'18), At Kathmandu
3. John JJ (2015) Evaluation and prediction of soiling loss on PV modules with artificially deposited dust. In: IEEE 42nd Photovoltaic Specialist Conference (PVSC)
4. Sobha Rani P (2018) Effect of Temperature and Irradiance on Solar Module Performance. IOSR J Electr Electron Eng (IOSR-JEEE) 13(2) Ver. III:36–40
5. Kiarashinejad Y (2019) Deep learning approach based on dimensionality reduction for designing electromagnetic nanostructures. ARXIS J

Chapter 32
Optimal Controller Design for DC–DC Buck Converter

Shubham Sharma and Kusum Lata Agarwal

1 Introduction

A DC–DC converter is used to convert one level of DC voltage to another. A linear DC–DC converter uses resistive voltage drop to regulate output voltage but this method is not efficient due to large power loss. To avoid this switch mode, DC–DC converters are used. Switch mode DC–DC converter uses power electronics devices and energy storing element which gives regulated dc output voltage and higher efficiency [1].

The switch-mode power electronics converters are some of the simplest power electronics converter which converts one level of electrical voltage into another level by switching action. These converters have received an increasing deal of interest in many areas. This is due to their wide application like power supplies for personal computers, office equipment, appliances control, telecommunication equipment, DC motor drives, automotive, aircraft, etc. They are also used in the satellites where DC buses at different voltage levels are supplied through these DC–DC converters [1].

The DC–DC converter is having low cost, small size and high efficiency but it contains large ripple in output and output voltage regulation is not good as compared with linear power supply so a good controller is required to improve these performance parameters. PID (proportional-integral-derivative controller) controller can be used to improve output voltage regulation and ripple in output. The Pulse width modulation (PWM) technique is used to generate the switching pulse [2]. The PID controller generates a signal that is given to pulse width modulator. Pulse width

S. Sharma (✉)
JIET-JODHPUR, Mogra Khurd, Rajasthan, India
e-mail: shubhamee81@gmail.com

K. L. Agarwal
Department of Electrical Engineering, JIET-JODHPUR, Mogra Khurd, Rajasthan, India
e-mail: kusum.agarwal@jietjodhpur.ac.in

M. Shorif Uddin et al. (eds.), *Intelligent Energy Management Technologies*, Algorithms for Intelligent Systems,
https://doi.org/10.1007/978-981-15-8820-4_32

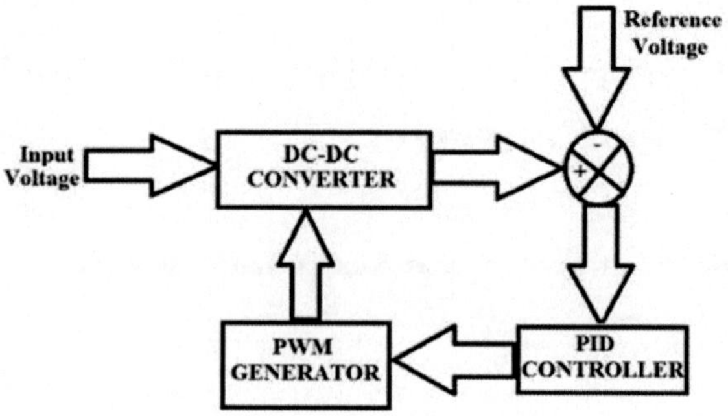

Fig. 1 Block diagram of DC–DC converter with PID controller

modulator adjusts the width of switching pulse such that the desired output voltage is achieved. The block diagram of DC–DC converter with PID controller is shown in Fig. 1.

The DC–DC converter controller can be designed for voltage mode control or current mode control. The DC–DC converter is a nonlinear converter so design of controller having lots of complexities. To design controller for nonlinear DC–DC converter linearized model of converter is required. Most important thing in design of a PID controller is the estimation of k_p, k_i, and k_d (Tuning of PID controller) for best performance of DC–DC converter [3]. In general, Trial and Error, Zeigler–Nichols and Internal mode control (IMC) methods are used for tuning of PID controller.

This method of tuning does not give the optimal operation of PID controller. For optimization of PID controller, (LQR) method is used for tuning of PID controller. LQR controller estimates k_p, k_i, and k_d to minimize cost function [2]. LQR controller adjusts weight matrices to get the desired performance parameters. The weight matrices can be determined using trial and error method, algebraic solution method and genetic algorithm etc. That means LQR tuning method tunes PID controller to minimize ripple in output and to improve output voltage regulation with minimum cost function. The LQR method also suppresses the disturbance occurring in the system with better performance parameters [2]. LQG controller is combination of LQR controller together with Linear quadratic Estimator (LQE) for good voltage regulation in output, for the rejection of small disturbances and for the elimination of noise. In this method output measurements are assumed to be disturbed by Gaussian noise.

The paper is organized as follows. In the second section, we have discussed briefly about the DC–DC buck converter and way to obtain the linearized state-space model of buck converter. In the third section, the PID tuning methodology using Zeigler-Nichols, LQR and LQG methods is discussed. Simulation results are discussed in the fourth section. Simulation is done in MATLAB Simulink environment.

2 DC–DC Buck Converter

DC–DC buck or step down regulator produces lower average output voltage as compared with input voltage. It is used in power supplies and speed control of motor. A DC–DC buck converter with resistive load is shown in Fig. 2.

DC–DC buck converter is low cost, small size and highly efficient converter but its output contains ripple and voltage regulation is poor as compared with linear converter. To improve ripple factor and voltage regulation, controller is required. There are many control schemes available. The operation of buck converter can be studied in two modes. Mode 1 means switch is in on position and mode 2 means switch is in off position. Dynamics of DC–DC converter can be understood by state-space representation of buck converter [3, 4].

Mode-1 Operation:

$$\dot{x} = A_1 x + B_1 u \tag{1}$$

$$y = C_1 x + D_1 u \tag{2}$$

where $A_1 = \begin{bmatrix} 0 & \frac{-1}{L} \\ \frac{1}{C} & \frac{-1}{RC} \end{bmatrix}$, $B_1 = \begin{bmatrix} \frac{1}{L} \\ 0 \end{bmatrix}$, $C_1 = \begin{bmatrix} 0 & 1 \end{bmatrix}$ and $D_1 = \begin{bmatrix} 0 \end{bmatrix}$

Mode-2 Operation:

$$\dot{x} = A_2 x + B_2 u \tag{3}$$

$$y = C_2 x + D_2 u \tag{4}$$

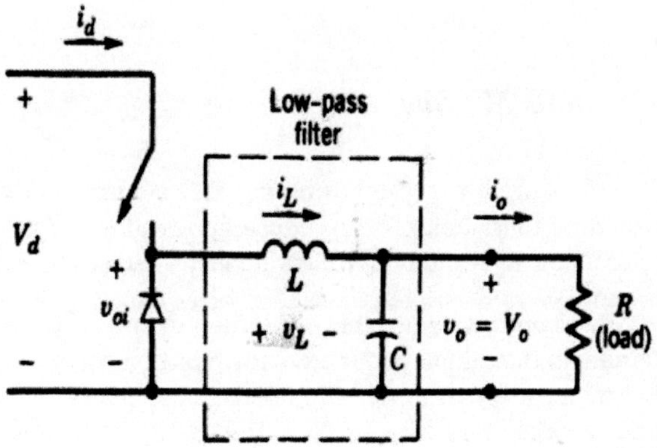

Fig. 2 DC–DC buck converter

where $A_2 = \begin{bmatrix} 0 & \frac{-1}{L} \\ \frac{1}{C} & \frac{-1}{RC} \end{bmatrix}$, $B_2 = \begin{bmatrix} 0 \\ 0 \end{bmatrix}$, $C_2 = [0\ 1]$, $D_2 = [0]$

From above, the average model of buck converter can be represented as

$$\dot{x} = Ax + Bu \tag{5}$$

$$y = Cx + Du \tag{6}$$

where $A = \begin{bmatrix} 0 & \frac{-1}{L} \\ \frac{1}{C} & \frac{-1}{RC} \end{bmatrix}$, $B = \begin{bmatrix} D/L \\ 0 \end{bmatrix}$, $C = [0\ 1]$, $D = [0]$

Using average model, transfer function of DC–DC buck converter can be written as:

$$G(S) = \frac{V_d}{LCS^2 + (L/R)S + 1} \tag{7}$$

3 PID Controller

PID controller performs proportional, integral and derivative action on error signal to get desired response. PID controllers are widely used in the industrial control applications. Error signal is basically a difference between the real value and desired value [5]. The most important task in the design of PID controller is to find proportional gain k_p, integral gain k_i and derivative gain k_d. The process of determining values of these gains is known is the tuning of PID controller [6]. The PID controller improves transient characteristics such as peak overshoot, rise time, settling time and reduces steady-state error [7]. The tuning methods used are explained below.

3.1 *Zeigler–Nichols Method*

Zeigler–Nichols is the simplest method of tuning PID controller. Z–N method is a heuristic tuning method that attempts to produce good values of k_p, k_i, and k_d. To tune PID controller using Z–N method, following steps need to be followed:

1. Make k_i and k_d zero until we get stable oscillation in the output. A gain at which stable oscillations in the output occur are known as control gain k_u.
2. After getting stable, oscillation determines the time period of one oscillation which is termed as T_u.
3. After measuring k_u and T_u, we can measure the PID controller gains k_p, k_i, and k_d are calculated by formula proposed by Zeigler–Nichols.

3.2 Linear Quadratic Regulator Method

The optimal control theory is the backbone of the modern control theory [8]. LQR is an optimal control method of tuning PID controller [3, 9]. LQR problem is defined as:

To Derive the State X of Linear Systems $\dot{X} = AX + BU$ to the Origin by Minimizing the Following Quadratic Performance Index

$$J = \frac{1}{2}\int_{0}^{\infty}(X^T QX + U^T RU)dt \tag{8}$$

Using closed-loop state feedback gain:

$$U(t) = -K.X(t) \tag{9}$$

where Q and R matrices are positive definite matrices. Matrices Q and R express relation between energy expense and error rate [10]. The LQR method gives the optimal value of weight matrices by solving algebraic Riccati equation.

$$A^T P + PA - PBR^{-1}B^T P + Q = 0 \tag{10}$$

The LQR weight matrices Q and R have been calculated from the following methods:

Trial and Error Method: In this method, we randomly select the values Q and R and determine k_p, k_i, and k_d. By doing some reasoning, right combination of matrices Q and R is obtained but this method is very time consuming [3, 4, 9].

Algebraic approach: The implementation of LQR controller requires minimization of cost function which is achieved by placing weights on control input and state variables. It is assumed that:

$$R = r \tag{11}$$

$$Q = \begin{bmatrix} q_{11} & 0 & 0 \\ 0 & q_{22} & 0 \\ 0 & 0 & q_{33} \end{bmatrix} \tag{12}$$

On solving Riccat equation, following values of parameter of q matrix were obtained

$$q_{11} = \frac{r}{B_{31}^2}[2\zeta\omega_n^3 - A_{31}^2] \tag{13}$$

$$q_{22} = \frac{r}{B_{31}^2} 2[A_{33} + 3\zeta\omega^n]A_{31} - [A_{32}^2 - (2\zeta^2 + 1)^2\omega_n^4] - [A_{31} + \zeta\omega_n^3]6\zeta\omega_n \quad (14)$$

$$q_{33} = \frac{r}{B_{31}^2}[-2(A_{32} + (2\zeta^2 + 1)\omega_n^2) - (A_{33}^2 - (3\zeta\omega)^2)] \quad (15)$$

So element of matrix Q can be obtained from above equations but for that we need to define damping ratio ζ and natural frequency ω_n. Steps followed for algebraic approach are:

1. Design mathematical model of system to be controlled.
2. Now define damping ratio ζ and natural frequency ω_n.
3. Obtain state gain matrix K in relation to p using ARE.
4. Now actual characteristic equation of system is determined.
5. Compare ARE with characteristics equation to obtain three elements of matrix p.
6. Fix the value of R and by using three elements of matrix p we need to determine values of elements of matrix Q.
7. Now calculate gain matrix K.
8. Obtain response in MATLAB simulink.

Genetic Algorithm Method: GA is a useful and power tool to solve the optimization problem with constrained and unconstrained condition. GA is basically based on natural selection process of biological evolution. GA plays an important role in optimal control problem. In this method, we find the values of gain matrices of LQR for different values of Q and R by optimizing cost function J. The genetic Algorithm problem is solved using following steps:

1. **Initialization**: The GA problem is solved by generating initial population. The population contains individuals. The individuals are basically set of chromosomes and each individual is having separate set of chromosome.
2. **Fitness Function**: The fitness function selection is very important procedure of genetic algorithm. A function which is needed to be optimally minimized is known as fitness function. In LQR problem, our cost function J is selected as fitness function.
3. **Selection**: Now fit chromosomes are selected from the population two create off-springs. This procedure is followed until we get chromosomes close to each other and less diversity is obtained. For selection, roulette wheel selection method is used.
4. **Crossover**: Crossover means reproduction. In this procedure, two parent chromosomes will crossover to create new child chromosome.
5. **Mutation**: As we know that in biological sense, child characteristics are not same as parents characteristics. There are some changes in genes at the time of growth of children which makes child different from parents. In GA, we change some bits of chromosomes to change characteristics of child this is known as mutation.

Algorithm for GA method for LQR problem is:

1. Start
2. The initial population of K is selected randomly.
3. Select the weight values on state variable and control input i.e. Q and R
4. Calculate the fitness function of each chromosome. If optimal J is obtained than go to step 7.
5. Now the reproduce next generation.
6. Change the bits of next generated chromosomes by applying crossover and mutation operations.
7. Determine best possible optimal value of gain matrix for given Q and R.
8. If result is not satisfying than go to step 3 otherwise go to next step.
9. End.

3.3 Linear Quadratic Gaussian

LQG controller is basically combination of LQR controller and Kalman filter (known as linear quadratic state estimator). LQG controller is applicable to linear system only and system is driven by the white Gaussian noise. So basically to design an LQG controller, we need to solve LQE problem than LQR problem. To design LQE, we assume that system is disturbed by noise. Our model is represented as:

$$\dot{x}(t) = Ax(t) + Bu(t) + \omega \tag{16}$$

$$y = Cx(t) + Du(t) + v \tag{17}$$

In this method, state X(t) is recovered by information known to us as output Y and control signal U. LQE tries to minimize the estimated error $e(t) = x(t) - \widehat{x}(t)$ where x(t) is unknown real state vector and $\widehat{x}(t)$ is a estimated vector. Our main aim is to minimize the cost function:

$$J_e = E[e_{est}^T(t).e_{est}(t)] \tag{18}$$

Now estimator will solve equation above and provide solution as

$$\widehat{x}(t) = (A - LC)\widehat{x}(t) + \begin{bmatrix} B & L \end{bmatrix}\begin{bmatrix} [u(t) \\ y(t)] \end{bmatrix} \tag{19}$$

where $L = PC^T V^{-1}$.

Table 1 Comparison of transient poerfomance

S.No	Controller	Rise time (ms)	Settling time (s)	Peak overshoot (%)	Steady-state error (V)
1	Z-N	160.969	0.018	26.52	0.02
2	LQR-TE	3.747	0.0025	26	0
3	LQR-AA	3.225	0.00030	22.72	0.08
4	LQR-GA	3.869	0.00034	26	0.09
5	LQG-TE	3.274	0.003	26	0
6	LQG-AA	26.39	0.024	25	−0.05
7	LQG-GA	3.281	0.00028	8	0

4 Results and Discussion

The tuning of PID controller was proposed in the previous sections. The conventional approach of designing a PID controller (Zeigler-Nichols) has been implemented but controller was not able to provide better transient performance and not able to compensate the change in input voltage and load. The LQR controller provided better transient performance and able to compensate the change in input voltage and load. The LQR weight matrices were obtained by trial error method, algebraic approach and genetic algorithm. To provide better performance with load disturbance, LQG controller is proposed. The LQG uses LQR controller with Kalman filter. Table 1 shows the comparison of transient parameters obtained by different PID tuning methods.

This comparison shows that LQG-GA method provides the best transient performance but if load disturbance of 160% was provided than LQR-GA not able to give good transient performance. The LQR-GA provides good transient performance till 300% load disturbance. The MATLAB simulink results for different methods are shown in figures below (Figs. 3, 4, 5, 6, 7, 8, 9 and 10).

5 Conclusion

In this paper, Zeigler–Nichols, LQR and LQG methods of tuning PID controller were introduced. The LQR and LQG methods provide optimal design of PID controller. These tuning methods have been successful to provide improved transient parameter like rise time, settling time, peak overshoot and steady-state error. The designed PID controller is capable of compensating the effect of change in input voltage and load disturbances.

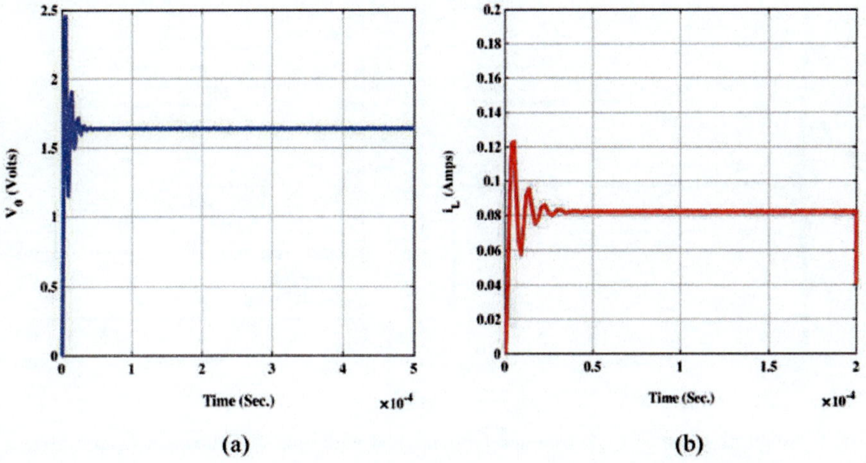

Fig. 3 Simulink result of P type Z-N controller **a** Output Voltage **b** Inductor Current

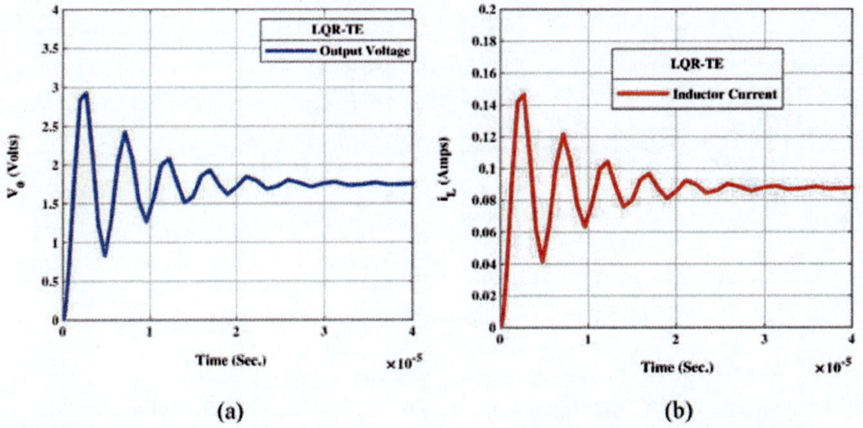

Fig. 4 Simulink result of LQR trial and error method **a** Output Voltage **b** Inductor Current

6 Future Scope

The optimal control methods LQR and LQG may also be used to improve transient performance of boost, buck-boost and Cuk converter. These methods will able to provide optimal solution of problem. The weight matrices of LQR state feedback system may also be obtained by using neural network and fuzzy controller. Future research can be done to design robust and discrete PID controller using optimal theory.

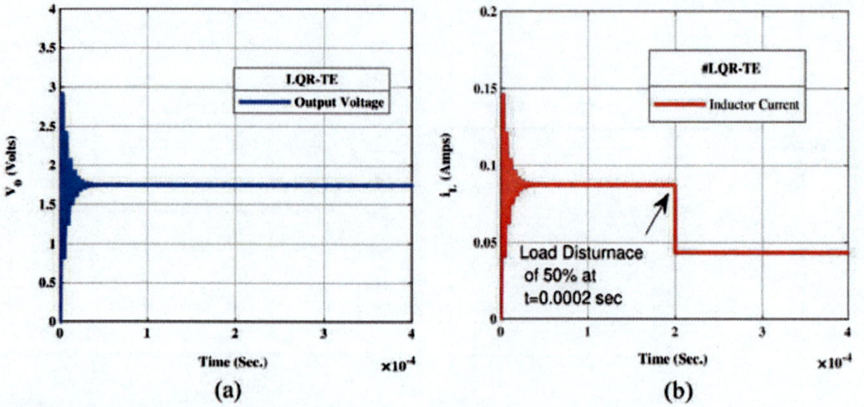

Fig. 5 Simulink result of LQR trial and error method with load disturbance **a** Output Voltage **b** Inductor Current

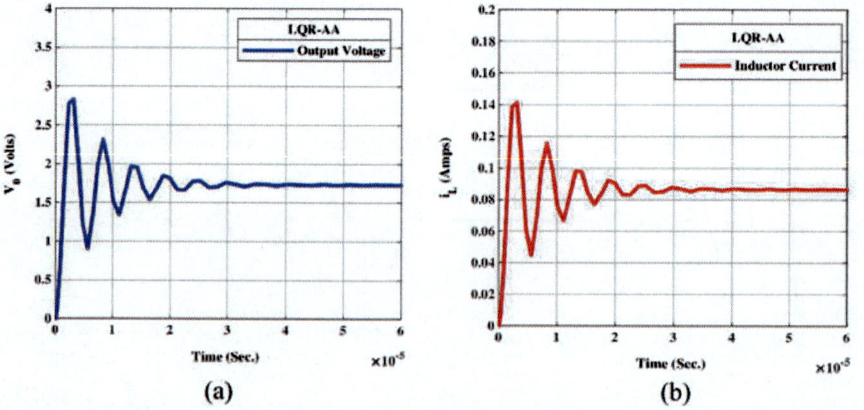

Fig. 6 Simulink result of LQR algebraic approach **a** Output Voltage **b** Inductor Current

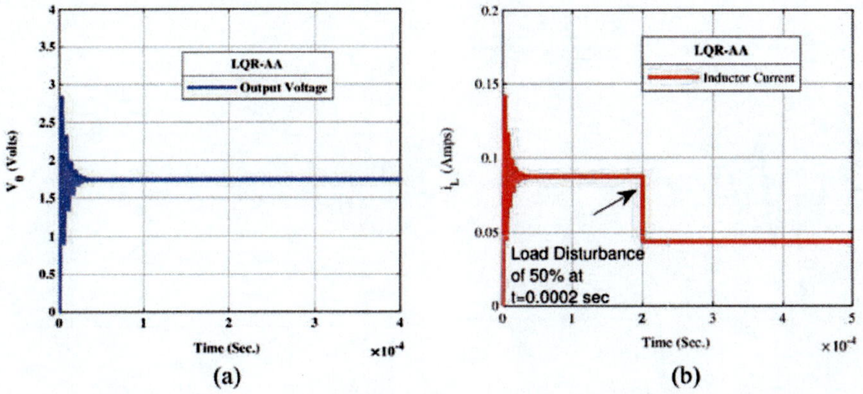

Fig. 7 Simulink result of LQR algebraic approach with load disturbance **a** Output Voltage **b** Inductor Current

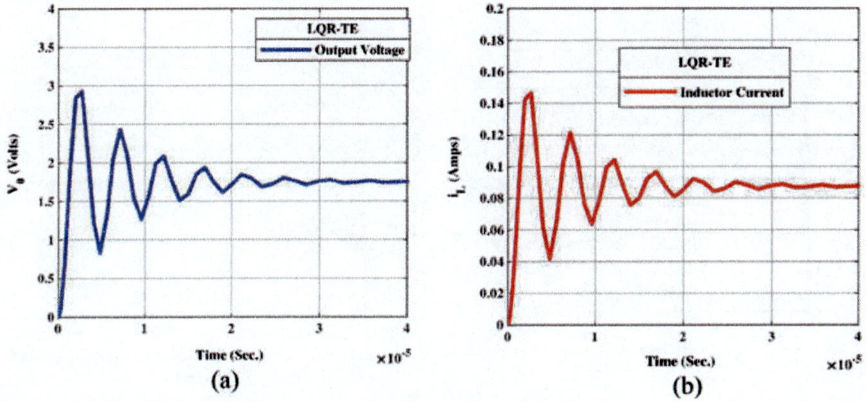

Fig. 8 Simulink result of LQR TE **a** Output Voltage **b** Inductor Current

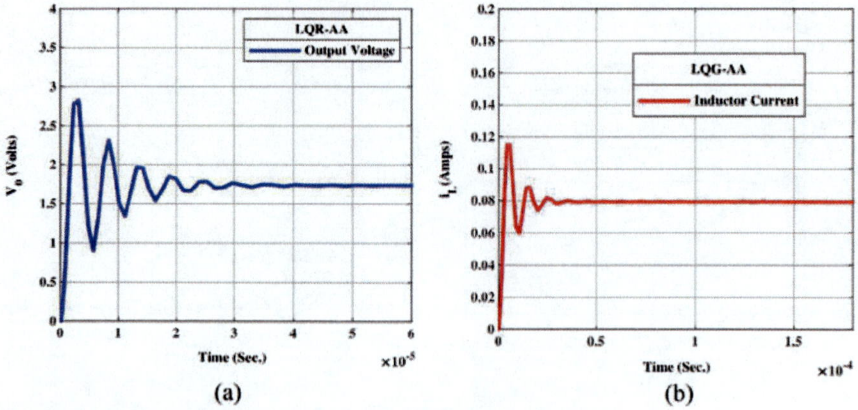

Fig. 9 Simulink result of LQG-AA **a** Output Voltage **b** Inductor Current

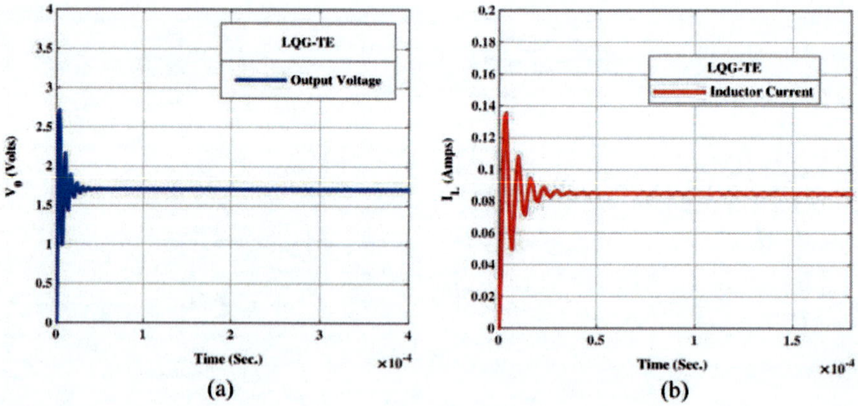

Fig. 10 Simulink result of LQG-GA **a** Output Voltage **b** Inductor Current

References

1. Mohan N, Undeland TM, Robbins WP (2003) Power electronics converters. John Wiley and Sons, Inc., Applications and Design
2. Giovanni Beccuti A, Papafotiou G, Morari M (2005) Optimal control of the boost dc-dc Converter. In: 14th IEEE conference on decision and control, and the European control conference. CDC-ECC
3. Srivastava S, Kumar Y, Misra A, Thakur SK, Pandit VS (2013) Optimum design ofbuck converter controller using LQR approach. In: 15th international conference onadvanced computing technologies, ICACT
4. Tan RH, Hoo LY (2015) DC-DC converter modeling and simulation using state spaceapproach. IEEE Conf Energy Convers CENCON 2015(2):42–47
5. Camacho-Solorio L, Sari˜nana-Toledo A (2014) I-LQG control of DC-DC boost converters. In: 11th International conference on electrical engineering, computing scienceand automatic

control. CCE

6. Li Y, Ang KH, Chong GCY (2006) Pid control system analysis and design. IEEE Control Syst Mag 26(1): 32–41
7. Li H, Ye X (2010) Sliding-mode PID control of DC-DC converter. In: Proceedings of the 2010 5thIEEE conference on industrial electronics and applications, ICIEA pp 730–734
8. Kumar C, Lal S, Patra N, Halder K, Reza M (2012) Optimal controller design for inverted pendulum system based on LQR method. IEEE Int Conf Adv Commun Control Comput Technol ICACCCTno. 978:259–326
9. Poodeh MB, Eshtehardiha S, Kiyoumarsi A, Ataei M (2007) Optimizing LQR and poleplacement to control buck converter by genetic algorithm. In: ICCAS internationalconference on control, automation and systems, pp 2195–2200
10. Tan GQ, Chen YH, Gu L (2014) LQR based optimal PID control for buck converter. Appl Mech Mater 687–691:3221–3226

Chapter 33
Comparative Analysis of Different Maximum Power Point Techniques for Solar Photovoltaic Systems

Shyam Lal Vishnoi and Kusumlata Agarwal

1 Introduction

The photovoltaic cell has nonlinear I–V characteristics and the direct extraction of energy is lower from the SPV module. To extract the maximum power from the SPV module, it is necessary to match the PV source to the load so that the operating point of PV module coincides with a maximum power point. This job will be done by DC–DC converter. To extract the maximum power from SPV module, there are so many different MPPT techniques from which output of the solar panel can be maximized. According to maximum power transfer theorem that power transferred will be maximum when the load resistance is equal to source resistance so in all techniques input resistances are so adjusted that it can be made equal to load resistance so that maximum power takes place. So to do the source resistance equal to load resistance duty cycle of the power semiconductor is adjusted by all these MPPT techniques. By adjusting the duty cycle power transferred can be maximized. So these MPPT techniques continuously track the peak power point and according to that, some action is being taken on duty cycle whether it has to increase or decrease. There are so many MPPT techniques available in literature out of which four different MPPT listed below will be discussed in this paper.

- Perturb and Observe (P&O) Method
- Incremental Conductance (INC) method
- Fractional open circuit voltage method (FOCV)
- Fractional short circuit current method (FSSC)

S. L. Vishnoi (✉) · K. Agarwal
Department of EE, Jodhpur Institute of Engineering and Technology, Jodhpur, Rajasthan, India
e-mail: shyamlal209e@gmail.com

K. Agarwal
e-mail: kusumagarwal@jietjodhpur.ac.in

M. Shorif Uddin et al. (eds.), *Intelligent Energy Management Technologies*, Algorithms for Intelligent Systems,
https://doi.org/10.1007/978-981-15-8820-4_33

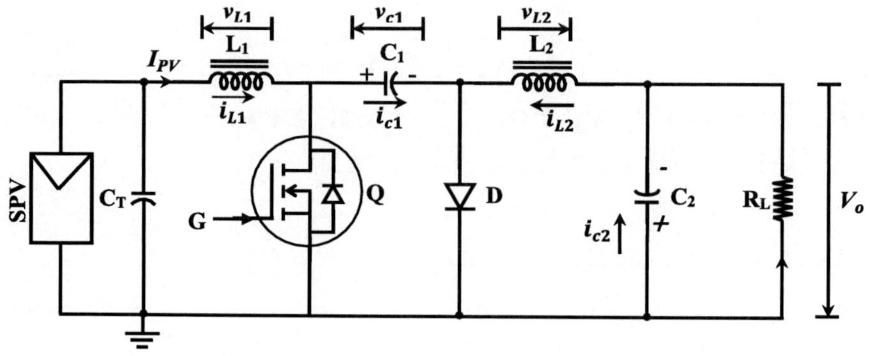

Fig. 1 Cuk converter interfacing a resistive load to an SPV panel

These methods are widely used although these methods are having inherent deficiencies despite the fact these methods are widely used. The researchers are consistently trying to improve the existing methods with the help of artificial neural network methods. This paper presents the comparison of different MPPT techniques with variable load in MATLAB SIMULINK environment and also hardware validation of two MPPT is also incorporated [1, 2].

2 DC–DC Cuk Converter

The proposed converter in this dissertation is Cuk converter. All the works have been carried out on this converter only. This converter has input side capacitance so that nonlinear output voltage will be nearly constant for DC–DC Cuk converter and a IRF840 MOSFET is used for switching operation and inductor is an integral part of this converter [3–5] (Fig. 1).

2.1 *Working of Cuk Converter*

See Fig. 2.

2.2 *Input–Output Relationship*

$V_o = V_{PV} \cdot \left(\frac{D}{1-D}\right)$ & $I_o = \frac{1-D}{D} I_{PV}$

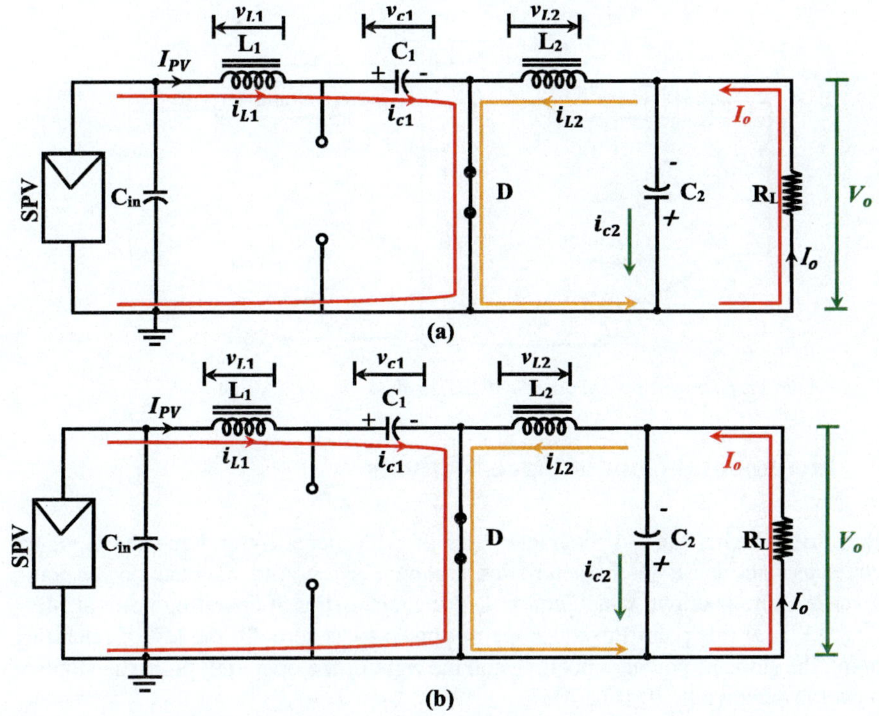

Fig. 2 Current roots during (a) DT_s time and (b) $(1-D)T_s$ time

2.3 Terminal Resistance

$$R_T = R_L\left(\frac{1-D}{D}\right)^2$$

3 Perturb and Observe (P&O) MPPT

The Perturb and Observe technique is very robust, very popular, very simple and efficient techniques generally used in most of the industries for maximum power point racking. So in this technique, voltage and current are sensed by voltage and current sensor respectively and then power is calculated based on the data available from sensor. So this power is compared with the previous stored data of power and according to this duty cycle is perturbed. Here the flow chart of P&O method is shown below for MPPT [6, 7] (Fig. 3).

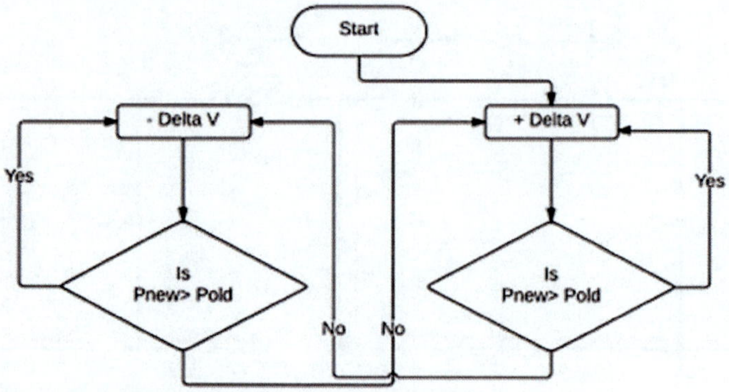

Fig. 3 Flow chart for perturb and observe (P&O) method

4 Incremental Conductance Methods

Here by referring the I–V characteristics of a PV panel by drawing the It v/s Vt characteristics and P–V characteristics as shown below. And at a peak power point, a vertical line is drawn which cut on I–V characteristics at operating point at MPP, i.e., at Pm, at this point Pm the slope of power $\frac{dP}{dV}$ is zero. To the left of operating point, the slope of power is positive and the right of the operating point the slope of power is negative [8, 9] (Fig. 4).

So in a nutshell it can be written as.

$(\frac{dI_t}{dV_t} + \frac{I_t}{V_t}) > 0$ Left of MPP.
$\frac{dI_t}{dV_t} + \frac{I_t}{V_t} = 0$ at MPP.
$\frac{dI_t}{dV_t} + \frac{I_t}{V_t} < 0$ Right of MPP.

5 Fractional Open Circuit Voltage MPPT Method

$$\frac{V_m}{V_{oc}} = k = 0.7$$

For V_{oc} to be at 100% then V_m will be approximately 70%. It is varying from 70 to 90% so once a particular PV panel is chosen then with the help of datasheet, the ratio of V_m to V_{oc} can be obtained which will be a constant [10, 11] (Figs. 5, 6).

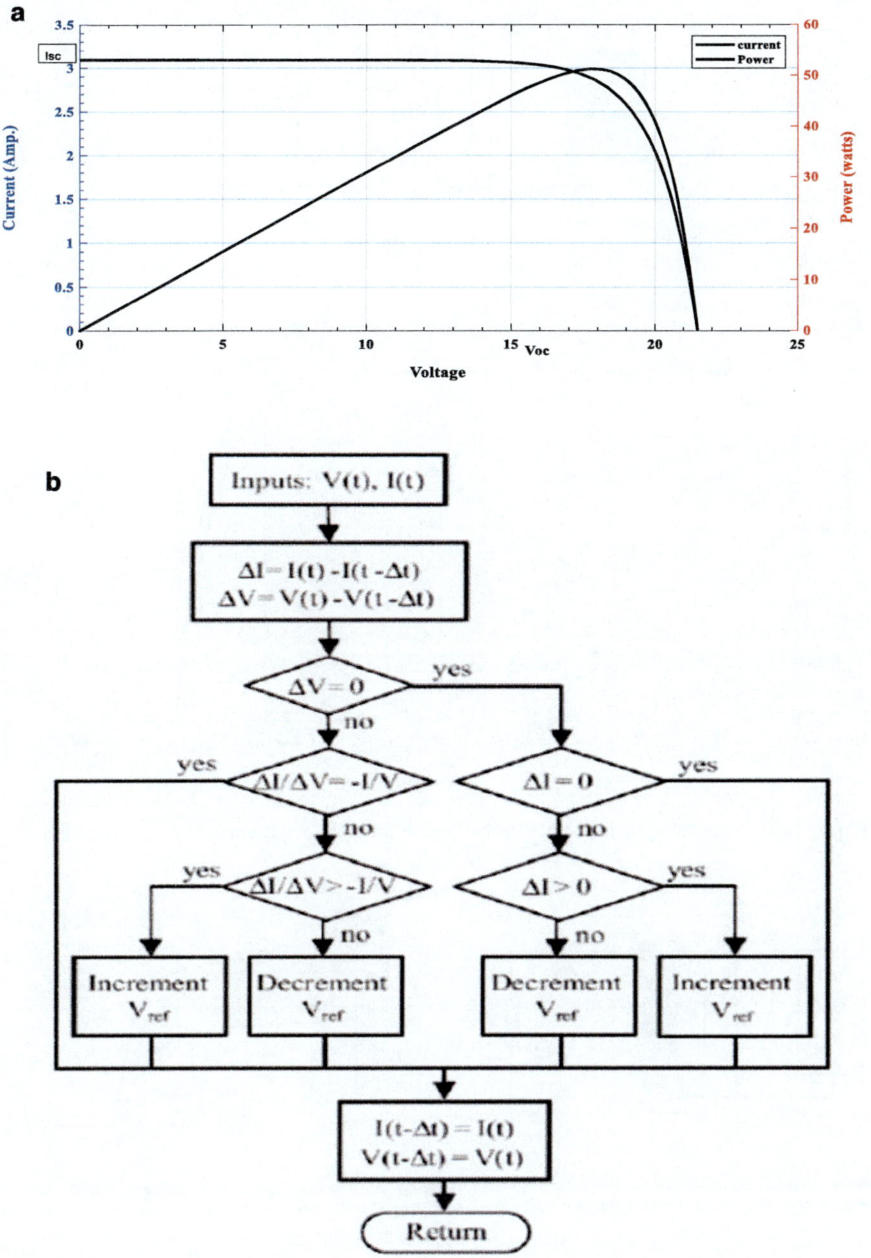

Fig. 4 **(a)** I–V & P–V characteristics of SPV Module, **(b)** Flow chart for incremental conductance method

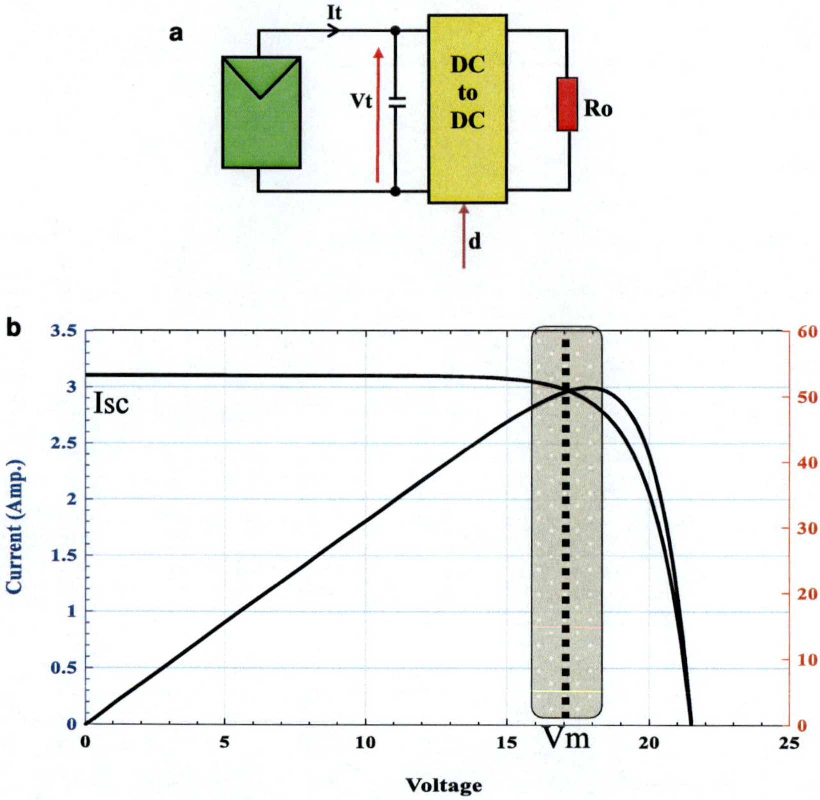

Fig. 5 **(a)** Typical PV system, **(b)** I–V & P–V curve for that PV system

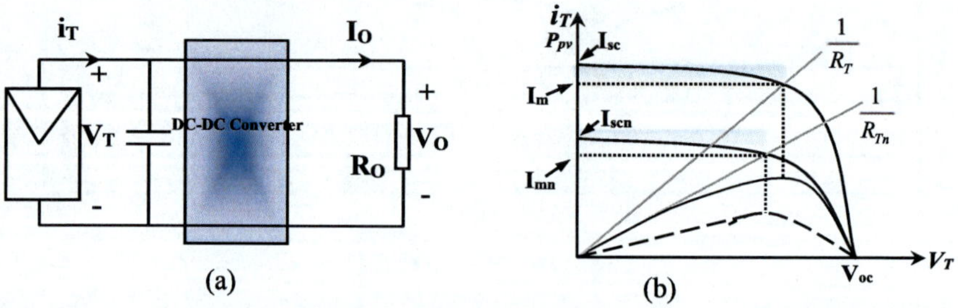

Fig. 6 **(a)** A typical PV interfacing system, **(b)** A typical I–V and P–V curve at different insolations

6 Fractional Short Circuit Current MPPT method-

The relationship between Isc and Im for a particular set or insolation so one can use the following assumption is much stronger than the voltage assumption that was made in the previous method so it can be written as

$$\frac{I_m}{I_{sc}} = \frac{I_m}{I_{sc}} = k = 0.9$$

7 Designed Parameters Value of Cuk Converter

8 PV System Design and Simulation in MATLAB/Simulink

See Fig. 7.

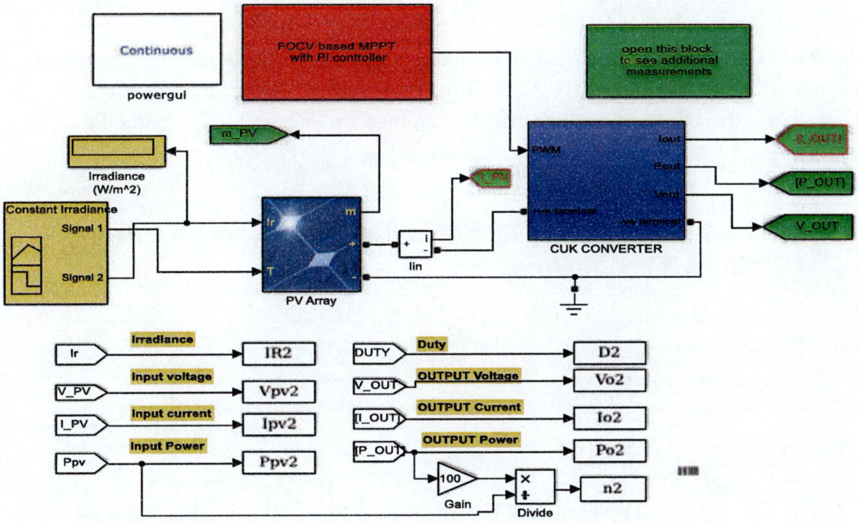

Fig. 7 Simulink model of PV system with MPPT algorithm

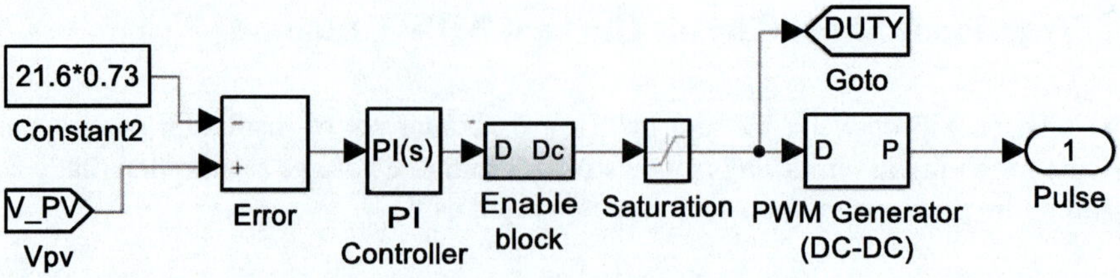

Fig. 8 Simulink block diagram of subsystem FOCV MPPT block

Table 1 Designed Parameters Value of Cuk Converter

Parameter	D	L_1	L_2	C_1	C_2
Theoretical Value	0.6663	8.35mH	16.68mH	177.59μF	2.50 μF
Practical value	*0.6663*	$L_1 = 8.53mH, r_1 = 0.62\Omega$	$L_2 = 18.766$mH, $r_2 = 0.285\Omega$	200 μF	47 μF

8.1 MATLAB Simulation of Fractional Open Circuit Voltage (FOCV) MPPT Technique

The MATLAB/Simulink of DC–DC Cuk converter with fractional open circuit voltage (FOCV) MPPT is shown in Fig. (8) below. The FOCV is applied after 0.1 s so that the effect of applied MPPT can be seen [12, 13] (Fig. 8 and Table 1).

8.2 MATLAB Simulation of Fractional Short Circuit Current (FSSC) MPPT Technique

See Fig. 9.

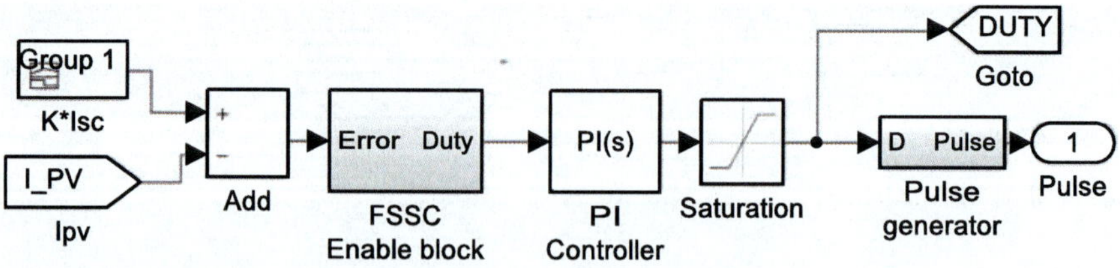

Fig. 9 Simulink block diagram of subsystem FSSC MPPT block

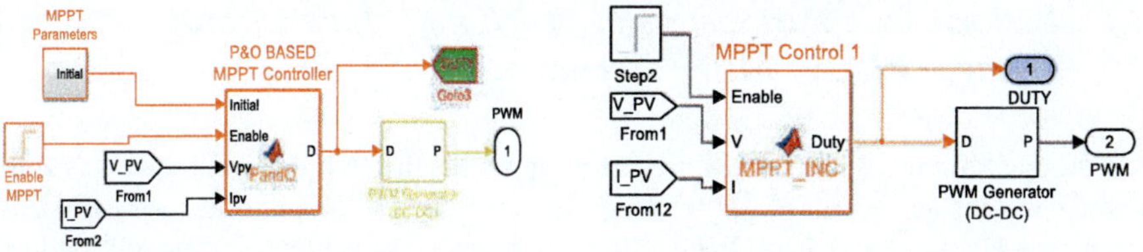

Fig. 10 Simulink block diagram of subsystem PO based MPPT controller

8.3 MATLAB Simulation of Perturb and Observe (P&O) MPPT Technique

See Fig. 10.

8.4 MATLAB Simulation of Incremental Conductance (INC) MPPT Technique

See Fig. 11.

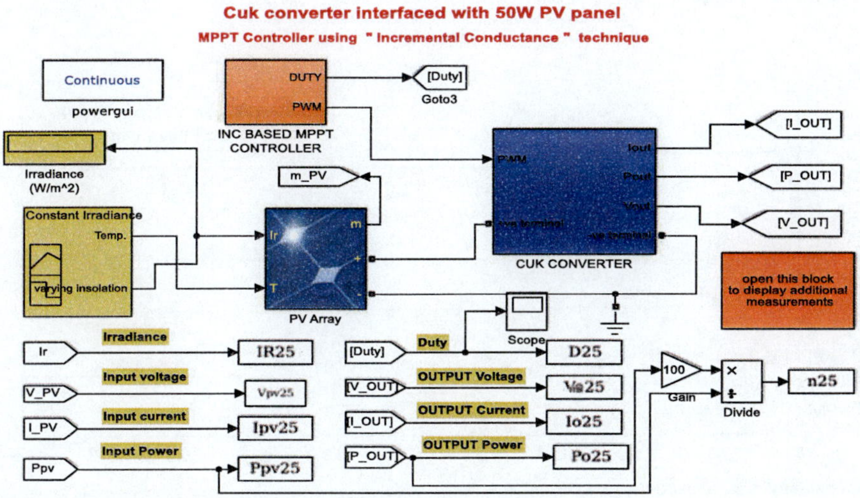

Fig. 11 Simulink block diagram of subsystem INC based MPPT controller

8.5 Performance Parameters of Fractional Open Circuit Voltage (FOCV) MPPT Technique

All the performance parameters are measured with the two different cases as one is for constant insolation of 1000 W/m^2 for variable load and another for different insolation levels like as 1000 W/m^2, 800 W/m^2 and 600 W/m^2 for constant load of 40Ω. FOCV is applied after 0.1 s so that the effect of MPPT can be seen [14, 15, 16] (Figs. 12 and 13).

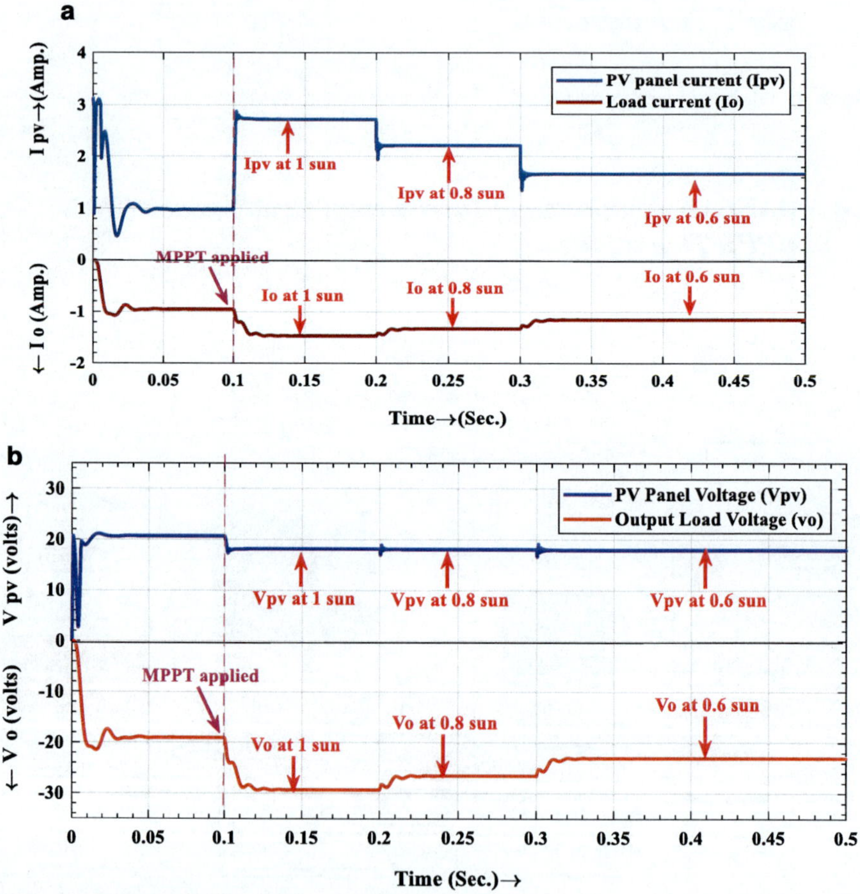

Fig. 12 **(a)** PV panel current (Ipv) and output load current (Io) with FOCV MPPT, **(b)** PV panel voltage (Vpv) output load voltage (Vo) with FOCV MPPT

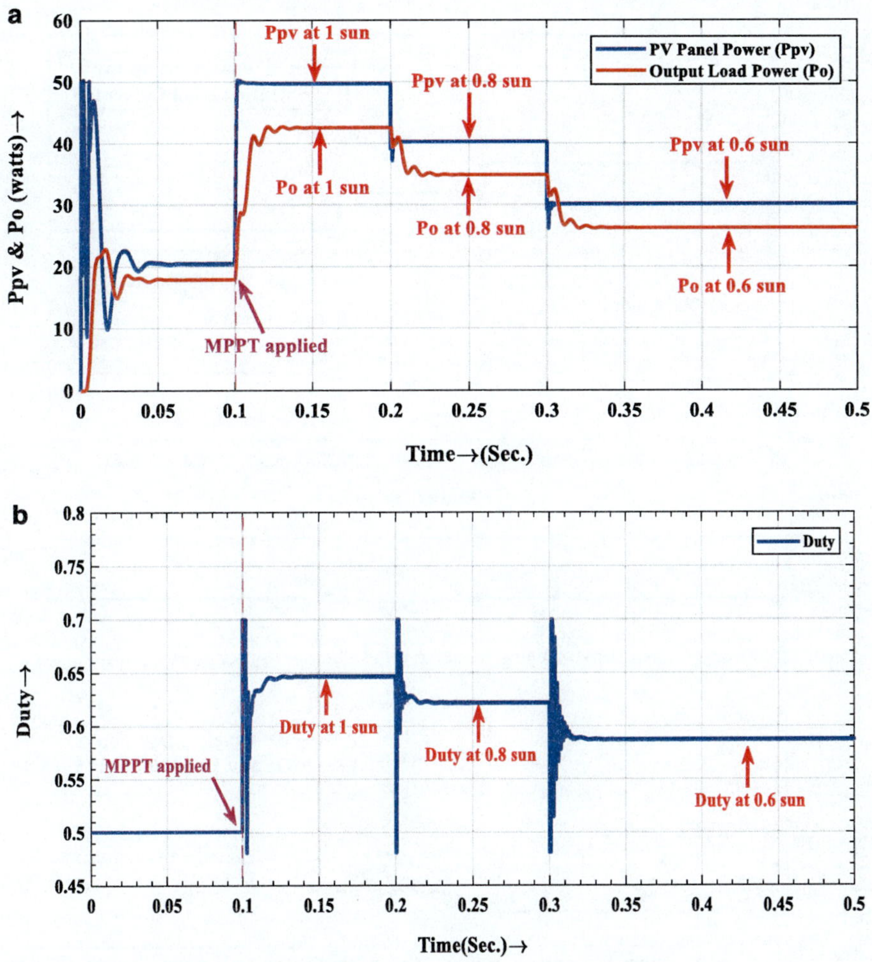

Fig. 13 **(a)** PV panel power (Ppv) output load power (Po) with FOCV MPPT, **(b)** Tracking of duty cycle waveform with FOCV MPPT

8.6 Performance Parameters of Fractional Short Circuit Current (FSSC) MPPT Technique

Figure 4.18 shows the PV panel current which is fairly constant at 2.82 at insolation of 1000w/m2 and at 2.3 A at 800w/m2 insolation. The PV panel voltage is fairly constant at 18 V which is very close to the peak power point voltage i.e. 17.7. The output voltage is 42 V at 1sun and 38 V at 0.8 sun and 33 V at 0.6 sun [17, 18] (Figs. 14 and 15).

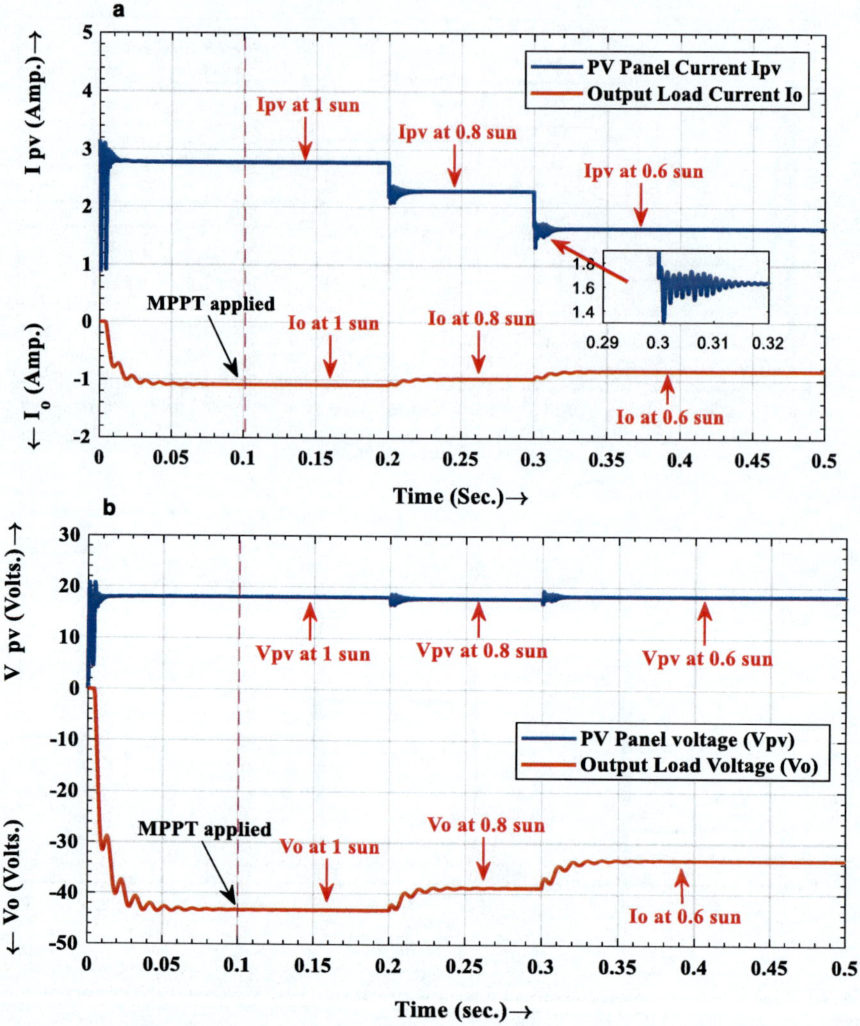

Fig. 14 **(a)** PV panel current (Ipv) output load current (Io) with FSSC MPPT, **(b)** PV panel voltage (Vpv) output load voltage (Vo) with FSSC MPPT

8.7 Performance Parameters of Perturb and Observe (P&O) MPPT Technique

See Figs. 16 and 17.

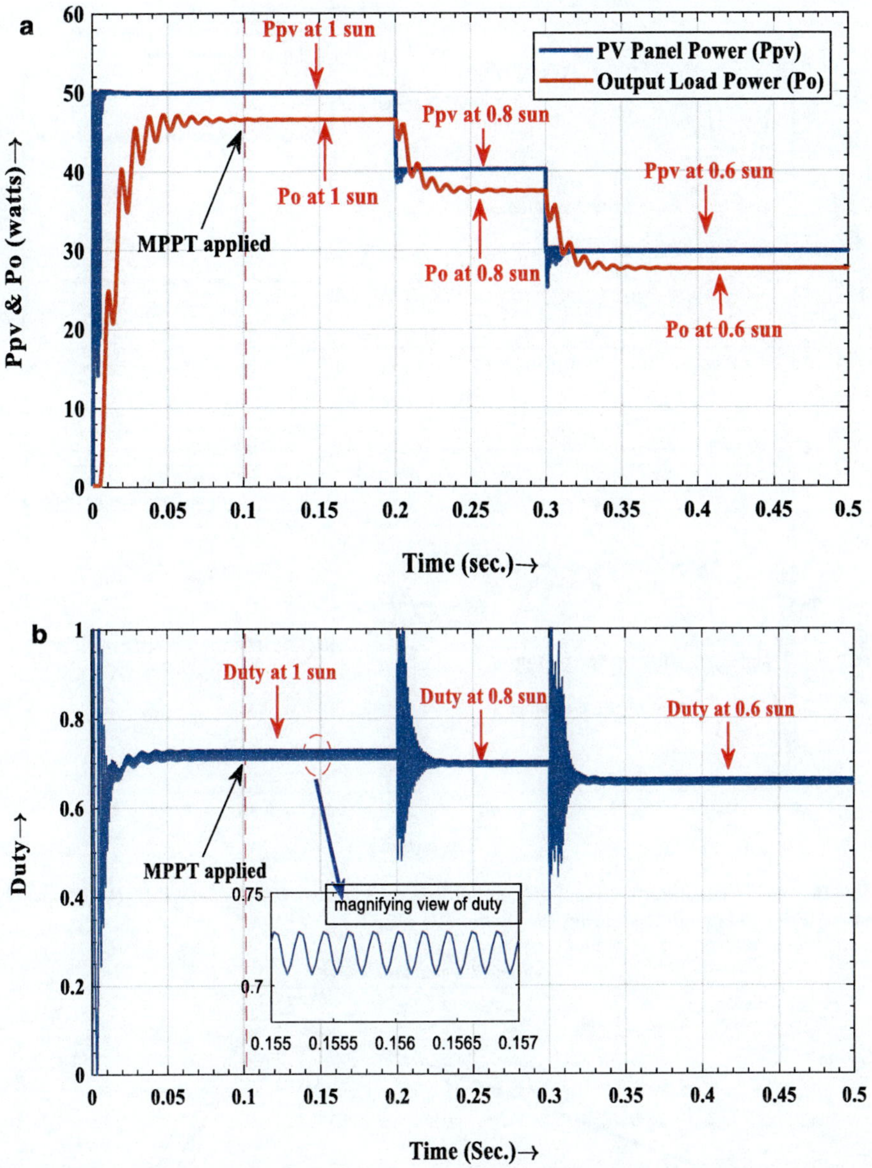

Fig. 15 **(a)** PV panel power (Ppv) output load power (Po) with FSSC MPPT, **(b)** Tracking of duty cycle waveform with FSSC MPPT

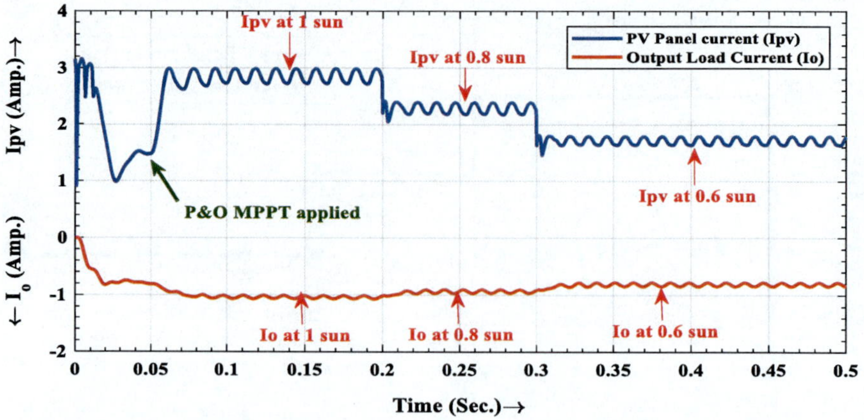

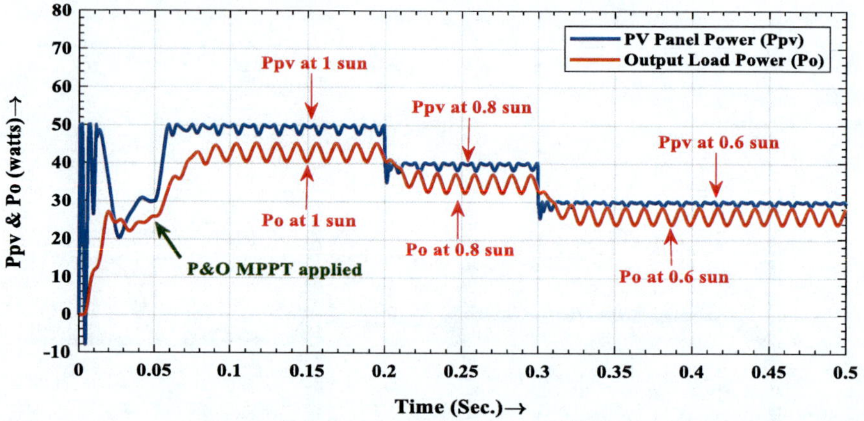

Fig. 16 **(a)** PV panel current (Ipv) and output load current (Io) with P&O MPPT, **(b)** PV panel power (Ppv) and output load power (Po) with P&O MPPT

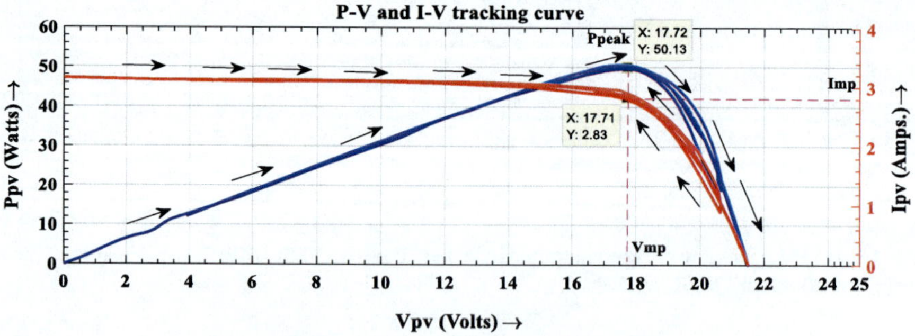

Fig. 17 P–V and I–V tracking curve with P&O MPPT applied

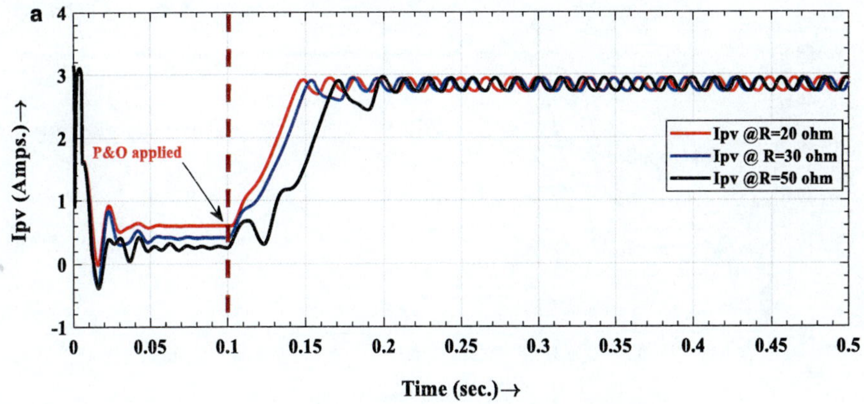

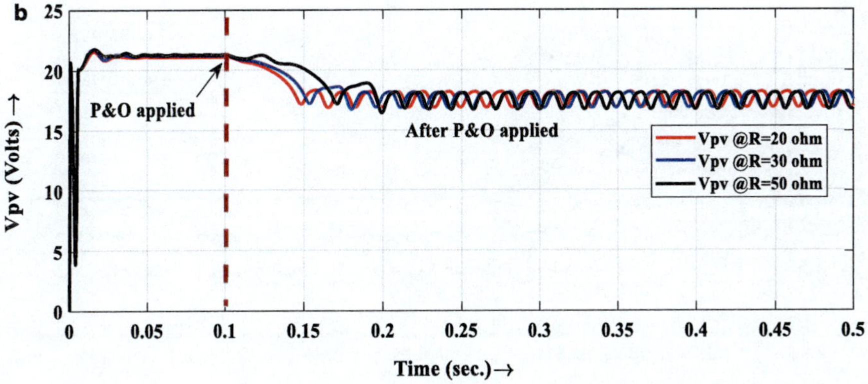

Fig. 18 **(a)** PV panel current (Ipv) of DC–DC Cuk converter with P&O MPPT for different loads, **(b)** PV panel voltage (Vpv) of DC–DC Cuk converter with P&O MPPT for different loads

8.8 Performance Parameters of P&O MPPT with Different Loads

See Figs. 18, 19 and 20.

8.9 Performance Parameters of INC MPPT with Different Loads

The figures below show the PV panel current and output current at different loads. Initially, MPPT has been disabled and then after some interval MPPT is being enabled so that effect of MPPT can be seen [19, 20] (Fig. 21).

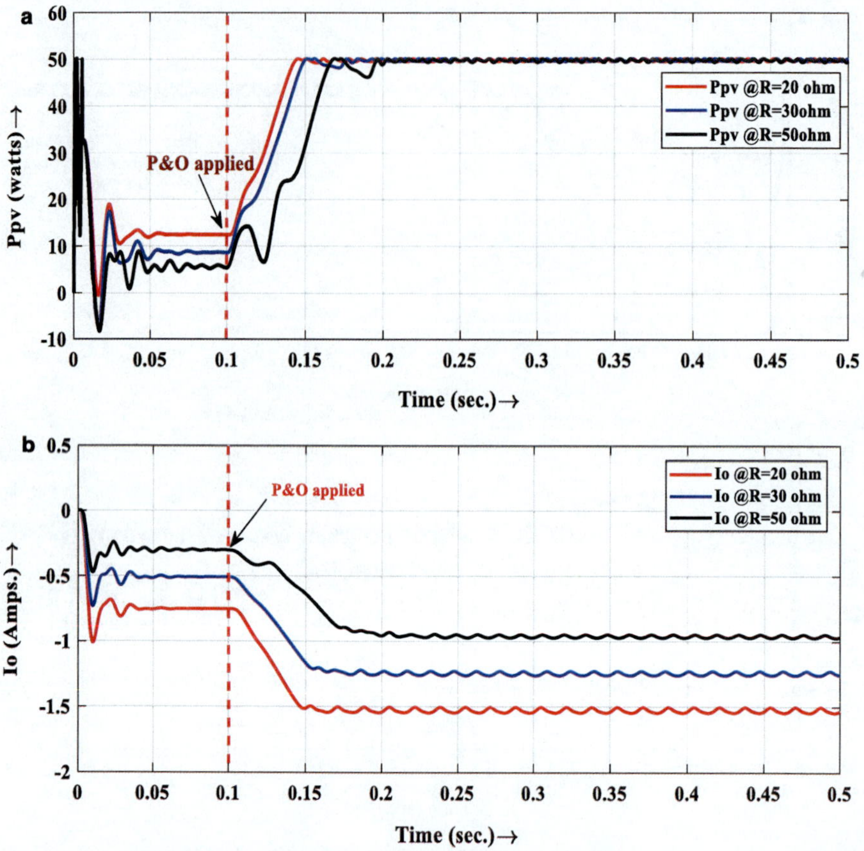

Fig. 19 **(a)** PV panel power (Ppv) of DC–DC Cuk converter with P&O MPPT for different loads, **(b)** Output current (Io) of DC–DC Cuk converter with P&O MPPT for different loads

9 Hardware Implementation of MPPT Techniques

There are so many MPPT techniques that can be implemented using hardware but in this thesis, two MPPT techniques are being implemented using hardware which is perturb and observe (P&O) MPPT and incremental conductance (INC) MPPT technique. For implementing these MPPT techniques, Arduino Uno Atmega320P [15] will be used as a microcontroller that will give the PWM signal to the DC–DC Cuk converter [21, 22, 23] (Fig. 22).

MOSFET driver circuit uses the IC TLP250 which is basically a MOSFET driver IC. Arduino is hardware cum software tool which is used as a controller just as microcontroller because it uses the microcontroller IC Atmega320P of ATMEL company and the programming is very easy in the Arduino. For sensing of data, voltage and current sensors are being used. One can use the ACS712-5A as a current sensors and

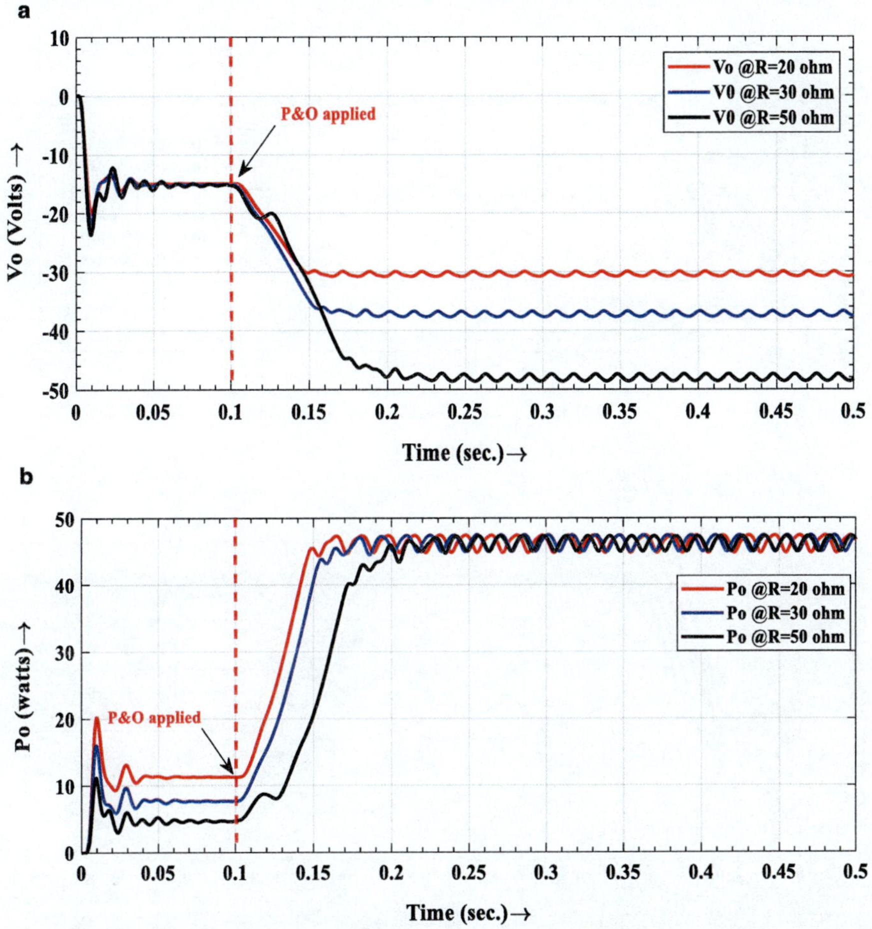

Fig. 20 **(a)** Output voltage (Vo) of DC–DC Cuk converter with P&O MPPT for different loads, **(b)** Output power (Po) of DC–DC Cuk converter with P&O MPPT for different loads

ACS712-25 V as voltage sensors [24, 25, 26, 27] (Figs. 23, 24 and 25) (Tables 2, 3, 4 and 5).

10 Comparative Analysis Experimental Results of Different MPPT Techniques

Table 6 shows the comparative analysis of the results obtained from the MPPT techniques applied to the DC–DC Cuk converter.

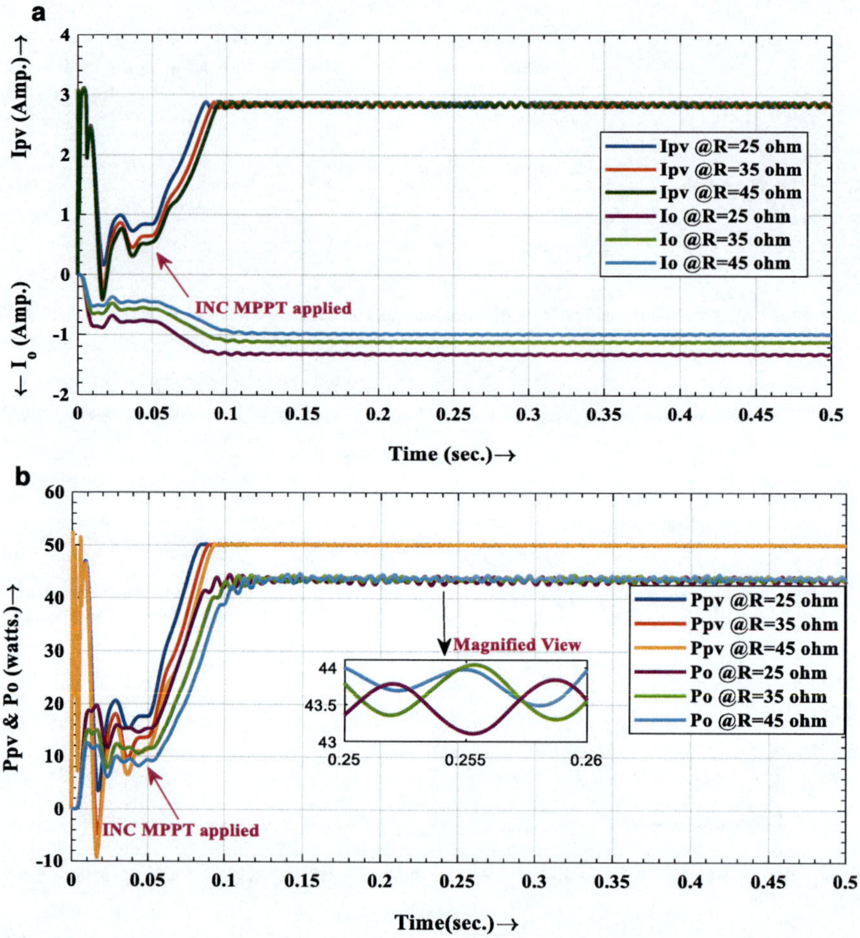

Fig. 21 **(a)** Input current (Ipv) and output current (Io) of DC–DC Cuk converter with INC MPPT for different loads, **(b)** Input power (Ppv) and output power (Po) of DC–DC Cuk converter with INC MPPT for different loads

The two MPPT techniques have been discussed and implemented on hardware and its results are obtained. Based on these results, a comparison has to be made between the results of these two MPPT techniques, i.e., P&O and INC MPPPT techniques applied to the DC–DC Cuk converter. Table 6 shows the comparative analysis of the results of P&O and INC MPPT applied to Cuk converter [26, 27, 28].

However, these output parameters are not constant, it is varying near these values. From the comparative analysis, it is seen that power output available in case of INC MPPT is more and smooth that can be seen from the graph of output power obtained [6, 6].

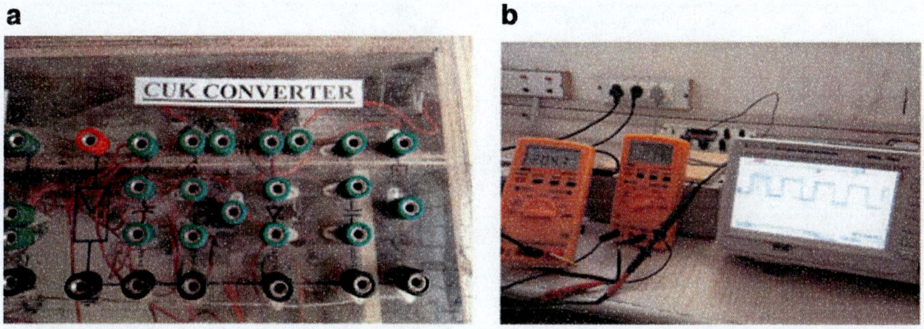

Fig. 22 **(a)** Duty cycle waveform generated by IC TLP250 on DSO, **(b)** Top view of hardware validation of DC–DC Cuk converter

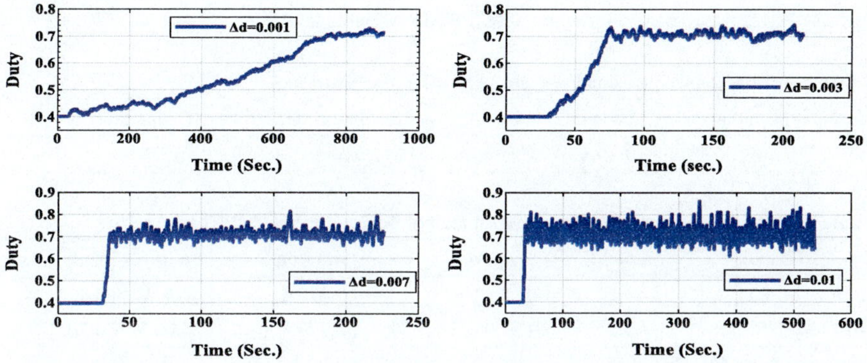

Fig. 23 The tracking of duty with the change in step size of duty with P&O MPPT

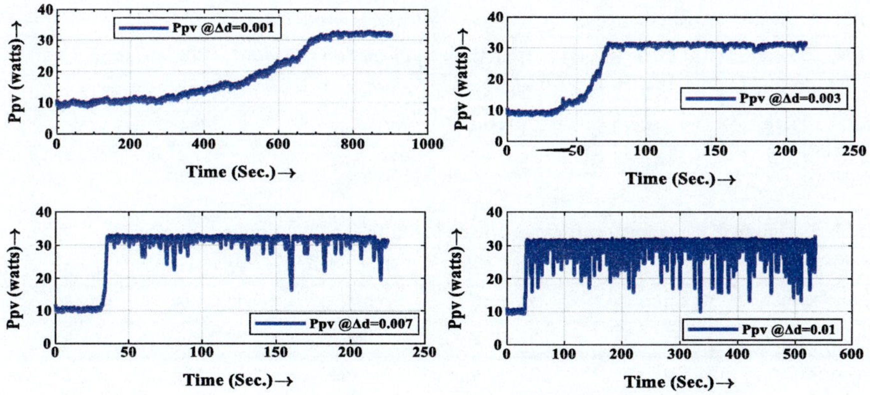

Fig. 24 The PV panel power with the change in step size of duty with P&O MPPT

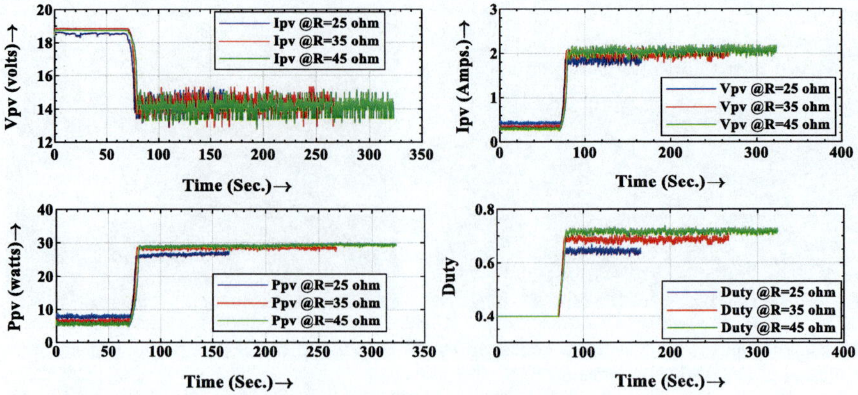

Fig. 25 Performance parameter with INC MPPT for variable load

Table 2 Available insolation and I_{SC} and VOC 40Ω load resistance

S. No	Load R (Ω)	Voc (Volts)	Isc (Amp)	Insolation (G) w/m^2	Temperature T (°C)
1	40Ω	19.30 V	2.645A	842 w/m^2	56 °C

Table 3 Output side parameters for constant load of 40Ω with P&O MPPT applied

MPPT Enable	Load R (Ω)	Vo (Volts)	Io (Amp)	Po (w)	Duty
Before MPPT	40Ω	12.82 V	0.325A	4.04 W	0.404
After MPPT	40Ω	31.08 V	0.773A	24.02 W	0.70
Variation After MPPT	40Ω	28.80 V–31.19 V	0.72A–0.78A	22 W–24 W	0.66–0.72

Table 4 Available insolation and I_{SC} and V_{OC} for variable load (INC)

S.No	Load R (Ω)	Voc (Volts)	Isc (Amp)	Insolation (G) w/m^2	Temperature T (ºC)
1	25Ω	19.68 V	2.033A	647 w/m^2	56 ºC
2	35Ω	19.70 V	2.197A	700 w/m^2	54 ºC
3	45Ω	19.64 V	2.277A	725 w/m^2	60 ºC

Table 5 Output side parameters for variable load with INC MPPT applied

MPPT	Load R (Ω)	Vo (Volts)	Io (Amp)	Po (w)	Duty
After MPPT	25Ω	23.90 V	0.958A	22.90 W	0.65
After MPPT	35Ω	28.43 V	0.807A	22.94watt	0.69
After MPPT	45Ω	32.18 V	0.716A	23.04 W	0.72

Table 6 Comparative analysis of experimental results of MPPT techniques

MPPT TECHNIQUE	MPP MODE	Insolation (W/m2)	Average Temperature (°C)	VOUT (Volt)	IOUT (Amp.)	POUT (Watts)	Average Duty (D)
P&O	Disable	850	56	12.82	0.325	4.04	0.404
	Enable	850	56	31.08	0.773	24.02	0.70
INC	Disable	850	56	12.45	0.309	3.84	0.404
	Enable	850	56	32.89	0.827	27.20	0.72

11 Conclusions

In this paper, four different MPPT techniques have been presented using MATLAB SIMULINK and two MPPT techniques have been experimentally realized. A comparative analysis of all four MPPT techniques has been made The simulation is done using MATLAB/SIMULINK to find out the performance of MPPT techniques under different weather conditions, i.e., steady-state and dynamic weather conditions. From the results obtained, it has been noticed that IC is found to be the best method for steady-state weather conditions as far as the efficiency and the power losses are concerned. However, the implementation of IC is complex because it requires two sensors for proper operation. FSSC and FOCV are the easiest to implement at software level but they require additional hardware arrangements for measuring the I_{sc} and V_{oc} PO is easy to implement in terms of both embedded software and hardware but is less efficient under both steady and dynamic weather conditions. The comparison presented in this paper will enrich the research community and will help in future research in this field [25,27–31].

References

1. Priyadarshini, Rai S (2014) Design, modelling and simulation of a PID controller for buck boost and cuk converter. Int J Sci Res (IJSR) 3(5)
2. Kushwaha BK, Narain A (2012) Controller design for cuk converter using model order reduction. In: International conference on power, control and embedded systems
3. Han BC, Kim M, Lee S, Lee JS (2015) Dynamic modeling and integral sliding mode controller design for cuk converter under load variation. In: 9th international conference on power electronics-ECCE Asia
4. Mokal BP, Vadirajacharya K (2017) Extensive modeling of DC-DC cuk converter operating in continuous conduction mode. In: International conference on circuits power and computing technologies [ICCPCT]
5. Rao P, Siraswar V, Pimple BB (2017) Efficient implementation of MPPT and comparison of converter for variable load Solar PV system. In: 2nd IEEE international conference on electrical, computer and communication technologies (ICECCT)
6. Safari A, Mekhilef S (2011) Simulation and hardware implementation of incremental conductance mppt with direct control method using cuk converter. IEEE Trans Industr Electron 58(4):1154–1161

7. Panigrahi BK, Thakura PR (2017) Implementation OF cuk converter with MPPT. In: 3rd international conference on advances in electrical, electronics,information, communication and bio-informatics (AEEICB17)
8. Samantara S, Roy B, Choudhury RSS, Jena B (2017) Modeling and simulation of integrated CUK converter for grid connected PV system with EPP MPPT hybridization. In: IEEE power, communication and information technology conference (PCITC) Siksha 'O' Anusandhan University, Bhubaneswar, India
9. Sahu TP, Dixit TV (2014) Modelling and analysis of perturb and observe and incremental conductance MPPT algorithm for pv array using cuk converter. In: IEEE student's conference on electrical, electronics and computer science
10. Kumar R, Choudhary A, Koundal G, Singh A, Yadav A (2017) Modelling/simulation of MPPT techniques for Photovoltaic systems using matlab. Int J Adv Res Comput Sci Softw Eng 7(4)
11. Hadeed Ahmed Sher, Ali Faisal Murtaza, Abdullah Noman, Khaled E. Addoweesh, Kamal Al-Haddad, "A New Sensorless Hybrid MPPT Algorithm Based on Fractional Short-Circuit Current Measurement and P&O MPPT", *IEEE transactions on sustainable energy, 2015*
12. Enslin JHR, Wolf MS, Snyman DB, Swiegers W (1997) Integrated photovoltaic maximum power point tracking converter. IEEE Trans Industr Electron 44(6):769–773
13. Elgendy MA, Atkinson DJ, Zahawi B (2016) Experimental investigation of the incremental conductance maximum power point tracking algorithm at high perturbation rates. IET Renew Power Gener 10(2):133–139
14. Tan RHG, Teow MYW (2014) A comprehensive modeling, simulation and computational implementation of buck converter using MATLAB/Simulink. In: IEEE conference on energy conversion (CENCON), pp 37–42
15. Makni W, Ben Hadj N, Samet H, Neji R (2016) Design simulation and realization of solar battery charge controller using Arduino Uno. In: 17th IEEE international conference on sciences and techniques of automatic control and computer engineering (STA), pp 635-639
16. Chy DK, Khaliluzzaman M, Karim R (2017) Analysing efficiency of DC-DC converters joined to PV system run by intelligent controller. In: IEEE international conference on electrical, computer and communication engineering (ECCE), pp 457–462
17. Halder T (2011) Charge controller of solar photo-voltaic panel fed (SPV) battery. In: India international conference on power electronics 2010 (IICPE2010)
18. Kollimalla SK, Mishra MK (2014) Variable perturbation size adaptive P&O MPPT algorithm for sudden changes in irradiance. IEEE Trans Sustain Energy 5(3):718–728
19. Singh B, Kumar R (2016) Solar PV array fed brushless DC motor driven water pump. In: IEEE 6th international conference on power systems (ICPS)
20. Popa D-L, Nicolae M-S, Nicolae P-M, Popescu M (2016) Design and simulation of a 10 MW photovoltaic power plant using MATLAB and Simulink. In: IEEE international power electronics and motion control conference (PEMC), pp 378-383
21. Sathya P, Natarajan R (2013) Design and implementation of 12V/24V closed loop boost converter for solar powered led lighting system. Int J Eng Technol (IJET) 5
22. Sahin ME, Okumus HI (2013) Small signal analyses and hardware implementation of a buck-boost converter for renewable energy applications. In: IEEE international conference on renewable energy research and applications, pp 330-335
23. Geethanjali MN, Sidram MH (2017) Performance evaluation and hardware implementation of MPPT based photovoltaic system using DC-DC converters. In: IEEE international conference on technological advancements in power and energy (TAP Energy)
24. Lin B-T, Lee Y-S (1997) Power-factor correction using Cuk converters in discontinuous-capacitor-voltage mode operation. IEEE Trans Industr Electron 44(5):648–653
25. Bandyopadhyay A, Parui S (2017) Dynamical behaviour of Cuk converter fed from a photovoltaic source. In: IEEE Calcutta Conference (CALCON), pp 397-402
26. Thiago A. Pereira , Thamires P. Horn , Walbermark M. dos Santos, Samir A. Mussa , Denizar C. Martins and Roberto F. Coelho, "Electrical characterizer of photovoltaic modules using the DC/DC Ćuk converter", IEEE International Conference on Industrial Technology (ICIT), pp. 954 - 959, February 2018

27. Singh SP, Gautam AK, Tripathi SP, Kumar B (2017) Performance comparison of MPPT techniques using Cuk converter for photovoltaic energy conversion system. In: IEEE 3rd international conference on computational intelligence and communication technology (CICT)
28. Sun Z, Yang Z (2017) Improved maximum power point tracking algorithm with Cuk converter for PV systems. J Eng 2017(13):1676–1681
29. Lina B-R, Huangb C-L, Tsaoa F-P (2009) Integrated Cuk-forward converter for photovoltaic-based LED lighting. Int J Electron 96(9):943–959
30. Ferrites and Accessories, [Online], Available: https://www.tdk-electronics.tdk.com/download/519704/069c210d0363d7b4682d9ff22c2ba503/ferrites-and-accessories-db-130501.pdf
31. Datasheet of TLP-250 Manufactured by TOSHIBA, [Online], Available: https://toshiba.semicon-storage.com/info/docget.jsp?did=16821

Chapter 34
MATLAB/Simulink-Based Tracking of Maximum Power Point in a Generalized Photovoltaic Module by Using DC-DC Boost Converter

Yogesh Joshi and Vinit Mehta

1 Introduction

Today, the whole world is facing an enormous drawback of electric power shortage, which is of great concern because of the conventional methods of generating electrical energy. In India, per capita electrical energy consumption is increasing fast during recent few years; it was around 347.5 kWh in 1992, which has reached to 1075 kWh in the year 2016 and around 1122 kWh in the year 2017 [1]. In general, from the year 2017, the contribution and shares of various power sectors to meet our energy demand and financial growth of India are; coal 59%, Hydro 14%, fossil fuels 17% including 8% of gas, and nuclear 2% [2]. The high dependency of coal, oil, and fossil fuels for energy generation results in greater energy insecurities in future. With the rapidly growing requirements of energy and as per the energy forecast, the electrical energy requirement for the year 2021–2022 will be around 1750 kWh and around 2300 kWh for the 2026–2027 year [3]. Therefore to meet out this demand, energy system needs to be more efficient and sustainable with minimum environmental and ecological impacts for harvesting more energy by using Renewable Energy Sources. Among all the Renewable Energy Sources, energy generation from photovoltaic effect is considered as more effective and sustainable resource, as it uses sun as a natural energy resource, which is eco-friendly, clean, inexhaustible, abundance in nature, free and energy supplied by it consistently. A major initiative has been taken up by the Government of India to promote and increase maximum utilization of solar energy by proposing "National Solar Mission" in the year 2008, which was retitled

Y. Joshi (✉)
Electrical Engineering Department, J.I.E.T, Jodhpur, Rajasthan, India
e-mail: yogi.jo89@gmail.com

V. Mehta
Electrical Engineering Department, J.N.V.U, Jodhpur, Rajasthan, India
e-mail: vinit741@gmail.com

M. Shorif Uddin et al. (eds.), *Intelligent Energy Management Technologies*, Algorithms for Intelligent Systems,
https://doi.org/10.1007/978-981-15-8820-4_34

Table 1 Phases and targets of (JNNSM)

Application	Phase-1 year (2010–2013)	Phase-2 year (2013–2017)	Phase-3 year (2017–2022)
Rooftop stand-alone solar plants	0.2 GW	1 GW	40 GW
Grid connected solar plants	1–2 GW	4–10 GW	60 GW
Total			100 GW

as "Jawaharlal Nehru National Solar Mission" (JNNSM) in the year 2010. The main aim of this operation is to deploy 20 Giga Watt (GW) of solar energy by the end of the year 2022. Three-phase approach is required for the execution of this operation. Different phases and goals of JNNSM are mentioned in Table 1 [4].

The energy obtained by the sun based upon the photovoltaic framework depends upon solar irradiance, cell temperature, and the voltage formed in the solar photovoltaic cells. Besides the fact that the solar photovoltaic system gives a new direction in the energy generation process, it also has some major drawbacks, which are.

1. Higher initial cost;
2. Low power conversion efficiency;
3. Electric power produced by sunlight-based photovoltaic framework is highly affected by the atmospheric conditions; variation in atmospheric conditions results in continuous change in electric power.

Figure 1 provides more clarity about the whole concept and working of how to separate most extreme power from the photovoltaic module using the MPPT system and DC-DC power converter.

As the climatic conditions for the whole duration of the day are not constant and certain parameters like solar irradiance and temperature are continuously varying all through the day, under such conditions it is difficult to find peak point where maximum power occurs. MPPT is an impedance matching and time-varying technique that interfaces changing source and possibly the fluctuating load. Among several MPPT algorithms, two more effective, fast and widely used MPPT methods

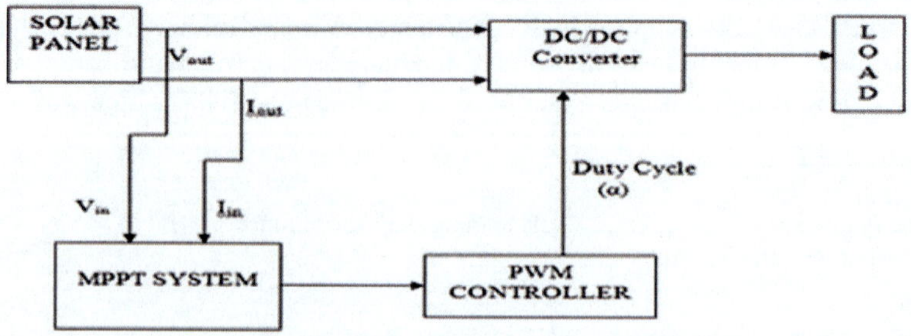

Fig. 1 Schematic diagram of maximum power tracking using DC-DC converter

that can work accurately and precisely in every single imaginable condition in order to get maximum power are.

1. Perturb and Observe Method (P&O Method),
2. Incremental Conductance Method (INC Method).

Perturb and Observe (P&O) and Incremental Conductance (INC) method are the two most optimizing and fast-tracking MPPT algorithms to track maximum power point by repeatedly updating the working voltage of the photovoltaic system. MPPT and PWM (Pulse-Width Modulation) controller helps in adjusting the duty cycle of the power DC-DC power converters with a fixed step size to transfer maximum power from the source to load.

Results obtained after simulation with conclusions are discussed in the later stage. Solar photovoltaic module, MPPT controller, and DC-DC power converter are designed and simulated using MATLAB/Simulink and simulated graphs such as I-V and P–V characteristics, graph of output power from DC-DC converter are obtained. Results of a simulated solar photovoltaic module are compared and validated with the actual results of the photovoltaic module.

2 State of the Art

Solar photovoltaic panels comprise of a number of solar photovoltaic modules which when connected with suitably designed power conditioning units such as Maximum Power Point Tracking (MPPT) controller and DC-DC converters produces maximum output power. Previous studies suggest that various investigations have been made on solar photovoltaic systems for different applications to gain as much benefits as possible from solar energy.

Some of the research works and papers presented on solar photovoltaics, simulation, and modeling of photovoltaic module in MATLAB/Simulink, interconnection of electrical load and photovoltaic module with a DC-DC power converter and implementation of Maximum Power Point Tracking (MPPT) strategies and methods are briefly conferred in this section.

Shyam. B and P. Kanakasabapathy had given an insight of various government policies, acts, plans, and strategies such as "Jawaharlal Lal Nehru Solar Mission" (JNNSM), JNNSM phases and targets initiated by Indian government agencies for country's economic growth and development. The approach toward development in the area of Smart Grid in India is additionally explained in their paper [1]. Marcelo Gradella Villalva, Jonas Rafael Gazoli, and Ernesto Ruppert Filho, had determined the three basic parameters of the photovoltaic model which are open-circuit voltage, short-circuit current, and maximum power point from the I-V characteristic curves. A photovoltaic array is designed and simulated in MATLAB/Simulink and using the simulated results mainly the output power of the photovoltaic array is matched and validated with the practically available photovoltaic array [5].

Huan-Liang Tsai, Ci-Siang Tu, and Yi-Jie Su had given an insight into how maximum power can be achieved through simulating and analyzing the proposed model by using power electronic devices [8]. Weidong Xiao, William. G. Dunford, and Antoine Capel had given a data-based approach for photovoltaic (PV) modeling. Nonlinear mathematical equations were solved for getting I-V characteristics of photovoltaic cells [9].

J. A. Cow and C. D. Manning had proposed a specific circuit-based simulated photovoltaic array model for the development of advanced solar photovoltaic (PV) conversion system [10]. S. Sheik Mohammed and D. Devaraj had presented a research study that provides a more comprehensive and detailed description on developing generalized photovoltaic module by means of simulation platform in MATLAB/Simulink and validating the simulated module with the monetarily accessible MSX 60 photovoltaic module [11].

Pandiarajan. N and Ranganath Muthu had portrayed about building up an exact power electronic circuit model necessary for designing and simulating photovoltaic integrated system [13]. Mei Shan Ngan, Chee Wei Tan have discussed two categories of MPPT algorithms, one is direct and another is indirect method [16].

Ting-Chung Yu and Yu-Cheng Lin had analyzed three most extensively used tracking methods specifically Perturb & Observe (P&O) algorithm, Incremental Conductance (INC) algorithm and Hill climbing algorithm. They described the complete development of photovoltaic system utilizing MPPT methods in MATLAB/Simulink by combining solar module and DC-DC converter with maximum power point tracking algorithms [17]. Trishan Esram and Patrick L. Chapman had given a comparative analysis of maximum power point tracking strategies on the photovoltaic array. They had discussed different methods for tracking maximum power point condition and their implementation [18].

In general, many simplifying assumptions have been made in the various proposed methods for maximum power point tracking. The aim of this paper is to develop a comprehensive computational methodology for building MATLAB/Simulink model to simulate with 'Perturb & Observe' and 'Incremental Conductance' algorithm incorporating DC-DC boost converter for getting maximum output power.

3 Proposed Methodology

The output power of the solar photovoltaic system is highly influenced by rapidly varying climatic conditions such as solar radiation and temperature. Solar photovoltaic system exhibits nonlinear I-V and P–V characteristics under different climatic conditions; operating point on P–V characteristics curve fluctuates continuously from its maximum position. The process of operating a photovoltaic array or module at or closer to maximum power point condition is known as Maximum Power Point Tracking (MPPT).

The method, which is to be applied for tracking maximum power point should be easy to implement, take lesser time to track, it should have adequate convergence speed, and price of implementation ought to be less. Various techniques have been proposed for tracking the Maximum Power Point (MPP), among these techniques, Perturb and Observe (P&O) and Incremental Conductance (INC) methods are most widely utilized techniques and are applied for tracking maximum power point.

(a) **Perturb and Observe (P&O) Method**:

P&O method is the most broadly utilized methodology for tracking maximum power due to its easy approach and very few parameters are required for the implementation of this method. The power generated by the photovoltaic module or array is perturbed, observed, and analyzed through iteration methods to locate maximum power point condition. Solar photovoltaic module or array voltage is perturbed periodically with increment and decrement in terminal voltage and comparing the output power received from photovoltaic with the previous perturbed cycle. The basis of this control algorithm is the slope (dP/dV) on the P–V characteristic curve as depicted in Fig. 2 [17].

Condition I: When the working voltage of photovoltaic module or array is perturbed in a particular direction and dP/dV > 0 (i.e., left aspect of the curve), then slope dP/dV is positive and direction of perturbation is toward Maximum Power Point (MPP) and P&O algorithm continues to perturb photovoltaic module or array voltage in the same direction.

Condition II: When dP/dV < 0, (i.e., right aspect of the curve), then slope dP/dV is negative and operating point moves far away from Maximum Power Point (MPP) and during this condition P&O algorithm reverses the direction of perturbation.

Condition III: At dP/dV = 0, condition of maximum power point is reached. Though maximum power can be obtained at dP/dV = 0, still some power is dissipated during tracking as the operating point does not remain stable and it oscillates around maximum power point. The basic operating procedure of P&O technique is given in Fig. 2.

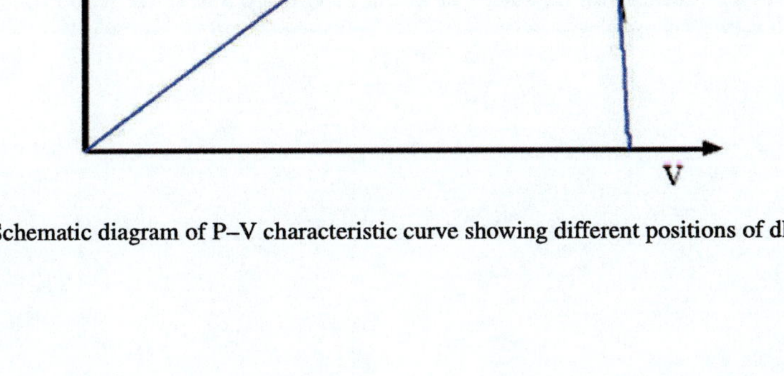

Fig. 2 Schematic diagram of P–V characteristic curve showing different positions of dP/dV

Advantages of P&O method:

1. Simple in structure and easy to implement.
2. Low computational demand.
3. Implementation cost is less.
4. Only one sensor, i.e., voltage sensor is required to sense the photovoltaic module or array voltage.

Disadvantages of P&O method:

1. Rapidly varying climatic conditions has a negative effect on the P&O method as algorithm starts tracking Maximum Power Point (MPP) in the wrong direction. The reason behind this is when atmospheric conditions changes, then simultaneously solar irradiance also changes rapidly due to which Maximum Power Point (MPP) deviates slightly from its position in either direction; P&O algorithm concludes this change as a change due to perturbation and accordingly in the successive iterations it starts perturbing in the wrong direction.
2. P&O method cannot find out exactly that when it has reached maximum power point (MPP) and operating point continuously oscillates around Maximum Power Point (MPP), which leads to some energy loss in the process.
3. Slow response speed at the event of changes in solar irradiance.

To overcome the limitations of the P&O algorithm, a new technique for tracking MPP has been established known as "Incremental Conductance methods"..

(b) **Incremental Conductance (INC) Method**:

The main concept of Incremental Conductance Method depends upon the way that slope of power curve at Maximum Power Point (MPP) is zero in the photovoltaic module; it is positive on left of Maximum Power Point (MPP) and negative on the right of the MPP, this can be explained as [18]

$$\begin{cases} \frac{dP}{dV} = 0, \text{ at MPP} \\ \frac{dP}{dV} > 0, \text{ left of MPP} \\ \frac{dP}{dV} < 0, \text{ right of MPP} \end{cases} \tag{1}$$

The above equation can be stated in terms of voltage and current as [16–18].

$$\begin{aligned} \frac{dP}{dV} &= \frac{d(IV)}{dV} \\ &= I + V\frac{dI}{dV} \end{aligned} \tag{2}$$

When the operating point is at MPP then dP/dV = 0; Eq. (2) can be rearranged as

$$I + V\frac{dI}{dV} = 0$$

Therefore,
$I + V\frac{dI}{dV} = 0$

$$\frac{dI}{dV} = -\frac{I}{V} \quad (3)$$

From the above Eq. (3) [18], it can be concluded that left side of the equation shows photovoltaic module's incremental conductance (dI/dV) and right side of the equation shows photovoltaic module's instantaneous conductance (I/V). Therefore, during the tracking process, operating point continue to perturb until incremental conductance value (dI/dV) equals the instantaneous conductance value or (I/V). The other two conditions when the slope of photovoltaic module power curve is on the left of the MPP (dP/dV > 0) and when it is on the right of the MPP (dP/dV < 0) can be expressed in terms of voltage and current by rearranging Eq. (2) as

$$\frac{dI}{dV} > -\frac{I}{V} \quad (4)$$

$$\frac{dI}{dV} < -\frac{I}{V} \quad (5)$$

Here, dI and dV represent change in current and voltage before and after the increment respectively, whereas I and V represent instantaneous values of voltage and current, respectively. The above two Eqs. (4) and (5) helps in determining the direction of perturbation so that operating point moves towards the direction of maximum power point. All these equations can be summarized as [18].

$$\begin{cases} \frac{dI}{dV} = -\frac{I}{V}; \text{ when } \frac{dP}{dV} = 0; \text{ at MPP} \\ \frac{dI}{dV} > -\frac{I}{V}; \text{ when } \frac{dP}{dV} > 0; \text{ at left of MPP} \\ \frac{dI}{dV} < -\frac{I}{V}; \text{ when } \frac{dP}{dV} < 0; \text{ at right of MPP} \end{cases} \quad (6)$$

Figure 3 shows how the operating point in the Incremental Conductance Method perturbs to reach Maximum Power Point (MPP). The pattern and approach of perturbation in the incremental conductance method is the same as that of perturb and observe method. From Fig. 3, it can be observed that when the operating point is in the left side of MPP, the direction of perturbation is positive whereas the direction of perturbation has to be changed (or reversed) when the operating point is at the right of MPP.

Condition I: When (dI = 0 and dV = 0) that means there is no change in voltage and current; it concludes that operating point is on the maximum position and MPPT

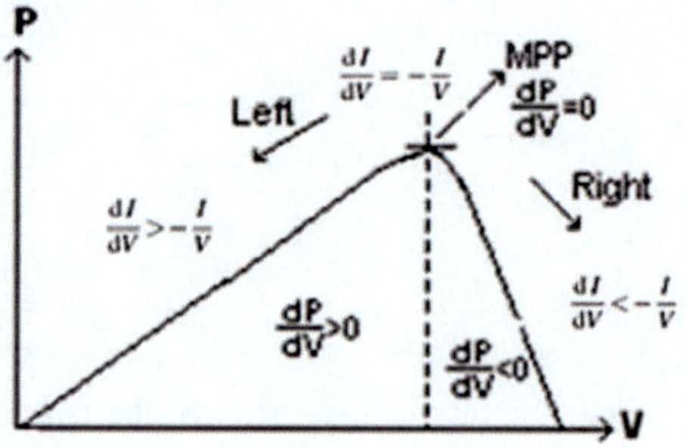

Fig. 3 Basic concept of incremental conductance method on P–V curve

still operates on maximum power point. Atmospheric conditions under this condition remains unchanged.

Condition II: When only change in current is accounted and there is no change in voltage; (dV = 0 and dI changes). Change in dI indicates a change in atmospheric conditions results in a change in solar irradiance, which leads to a change in maximum power point. When dI > 0 then there is a rise in the amount of solar irradiance which results in voltage of maximum power point.

Condition III: When current and voltage both changes and both do not equal to zero (dI/dV $\neq$ 0). In case, if both voltage and current are changing then Eqs. (4) and (5) helps in determining the direction of perturbation to reach maximum power point. If dI/dV > -I/V, then dP/dV > 0, this indicates that operating point on P–V curve is toward the left of the MPP.

Figure 4 explains the whole process, about how MPPT in combination with DC-DC converters is used to extract maximum power from the solar photovoltaic module or array.

Figure 4 shows that voltage V_{pv} and current I_{pv} obtained from solar panel are input to control algorithm. Control algorithm may be P&O method or Incremental Conductance method, implemented to track maximum power point condition by perturbing duty cycle. PWM (Pulse-Width Modulation) controller helps in adjusting and monitoring duty cycle (α) used for triggering purpose of power electronic switch

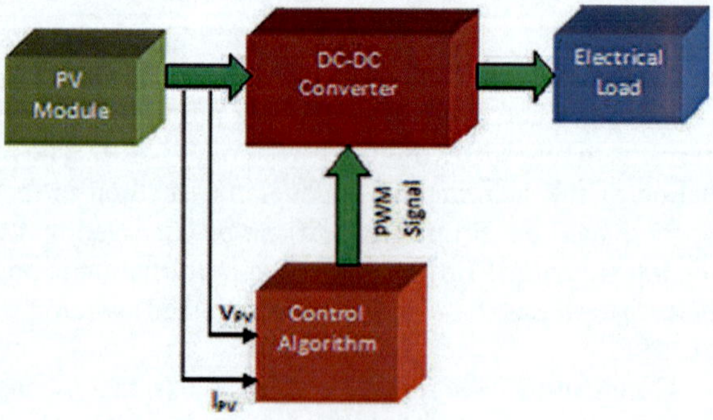

Fig. 4 Interconnection of solar photovoltaic module and load through DC-DC converter using control mechanism

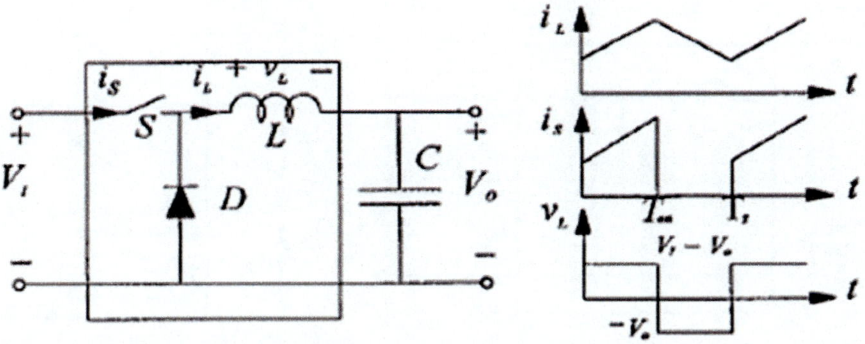

Fig. 5 Schematic diagram of Buck converter and waveforms

(i.e., MOSFET) of DC-DC Converter. Duty cycle (α) of the converter is controlled in such a way that the source will send maximum power to the load.

(c) **Types of DC-DC Converters**

1. Buck Converter,
2. Boost Converter, and
3. Buck–Boost Converter.

c1. **Buck Converter**

Buck Converter is generally used when photovoltaic module or array voltage (input voltage) is high and battery voltage (output voltage) is low. It is also termed as step-down converter (Fig. 5).

In Buck Converter, switch S operates at a higher frequency to produce a chopped output voltage. By adjusting on/off duty cycle of the switching, power flow is controlled. The average output voltage is given by [19]

$$\frac{V_o}{V_i} = \frac{t_{on}}{T} = D \tag{7}$$

c2. **Boost Converter**

Boost Converter is generally used when a photovoltaic module or array voltage (input voltage) is low and battery voltage (output voltage) is high. It is also known as a step-up converter (Fig. 6).

The average output voltage for boost converter is given as [20]

$$\frac{V_o}{V_i} = \frac{T}{T_{off}} = \frac{1}{1 - D} \tag{8}$$

c3. **Buck–Boost Converter**

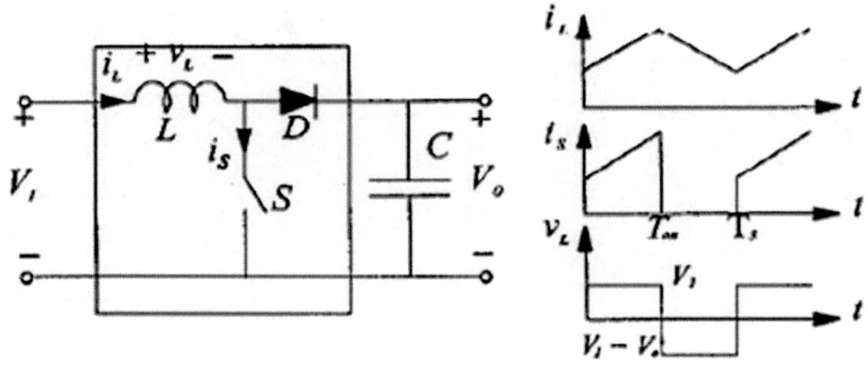

Fig. 6 Schematic diagram of boost converter and waveforms

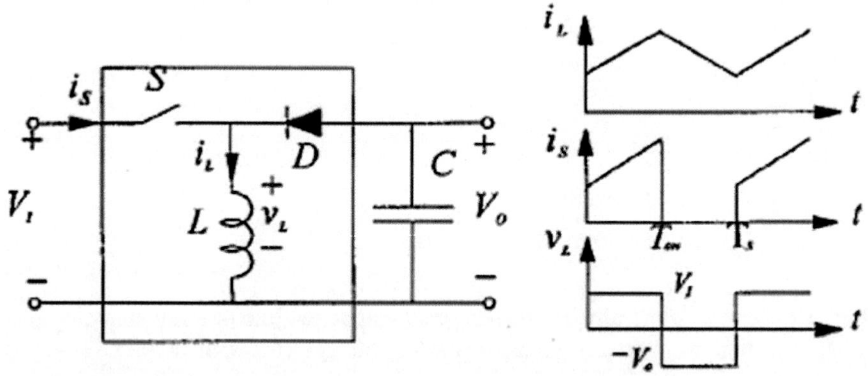

Fig. 7 Schematic diagram of Buck–Boost converter and waveforms

Buck–Boost Converter generally used when battery voltage (output voltage) is either high or low than the input voltage. It can be used as a step-up or step-down converter (Fig. 7).

The average output voltage for Buck–Boost converter is given by [21]

$$\frac{V_o}{V_i} = \frac{t_{on}}{t_{off}} = \frac{D}{1 - D} \tag{9}$$

4 Results and Discussion

The MATLAB/Simulink Models of Photovoltaic Module and Implementation of MPPT Methods are as follows:

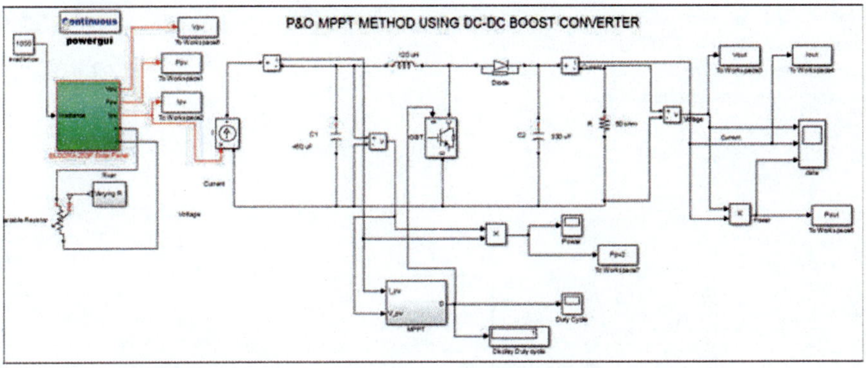

Fig. 8 MATLAB/Simulink Blocks of Perturb and Observe (P&O) MPPT Method

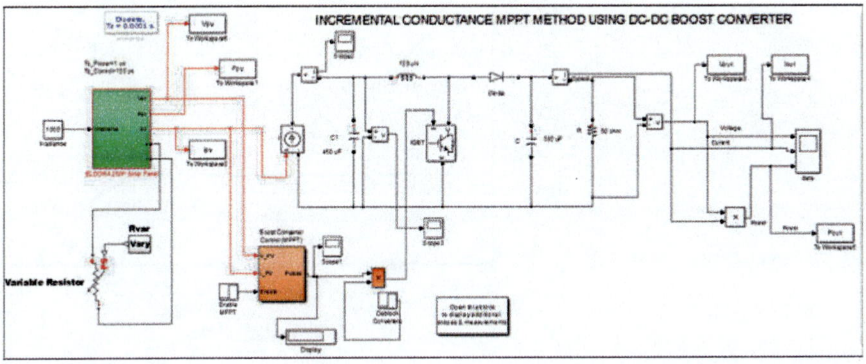

Fig. 9 MATLAB/Simulink Blocks of Incremental Conductance (INC) MPPT Method with DC-DC boost converter

ELDORA-250P, 60 cell polycrystalline solar photovoltaic module manufactured by Vikram Solar Pvt. Ltd., is proposed as a reference model for evaluating and validating the simulated photovoltaic model designed in MATLAB/Simulink. ELDORA-250P is a highly efficient solar photovoltaic module comprising 60 polycrystalline solar cells, ideally used in rooftop and ground-mounted applications.

Simulated results are compared with the practically and commercially available ELDORA-250P module for validating the proposed simulated module. An ideal ELDORA-250P solar photovoltaic module is developed in MATLAB/Simulink environment by considering series resistance $R_s = 0\ \Omega$ and shunt resistance $R_{sh} = \infty$. I-V Characteristics of the proposed model ELDORA-250P at different solar irradiance and at constant temperature of 25 °C are shown in Fig. 10, which are used as a reference for validating the characteristics to be obtained from the photovoltaic module designed in this thesis.

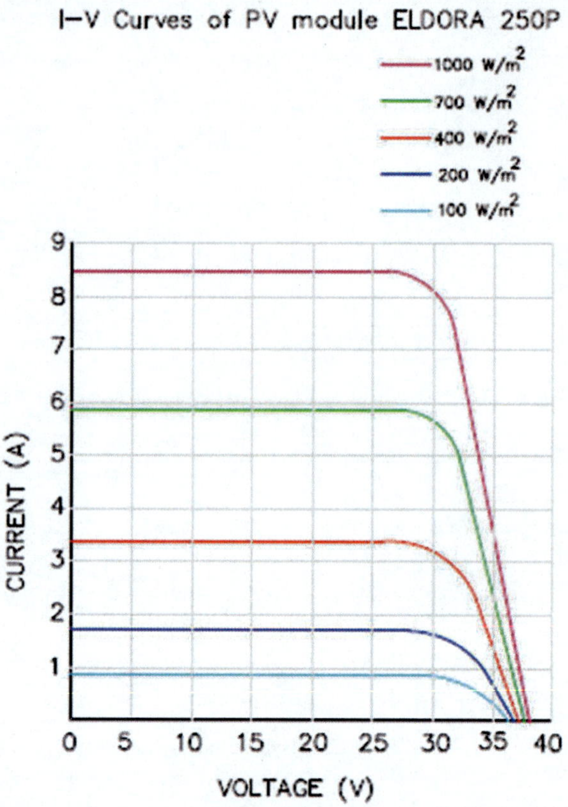

Fig. 10 V-I characteristic curves of ELDORA-250P at constant temperature 25 °C and varying solar irradiance

According to the specification of ELDORA-250P solar photovoltaic module, basic parameters, which are required to design a photovoltaic module at Standard Test Condition (STC), are given in Table 2 of the manufacturer's datasheet.

The output power of a solar photovoltaic module is affected mainly due to change in atmospheric conditions and the two major environmental factors, responsible for influencing output power of solar photovoltaic module, are power density of sunlight,

Table 2 Commercial photovoltaic module ELDORA-250P datasheet at STC with solar irradiance 1000 W/m^2 and temperature 25 °C

Module parameters	ELDORA-250P
Maximum power, P_m (W)	250
Maximum voltage, V_m (V)	30.58
Open circuit voltage, V_{oc} (V)	37.55
Short circuit current, I_{sc} (A)	8.71
NOCT	45 °C ± 85 °C

i.e., solar irradiance (W/m^2) and working temperature (°C). Simulated Model of ELDORA-250P is tested for two different conditions using MATLAB/Simulink, which are.

1. Constant temperature and varying solar irradiance,
2. Constant solar irradiance and varying module temperature.

Parameters obtained after simulation at constant solar irradiance (1000 W/m^2) and varying module temperature (25 °C, 35 °C, 45 °C, 55 °C, 65 °C, and 85 °C) are given in Table 3.

Simulated results of photovoltaic module parameters obtained from simulated I-V and P–V characteristic graphs are compared against corresponding actual results of commercially available ELDORA-250P module parameters which are given in Table 4.

Table 3 Module parameters extracted after simulation at different temperature level and at constant solar irradiance of 1000 W/m^2

Module parameters	Temperature (°C)					
	25 °C	35 °C	45 °C	55 °C	65 °C	85 °C
Maximum power, P_m (W)	257.6	246.6	238.3	228.1	216.8	198.1
Maximum voltage, V_m (V)	32.1	29.38	28.88	28.26	27.55	24.38
Open-circuit voltage, V_{oc} (V)	37.2	36.04	34.88	33.72	32.55	30.2
Short-circuit current, I_{sc} (A)	8.71	8.74	8.78	8.81	8.84	8.91

Table 4 ELDORA-250P commercial module validation with the simulated module at STC with solar irradiance of 1000 W/m^2 and at 25 °C temperature

Module parameters	ELDORA-250P (Actual Results)	ELDORA-250P (Simulated Results)	% Error = (Actual Results-Simulated Results)/Actual Results × 100%
Maximum power, P_m (W)	250	257.6	3.04%
Maximum voltage, V_m (V)	30.58	32.1	4.97%
Open-circuit voltage, V_{oc} (V)	37.55	37.2	0.93%
Short-circuit current, I_{sc} (A)	8.71	8.71	0.00%
NOCT	45 °C ± 85 °C		

Table 5 Comparison of simulated results of P&O and INC MPPT methods

MPPT method	Output Power P_{out} (Watt)	Output voltage V_{out} (Volts)	Output current I_{out} (Ampere)
(P&O) method	797.47	199.68	3.99
INC method	897.75	211.86	4.23

From Table 4, it can be easily understood that simulated results obtained at STC are approximately nearer to the actual results with small deviations, which may be considered as negligible and are articulated in the form of % error.

DC-DC boost converter is modeled in MATLAB/Simulink for enhancing power to be received from ELDORA-250P simulated photovoltaic module using Perturb & Observe (P&O) and Incremental Conductance (INC) MPPT controller; both MPPT controllers are also designed in MATLAB/Simulink controlling the switching action of MOSFET. Both the MPPT controllers control duty cycle used for triggering MOSFET of the DC-DC boost converter. Load resistance of 50 Ω is connected across Simulink model of the DC-DC boost converter.

DC-DC boost converter designed in MATLAB/Simulink consists of two capacitors, one with a rating of 450 μF and another with a rating of 330 μF, one inductor of 120 μH, one resistive load of 50 Ω, one power diode, and one MOSFET. Both the MPPT methods are tested using simulated ELDORA-250P photovoltaic module at standard test condition (STC) having solar irradiance of 1000 W/m^2 and temperature of 25 °C.

Results obtained after simulation are as follows:

Simulated results achieved from both the MPPT methods after simulation are analyzed in tabulated form in Table 5.

5 Conclusion

The proposed photovoltaic module built in MATLAB/Simulink illustrates and verifies nonlinear I-V and P–V characteristic curves of the photovoltaic module. Essential parameters are obtained from simulated I-V and P–V graphs which are mentioned in Table 2. After analyzing the simulated I-V and P–V characteristics, it is concluded that with the increase in solar irradiance, both open-circuit voltage V_{oc} and short-circuit current I_{sc} of photovoltaic module increases, thereby increasing the output power of the photovoltaic module. The reason behind the increase in output power is the logarithmic dependence of open-circuit voltage V_{oc} on solar irradiance whereas short-circuit current I_{sc} is directly proportional to the solar irradiance.

When photovoltaic module is built in MATLAB/Simulink under different temperature levels and at constant solar irradiance and simulated I-V and P–V characteristic graphs are generated. Essential parameters are obtained from simulated I-V and P–V graphs which are mentioned in Table 3. After analyzing the simulated

I-V and P–V characteristics, it is concluded that with the increase in module's working temperature, short-circuit current I_{sc} of photovoltaic module increases marginally whereas open-circuit voltage V_{oc} of photovoltaic module decreases drastically, thereby decreasing the output power of the photovoltaic module. Hence, it is concluded that temperature has a significant effect in determining solar photovoltaic module's efficiency and performance of the photovoltaic module is highly affected by the increase in the module's working temperature.

Advantage of using DC-DC boost converter amongst the other converters is that in case if MPPT circuit fails to operate and switching action of MOSFET will not take place, therefore in this condition load is directly connected to the panel and panel will still supply power to the load but at less efficiency. Other DC-DC converters cannot fulfil this condition. This paper describes only about implementation of boost converters.

References

1. Shyam B, Kanakasabapathy P, Renewable energy utilization in india-policies, opportunities and challenges. In: 2017 IEEE international conference on technological advancements in power and energy (TAP Energy).
2. Central Electricity Authority of India, "Draft-National Electricity Plan (2016),[Online], https://www.cea.nic.in/reports/committee/nep/nep_dec.pdf. [Accessed: July 6, 2017].
3. Central Electricity Authority of India, National Electricity Plan (2018). https://www.cea.nic.in/reports/committee/nep/nep_jan_2018.
4. Jawaharlal Nehru National Solar Mission Document
5. Villalva MG, Gazoli JR, Filho ER (2009) Comprehensive approach to modeling and simulation of photovoltaic arrays. IEEE Trans Power Electron 24(5)
6. Lynn PA (2010) Electricity from sunlight: an introduction to photovoltaics. Wiley, New York, p 238
7. Markvart T (1994) Solar electricity. Wiley, New York
8. Tsai H-L, Tu C-S, Su Y-J (2008) Member, IAENG, Development of generalized photovoltaic model using Matlab/Simulink. In: WCECS 2008. San Francisco, USA
9. Xiao W, Dunford WG, Capel A (2004) A novel modeling method for photovoltaic cells. In: 35th annual IEEE power electronics specialists conference. Aachen, Germany
10. Cow JA, Manning CD (1996) Development of model for photovoltaic arrays suitable for use in simulation studies of solar energy conversion systems. In: Power electronics and variable speed drives, Conference Publication No. 429, ©IEE
11. Sheik Mohammed S, Devaraj D (2014) Simulation and analysis of stand-alone photovoltaic system with boost converter using MATLAB/Simulink. In: International conference on circuit, power and computing technologies [ICCPCT]
12. Sheik Mohammed S (2011) Modeling and simulation of photovoltaic module using MATLAB/Simulink. Int J Chem Environ Eng 2(5)
13. Pandiarajan N, Muthu R (2011) Development of power electronic circuit oriented model of photovoltaic module. Int J Adv Eng Technol IJAET II(IV):118–127
14. Pandiarajan N, Ramaprabha R, Muthu R, Application of circuit model for photovoltaic energy conversion system. (Manuscript Title—Revised), Research Article, https://www.hindawi.com/journals/ijp/2012/410401.
15. Sridhar R, Jeevananathan, Selvan NT, Banerjee S (2010) Modeling of PV array and performance enhancement by MPPT algorithm. Int J Comput Appl (0975–8887) 7(5)

16. Ngan MS, Tan CW (2011) A study of maximum power point tracking algorithms for stand-alone photovoltaic systems. IEEE Appl Power Electron Colloquium (IAPEC)
17. Yu T-C, Lin Y-C, A study on maximum power point tracking algorithms for photovoltaic systems
18. Esram T, Chapman PL (2007) Comparison of photovoltaic array maximum power point tracking techniques. IEEE Trans Energy Convers 22(2)
19. https://en.wikipedia.org/wiki/Buck_converter (Access on Date 2 June 2019, Time 07:45 p.m.)
20. https://en.wikipedia.org/wiki/Boost_converter (Access on Date 2 June 2019, Time 07:45 p.m.)
21. https://en.wikipedia.org/wiki/Buck–boost_converter (Access on Date 2 June 2019, Time 07:45 p.m.)

Chapter 35
Design and Simulation of MPPT-Operated DC-DC Flyback Converter Used for Solar Photovoltaic System

Ranjana Choudhary and Shrawan Ram Patel

1 Introduction

One of the major problems faced by the whole world is related to the environment like pollution, global warming, waste disposal, greenhouse gases, CO_2 emissions and many but the reason behind these problems are basic needs of humans. Causes of co_2 emission are natural as well as human sources. Natural sources like decomposition, respiration, ocean release, and the human sources are like production of cement, deforestation, burning of fossil fuels. The demand for electricity has increased the use of fossil fuel like coal, lignite, oil, gas. To fulfil the requirement of electricity, power plants are necessary. In India, almost 60% of the electricity is generated from thermal power plants which use coal as a fuel. There are some other power plants also in which gas and oil are used to burn and generate electricity but it causes Co_2 emission and resultantly, harming our environment. So it is necessary to switch to some other sources to generate electricity which can be environment friendly. The need for human to generate electricity using environment-friendly sources has brought this attention toward natural sources like sun, water, air. Electricity generation using these sources requires a technique which should be the most efficient. For bulk generation as well as for small-scale generation or domestic-level generation, solar energy is the most suitable form of energy. Being particular about solar energy, there are many techniques to extract solar energy but the demand is for the most efficient and economic technique which is the key objective of the paper. In literature, it was found that the Maximum Power Point Tracking Method is the most efficient and economic

R. Choudhary (✉)
EE Department, JIET-Jodhpur, Mogra Khurd, Rajasthan, India
e-mail: ranjana.choudhary2017@gmail.com

S. R. Patel
Department of Electrical Engineering, JIET-Jodhpur, Mogra Khurd, Rajasthan, India
e-mail: shrawanram.patel@jietjodhpur.ac.in

M. Shorif Uddin et al. (eds.), *Intelligent Energy Management Technologies*,
Algorithms for Intelligent Systems,
https://doi.org/10.1007/978-981-15-8820-4_35

to harness the solar energy using PV Panel. PV modules, MPP Techniques along with net metering can be used at consumer level so that the surplus electrical energy can be fed to the AC Grid. This prototype will require DC-DC converter, DC-AC converter, etc. DC-DC conversion with solar PV power has become very noteworthy for efficient and economical energy production to exploit the maximum solar power possible at every moment, increasing profits and sustainability. The objective of this paper is "design and simulation of MPPT guided an isolated dc-DC Flyback converter used for solar PV system".

There are several methods for exploiting PV solar energy and many existing conventional techniques had incorporated the use of DC-DC converter along with DC-AC converter (inverter) for AC voltage applications. This type of converter topologies has two foremost features as reported in the literature. The first one is where the transformerless inverter is used which has several advantages like abridged size and low cost with high working efficiency. In this topology, the main drawback is that voltage boosting is not achieved as per the need of AC Voltage application or grid applications of voltage range 85–265 V AC. There is another problem of isolation between low voltage circuit (PV side) and high voltage circuit (AC side) and also the problem of grounding in the solar cell side is serious in case of transformerless topologies. The second feature is one where the transformer or coupled inductor system is used and the operating frequency is high. This feature will mitigate the problem of grounding of PV Panel and paved a way of boosting the voltage. Also, the problem of earth parasitic capacitance which will become a source of leakage current is avoided by using an isolation transformer. Since the cost of a line frequency transformer is very high and due to its large volume and weight, the high frequency (say 100 kHz) transformer is preferred [1, 2]. In line with this dissertation work, several authors have worked on Flyback converter with different aspects as follows.

Xiong et al. [3] have implemented the improved maximum or peak power point tracking in a PV system based on Flyback converter. A versatile simulation using MATLAB-based model is constructed for the output voltage behaviours or characteristics of a particular PV system varies nonlinearly with different solar insolation (G W/m^2), panel temperature.

Daniel Calheiros de Lemos [4] has presented MPPT-guided DC-DC Flyback Converter, which fed the solar energy into the grid. The converter used is situated within the family of DC-DC boost/buck power electronic converters. The algorithm presented here is based on Perturb and Observe technique, where it only controls the power generated by the panel, but the algorithm presented controls the power and the current of the panel. It has been found from the literature survey that several assumptions have been made in order to simulate and to realize the performance of DC-DC Flyback converter using different MPPT techniques in the various proposed methods. Many authors have tried to deal with the DC-DC Flyback converters interfacing a load with either Solar PV panel/module/array or rectified DC supply in order to extract the maximum power from the given source. The goal of this paper is to build the software simulation of Flyback converter, being operated by InC MPPT technique, and obtain the performance of converter with varying insolation, temperature and load impedance.

Patil and Kumar [5] have been presented the analysis of open- and closed-loop control of DC-DC Flyback converter. The main objective is to get the voltage mode control in the closed-loop model using a robust PID controller by which the working efficiency is augmented by 9%. The designed models of the open- and closed-loop DC/DC Flyback converter are verified using a MATLAB simulation.

Prabhakaran and Prasad [6] have designed and analysed an AC-DC and DC-AC isolated Flyback converter without the practice of bridge circuit. The proposed Flyback inverter is driven in continuous conduction (where inductor current is continuous) Modes with low apex current and superior efficiencies. The comparative investigation of the proposed circuit/topology with conventional circuit consists of AC-DC and DC-AC converter is addressed in the simulation of the proposed converter and also implemented using MATLAB Simulink.

2 Basics of Flyback Converter

The Flyback converter is derived from the buck–boost converter with the coupling inductor. The advantage of Flyback converter is that it can provide voltage multiplication by varying the number of turns of the coupled inductor and can provide multiple output DC/DC converters at power levels of 150 W or less. The design parameters of a Flyback converter is discussed in this chapter.

3 Working of Flyback Converter

The Flyback converter is similar to buck–boost converter in design and working. The only difference in construction is of coupling inductor. During ON state, when the MOSFET switch is Turn-ON or closed, the primary winding of the coupled inductor is connected directly to the PV module or source. The current flowing in the primary winding of coupled inductor increases and magnetic flux in the coupled inductor also increases. In this duration, energy is stored in the primary winding of the coupled inductor and the voltage which was induced in the secondary winding of the inductor is inverted. Hence, the diode is in the blocked stage or reversed-biased. During ON state, the capacitor connected at the output side delivers electrical energy to the load.

During OFF state, when the MOSFET switch is Turn-OFF, the current of primary and magnetic flux decreases. The voltage of secondary inductor is positive and the diode comes into conduction state which allows the current to flow to the load circuit. The energy from the secondary coupled inductor is supplied to the capacitor and then supplied to the load and this cycle is continued. The operation of Flyback converter can be continuous conduction mode or discontinuous mode but in this paper, the continuous conduction mode of Flyback converter is chosen so that maximum energy can be harnessed from solar PV module/array (Fig. 1).

3.1 Continuous Conduction Mode of Flyback Converter

A DC-DC converter that provides isolation between the input side and output side is shown in Fig. 2a. In Fig. 2b, it uses the transformer model which includes the magnetizing inductance Lm. To understand the operation of the circuit, a simplified transformer model is considered.

The following additional assumptions for the analysis are made:

- The output capacitor Co is very large and it is providing a constant output voltage Vo.
- The circuit is in steady-state operation which means that all the voltages and currents are cyclic in nature, i.e. they are starting and ending at the same points over one cycle/switching period.

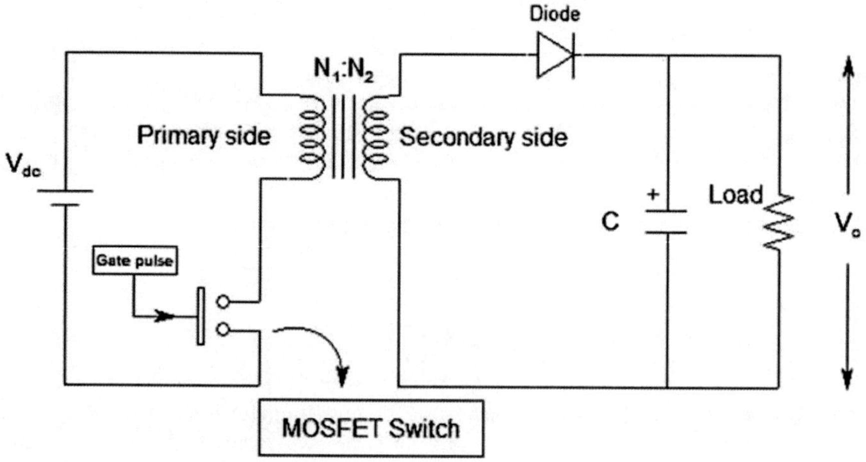

Fig. 1 Circuit diagram of a DC-DC flyback converter

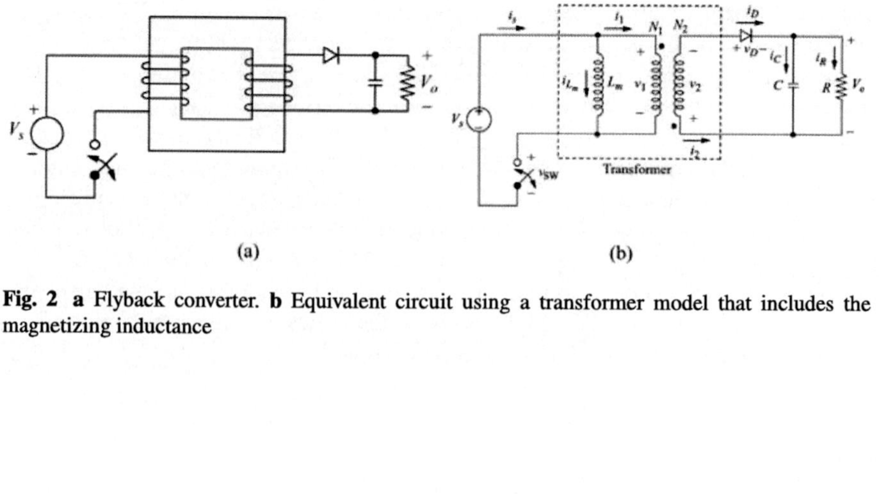

Fig. 2 **a** Flyback converter. **b** Equivalent circuit using a transformer model that includes the magnetizing inductance

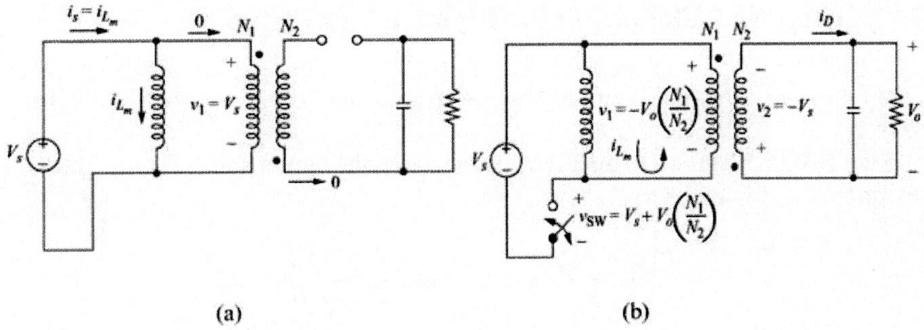

Fig. 3 Flyback Converter circuit during **a** T_{on} **b** T_{off}

- The duty cycle of the switch is D, ON-time period is of DT and OFF-time period is of (1-D)T.
- The semiconductor switch and power diode are ideal, giving a zero conduction loss. The basic operation of the Flyback converter is the same as that of the buck–boost converter.
- When the switch is closed, energy is stored in Lm. When the switch is open, it is transferred to the load. This circuit is analysed for both closed and open switch positions to determine the relationship between input and output.

Various waveforms of Flyback converter are shown in Sect. 3.13. The following relationships are required to design the Flyback converter (Fig. 3):

- Output Voltage (V_o)

$$V_o = V_{pv}\left(\frac{D}{1-D}\right)\frac{N_2}{N_1} \tag{1}$$

- Inductor Current

$$I_{L_m} = \frac{V_o}{R_L}\left(\frac{1}{1-D}\right)\frac{N_2}{N_1} \tag{2}$$

- Minimum Value of Inductance (L_m)

$$L_{m\_min} = \frac{(1-D)^2 R}{2f}\left(\frac{N_1}{N_2}\right)^2 \tag{3}$$

- Value of Output Capacitor

$$C_o = \frac{D}{R_L f.\frac{\Delta V_o}{V_o}} \tag{4}$$

4 Parameter Calculation for Flyback Converter

The following steps are required to determine the parameters of the Flyback converter:

Step 1: With no losses in the Flyback converter, the output voltage across the load resistance of 40Ω will be

$$P_{in} = P_{out} \Rightarrow V_{mpp}\, I_{mpp} = \frac{V_o^2}{R_L}$$

$$50 = \frac{V_o^2}{40} \Rightarrow V_o = 44.7213\ volt$$

If the turn ratio is $N_1/N_2 = ½$, then the duty of Flyback converter will be

$$V_o = V_{in}\left(\frac{D}{1-D}\right)\frac{N_2}{N_1} \Rightarrow 44.7213 = 17.7 \times \frac{D}{1-D} \times \frac{2}{1}$$

D = 0.5581 or 55.81%

Step 2: The average inductor current through the inductor L_m is given by

$$I_{Lm} = \frac{V_o^2}{V_{pv}\, D\, R_L} = \frac{44.7213^2}{17.7 \times 0.5581 \times 40} = 5.06155A$$

Step 3: The minimum value of inductance is

$$L_{M\_\min} = \frac{(1-D)^2}{2f} R \left(\frac{N_1}{N_2}\right)^2 = \frac{(1-0.5581)^2}{2 \times 50000} 40 \times \left(\frac{1}{2}\right)^2 = 19.52\,mH$$

Step 4: The output capacitance C_o is given by

$$C_o = \frac{D}{R_L f.\frac{\Delta V_o}{V_o}} = \frac{0.5581}{40 \times 50000 \times 0.01} = 27.05\mu F$$

5 Subsystem Configuration of Flyback Converter

Using a Flyback converter, the output voltage can be done step-up or down depending upon the duty ratio of the converter. In this thesis, the proposed Flyback converter is used as an interface between the load with a PV panel of 50 Wp. The controlling parameter of the Flyback converter for a fixed value of load resistance and operating frequency are provided to 'Pulse Generator' block to produce the pulse signal. This pulse signal is given to IGBT gate terminal for its ON/OFF operation. Figures 4

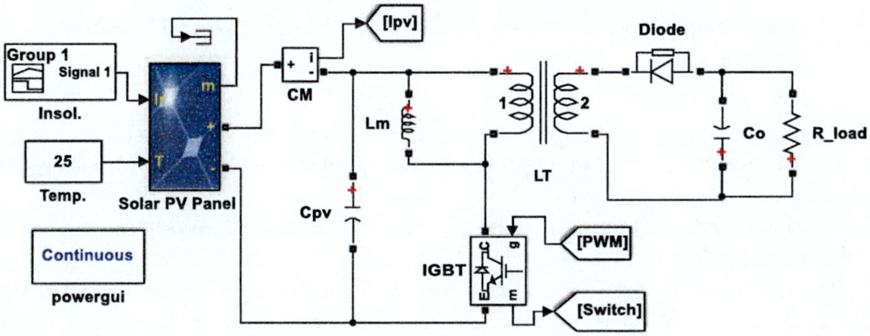

Fig. 4 Simulation circuit

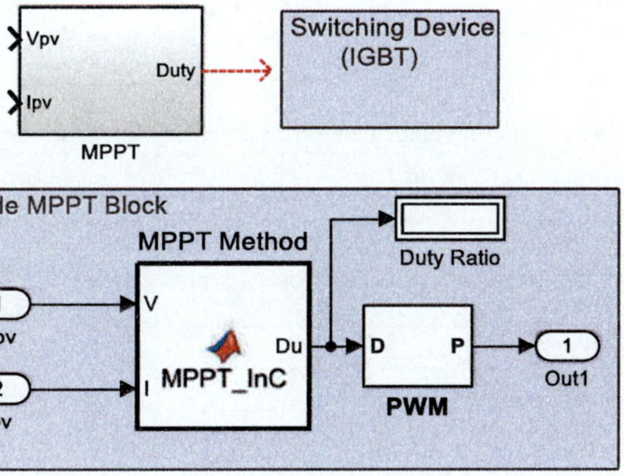

Fig. 5 Maximum power point tracking block

and 5 show the simulation diagram of Flyback converter interfacing the load to SPV panel with MPP technique. The PV module is subjected to variable insolation levels (200 W/m^2 to 1000 W/m^2) and panel temperature is kept constant to observe the effect of insolation on the performance of the PV module.

6 Experimental Results

The experimental setup for the proposed Flyback DC-DC converter is simulated in MATLAB, built and tested for effectiveness under varying load impedance, panel temperature and solar insolation.

6.1 Performance of Flyback Converter at Different Insolation Levels

The operating voltage of the PV panel and terminal current is varying with insolation levels. The panel current is increasing with insolation and decreasing with a decrease in insolation as shown in Table 1.

The input and output power of the Flyback converter with varying insolations is depicted in Fig. 6.

The efficiency of the Flyback converter with insolation is shown in Fig. 6. At standard condition (G = 1000 W/m^2 and Panel Temperature = 25 °C), the efficiency of converter is around 93%. From the efficiency curve, it can be stated that the conversion efficiency of the DC/DC Flyback converter is increasing with increasing solar insolation and temperature remains constant. But in real world, the temperature of the PV panel is also increasing with insolation and this effect is highlighted in the next session.

Table 1 Photovoltaic panel current with different insolations at constant temperature

Insolation (G)W/m^2	Current (I_{pv}) A
200	0.5527
400	1.127
600	1.687
800	2.276
1000	2.849

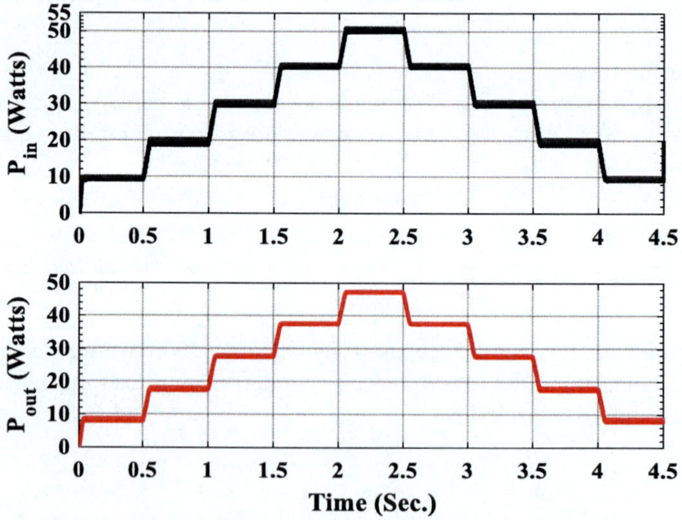

Fig. 6 Input and output power of Flyback converter with varying solar insolation

6.2 *Performance of Flyback Converter at Different Temperature Levels*

When the PV panel is subjected to both different insolations and temperatures, the output power is slightly lowered compared to the previous one (when the temperature is constant) as shown in Table 2.

The compared power of Flyback converter is shown in Fig. 7. From Fig. 7, one can say that as insolation is increasing the temperature of the PV module is increasing and this will cause of alleviation of the corresponding output power of the PV module. And also, with increased insolation, the photogenerated current is increased. With increased panel current the switching losses and the conduction losses are increased.

Table 2 Output power of flyback converter with different insolation levels and different temperature levels

Insolation (G) (W/m^2)	Panel temperature (°C)	Output power (P_o) Watts
200	25	8.441
400	30	17.43
600	40	26.23
800	50	34.16
1000	60	41.26

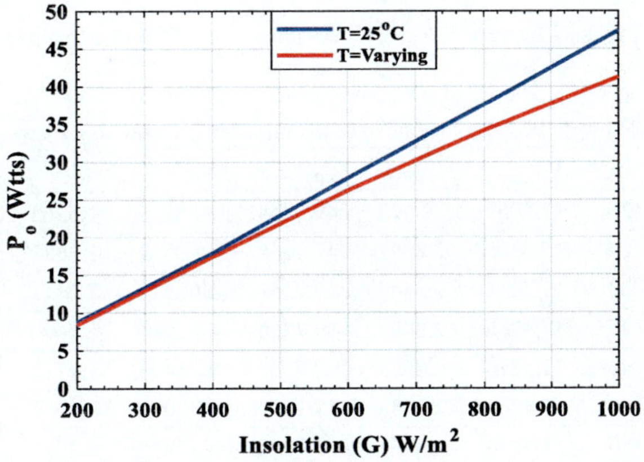

Fig. 7 Output power for different temperatures and different insolations

6.3 Performance of Flyback Converter with Varying Load Conditions

The load resistance, in this paper, was varied from 20Ω to 50Ω and the output power is slightly decreasing as shown in Fig. 8.

The MPPT technique is trying to extract the maximum power from the PV module for any value of load resistance. Here, the load resistance is increasing and to match this increased load resistance with characteristics impedance of solar PV panel, the duty should be increased. If the switching device is Turn-ON for a lonng time then PV panel can't be able to feed the maximum available power to the load. The comparative analysis of the Flyback converter with varying insolation is depicted in Table 3.

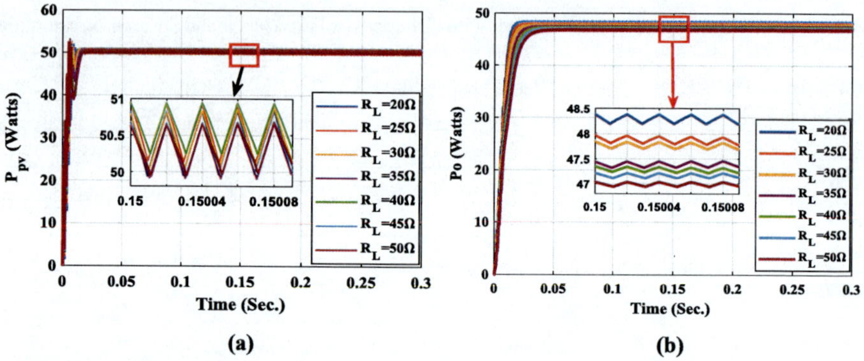

Fig. 8 **a** Input power and **b** output power of Flyback converter for varying loads

Table 3 Performance of flyback converter with variable loads at standard conditions

R_L	V_{pv}	I_{pv}	P_{pv}	V_o	I_o	P_o	η	Duty
20	17.4057	2.8527	49.5854	30.5296	1.5265	47.0289	94.84%	0.4669
25	17.4348	2.8522	49.6668	34.0519	1.3621	46.8063	94.24%	0.4933
30	17.5034	2.8443	49.7361	37.3515	1.2451	46.8081	94.11%	0.5142
35	17.4766	2.8498	49.7566	40.216	1.1490	46.5322	93.52%	0.5326
40	17.5276	2.8405	49.7322	42.8496	1.0712	46.239	92.98%	0.5473
45	17.4783	2.8483	49.7269	45.3022	1.0067	45.9609	92.43%	0.5615
50	17.5384	2.8376	49.7046	47.5855	0.9517	45.6603	91.86%	0.5726

6.4 Performance of Flyback Converter Under Dynamic Weather Conditions

The output power of the Flyback converter is also increasing by increasing the solar insolation. The output power is a combined effect of temperature and solar insolation because the photogenerated current is directly proportional to the insolation but terminal voltage falls logarithmically with an increase in panel temperature. The power of Flyback converter is obtained for different insolations at different temperatures as shown in Fig. 9.

The comparative efficiency of Flyback converter with varying insolation and temperature is depicted in Fig. 9b. The efficiency of the Flyback converter is high for lower temperature for the same insolation level. The efficiency of the Flyback converter with variable load is depicted in Fig. 10.

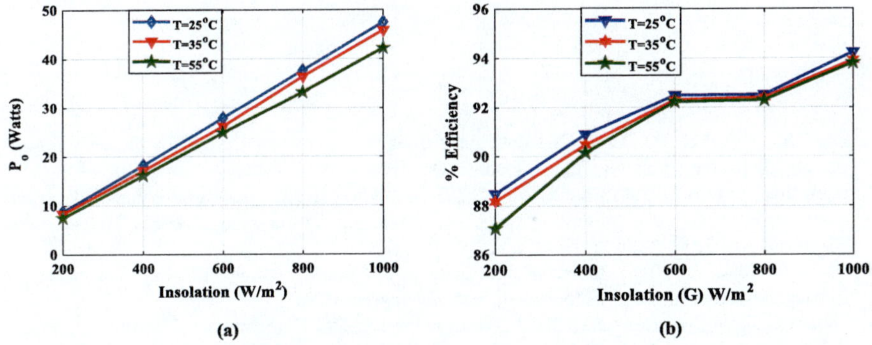

Fig. 9 **a** Output Power with varying insolations at constant temperature. **b** Output Power with varying insolations at varying temperatures

Fig. 10 Efficiency and power of Flyback converter versus duty ratio

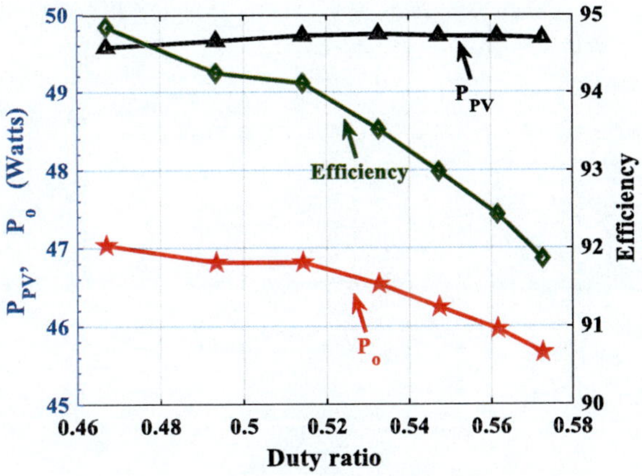

7 Conclusions

The proposed Flyback DC-DC converter is realized using MATLAB for different temperature levels, insolation levels and for varying insolation levels. Following main points are the conclusion for this paper.

(1) With increasing temperature level of solar PV panel, the efficiency of the system falls down.
(2) With increasing insolation, the efficiency of DC-DC converter is increasing when the temperature remains constant but in practice it is not possible. Therefore, the efficiency of the converter will slightly fall with increasing insolation level.
(3) With increasing load impedance, the duty of converter is increased and this results in large conduction loss in switch and hence efficiency will decrease.

References

1. Zhao S, Li Q, Lee FC, Li B (2018) High frequency transformer design for modular power conversion from medium voltage AC to 400V DC. IEEE Trans Power Electron 33(9):7545–7557
2. Shen J-M, Jou H-L, Jinn-Chang Wu (2012) Novel transformer less grid-connected power converter with negative grounding for photovoltaic generation system. IEEE Trans Power Electron 27(4):1818–1829
3. Xiong X, Shen A (2016) Improved maximum power point tracking in PV system based on flyback converter. In: 2015 Chinese automation congress (CAC)
4. Daniel Calheiros de Lemos,MPPT Algorithm for DC-DC Converter for Photovoltaic Panel Applications, [online available] https://fenix.tecnico.ulisboa.pt/ArtigoDaniel_Lemos_N77168.pdf
5. Patil B, Kumar P (2016) Performance analysis of a flyback converter. Int J Adv Res Electr Electron Instrum Eng 5(9)
6. Prabhakaran J, Prasad MB (2017) Design of novel bridgeless AC-DC-AC fly back convertor using Matlab. Int J Res Appl Sci Eng Technol (IJRASET) 5(III)

Chapter 36
To Improve Power Transfer Capacity Using TCSC FACTS Controller

Kishore Singh Gehlot and Shrawan Ram

1 Introduction

Electrical networks are interconnected to different generating stations and load centers according to the existing plan. But load demands on the system are not constant. With the increase of industrial growth and domestic load, more power is consumed by the different loads. To fulfill the load demand, either electrical system network has to be re-evaluated or the power carrying capability of the transmission line has to be increased. From economic point of view, modification or alteration of the electric network is costly. Thus the aim is to increase the power carrying capability of the transmission line.

How to Improve Power Flow:

The various parameters involved in power flow are.

(a) Load angle,
(b) Transmission line impedance, and
(c) Operating variables such as voltage and current.

To control the power flow from one bus to another bus, either of three parameters to be controlled. Power systems of today are mechanically controlled. Mechanical switching action is slow and power flow control is not fast enough according to the load variation. Another problem of mechanical control is that it cannot be initiated frequently as it leads to wear and tear.

To maintain both dynamic and steady-state operation, the new technology, i.e., FACTS (Flexible AC Transmission System) is used which is a power–electronics-based system. Its main role is to enhance controllability and power transfer capability

K. S. Gehlot (✉) · S. Ram
Department of EE, Jodhpur Institute of Engineering and Technology, Jodhpur, Rajasthan, India
e-mail: kishoresinghgehlot7@gmail.com

M. Shorif Uddin et al. (eds.), *Intelligent Energy Management Technologies*,
Algorithms for Intelligent Systems,
https://doi.org/10.1007/978-981-15-8820-4_36

in AC systems. FACTS technology uses switching power electronics to control power flow in the range of tens to hundreds of megawatts.

The various FACTS controller are capable of controlling the interrelated line parameters and other operating variables as mentioned in this paragraph. Thus FACTS technology governs the operation of the transmission line by providing shunt impedance, series impedance, line current, phase angle, voltage, and damping of oscillations at various frequencies below the rated frequency. By providing added flexibility, FACTS controllers can enable a transmission line to carry power:

- By maintaining proper insulation of transmission line without over-voltage,
- Up to its thermal rating,
- By maintaining stability in the system,

Thus, the use of FACTS technology increases the power carrying capability of the existing transmission network which is more economical.

Various FACTS Controllers:

In general, FACTS controllers may be divided into four categories:

- Shunt controller,
- Series controller,
- Combined series-shunt controllers, and
- Combined series-series controllers.

In this paper, it is considered about one of the Series Controller that is Thyristor-Controlled Series Capacitor (TCSC) to control the power flow in the transmission line. TCSC is one of the FACTS devices, which consist of a series capacitor bank shunted by a Thyristor-Controlled Reactor (TCR) in order to provide a smoothly variable series capacitive reactance by varying the firing angle of Thyristor-Controlled Reactor (TCR).

2 Power Flow Solution of Large Electrical Network

The main objective of a power flow study is to do determination the steady-state operating condition of the electrical power network. The steady state may be determined by finding out the flow of active and reactive power throughout the power system and the voltage magnitude and phase angles at all nodes of the network.

The planning and daily operation of current power systems leads to many power flow studies. Such information is used to carry out security evaluation analysis, where the nodal voltage magnitudes and active and reactive power flows in transmission lines and transformers are carefully observed to assess whether or not they are within the given operating limits. If the power flow study indicates that there are voltage magnitudes outside limits at certain points in the network, then appropriate control actions become necessary to regulate the voltage magnitude. Similarly, if the study presumes that the power flow in a given transmission line is out of the power carrying capacity of the line then control action will be taken.

Generalized Power Flow Solution For 5-Bus Network: A 5-bus network is given in the next page in Fig. to analyze the power flow solution and to determine the active power flow and reactive power flow from each bus. Also, nodal voltage magnitude and nodal phase angle is determined where these quantities are unknown. For this reason, a computer program is used to solve the power flow solution.

Data for 5-bus Network

Transmission Line Data

Transmission line	From	To	Resistance in p.u	Reactance in p.u	Susceptance in p.u
Tline-1	Giral	Merta	0.02	0.06	0.06
Tline-2	Giral	Sangariya	0.08	0.24	0.05
Tline-3	Merta	Sangariya	0.06	0.18	0.04
Tline-4	Merta	Piparcity	0.06	0.18	0.04
Tline-5	Merta	Jodhpur	0.04	0.12	0.03
Tline-6	Sangariya	Piparcity	0.01	0.03	0.02
Tline-7	Piparcity	Jodhpur	0.08	0.24	0.05

Generator Bus Data

Bus no	Bus type	Nodal voltage in p.u	Nodal phase angle in p.u	Active power injected in p.u	Reactive power injected in p.u	Generators reactive power upper limits in p.u	Generators reactive power lower limit in p.u
Giral-1	Slack Bus	1.06	0	Unknown	Unknown	5	−5
Merta-2	Generator PV Bus	1.00	To be calculated	0.4	To be calculated	3	−3

Load Bus Data

Bus no	Bus type	Nodal voltage in p.u	Nodal phase angle in p.u	Active power drawn in p.u	Reactive power drawn in p.u
Merta-2	LoadPQ Bus	To be calculated	To be calculated	0.20	0.10
Sangariya-3	LoadPQ Bus	To be calculated	To be calculated	0.45	0.15
Piparcity-4	LoadPQ Bus	To be calculated	To be calculated	0.40	0.05
Jodhpur-5	LoadPQ Bus	To be calculated	To be calculated	0.60	0.1

General Parameters: Maximum Iteration = 100, Tolerance = 1e − 12.

Computer Program for Power Flow Solution Using Newton–raphson Method:

For the 5-bus network shown in Fig. 4, a computer program is written using **MATLAB code** to solve the power flow equation (Figs. 1, 2, 3 and 4).

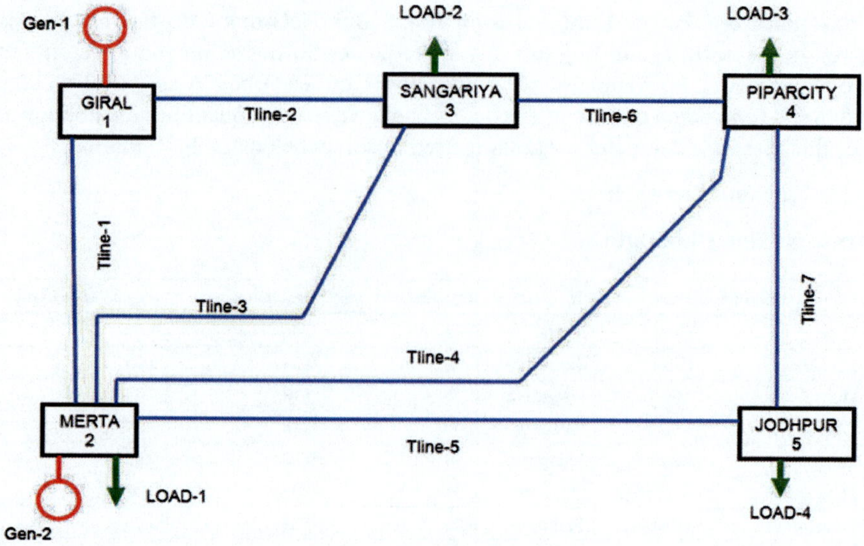

Fig. 1 A 5-Bus network for analyzing the power flow solution

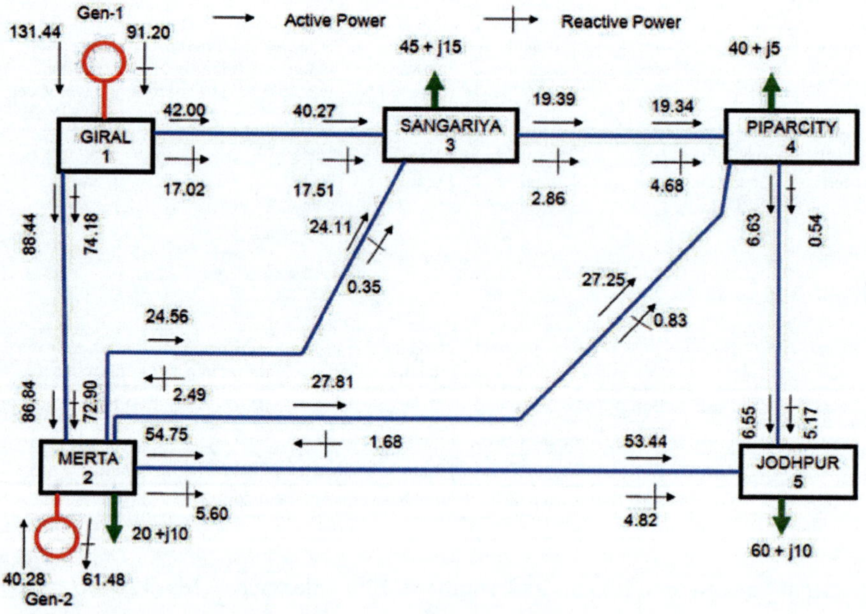

Fig. 2 Active and reactive power flow shown at each bus for 5-bus network

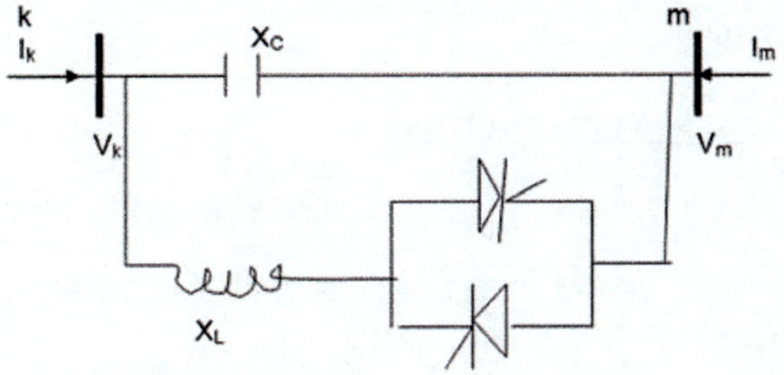

Fig. 3 TCSC connected between to bases k and m

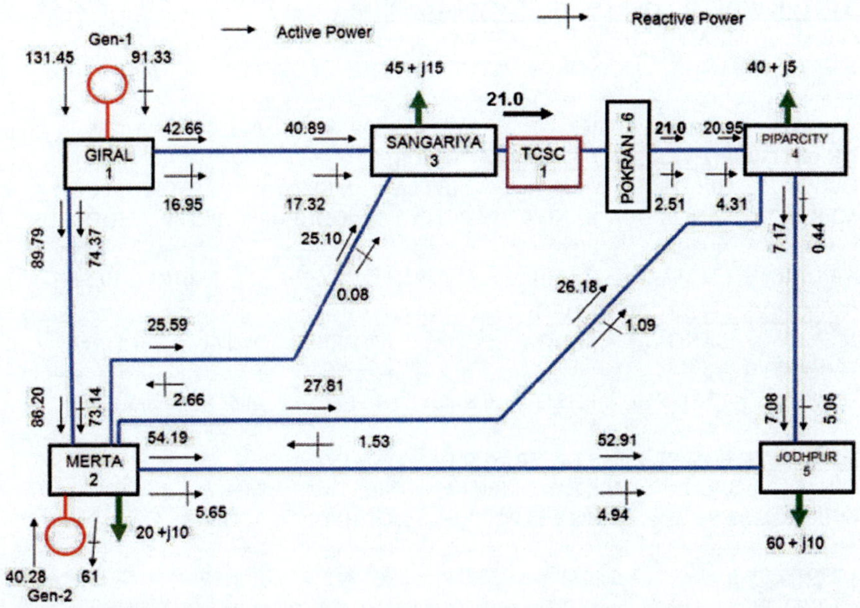

Fig. 4 Active and reactive power flow shown at each bus for 6-bus network where one TCSC is used to control the power flow

Summary of Outputs:

ACTIVE POWER SENT FROM EACH BUS:

GIRAL MERTA	GIRAL SANGA	MERTA SANGA	MERTA PCITY	MERTA JODHPUR	SANGA PCITY	PCITY JODHPUR
89.4458	42.0023	24.5614	27.8124	54.7540	19.3916	6.6320

ACTIVE POWER RECEIVED AT EACH BUS:

MERTA	SANGA	SANGA	PCITY	JODHPUR	PCITY	JODHPR
-86.8455	-40.2730	-24.1132	-27.2521	-53.4448	-19.3461	-6.5552

ACTIVE POWER LOSS ON TRANSMISSION LINE:

GIRAL MERTA	GIRAL SANGA	MERTA SANGA	MERTA PCITY	MERTA JODHPUR	SANGA PCITY	PCITY JODHPUR
2.6002	1.7293	0.4482	0.5603	1.3092	0.0455	0.0768

REACTIVE POWER SENT FROM EACH BUS:

GIRAL MERTA	GIRAL SANGA	MERTA SANGA	MERTA PCITY	MERTA JODHPUR	SANGA PCITY	PCITY JODHPUR
74.1880	17.0217	-2.4910	-1.6897	5.6034	2.8679	0.5491

REACTIVE POWER RECEIVED AT EACH BUS:

MERTA	SANGA	SANGA	PCITY	JODHPUR	PCITY	JODHPR
-72.9084	-17.5125	-0.3523	-0.8306	-4.8292	-4.6878	-5.1708

REACTIVE POWER LOSS ON TRANSMISSION LINE:

GIRAL MERTA	GIRAL SANGA	MERTA SANGA	MERTA PCITY	MERTA JODHPUR	SANGA PCITY	PCITY JODHPUR
1.2797	-0.4908	-2.8433	-2.5202	0.7742	-1.8199	-4.6216

NET ACTIVE POWER INJECTED/OUTAGE AT GENERATOR/LOAD BUS:

GIRAL	MERTA	MERTA	SANGA	PCITY	JODHPUR
131.4481	40.2823	-20.0000	-44.9946	-39.9663	-60.0000

NET REACTIVE POWER INJECTED/OUTAGE AT GENERATOR/LOAD BUS:

GIRAL	MERTA	MERTA	SANGA	PCITY	JODHPUR
91.2097	-61.4856	-10.0000	-14.9969	-4.9692	-10.0000

TOTALGENERATION_ACTIVEPOWER = 171.7304

TOTALGENERATION_REACTIVEPOWER = **29.7241**

TOTAL_ACTIVE_LOAD = **-165**

TOTAL_REACTIVE_LOAD = **-40**

TOTAL_ACTIVEPOWER_LOSS = **6.7304**

TOTAL_REACTIVEPOWER_LOSS = **-10.2759**

TOTALITERATION = **6**

Conclusion:

The summary of the result is superimposed on the given 5-bus network shown in Fig-6.5 so that at each bus we can find out how much power flows from each bus and the losses on the transmission line are also mentioned. The summary of output shows that at each bus, power mismatch equation is satisfied. Also, the power flow solution converges to a prescribed tolerance of, i.e., 12 within six iterations.

3 Power Flow Control Using Tcsc

As Thyristor-Controlled Series Capacitor (TCSC) will control the power flow in the transmission line of a large electrical network, here we will model the Thyristor-Controlled Series Capacitor (TCSC) as a variable reactance which varies in terms of firing angle of the thyristor.

Power Flow Solution for 6-Bus Network:

Power flow solution for the network where Thyristor-Controlled Series Capacitor (TCSC) is used in the network is solved in a similar way using Newton–Raphson (NR) method. In the next, a 6-Bus network is drawn, in which a TCSC is used in between Sangriya to Pokran. By writing a MATLAB code, we study how the Thyristor-Controlled Series Capacitor is effective for controlling the specified amount of active power in between two buses.

Data for 6-bus Network in Which Tcsc-1 is Connected:

Transmission Line Data for 7 Lines

Transmission line	Fm	To	Resistance in p.u	Reactance in p.u	Susceptance in p.u
Tline-1	Giral	Merta	0.02	0.06	0.06
Tline-2	Giral	Sangariya	0.08	0.24	0.05
Tline-3	Merta	Sangariya	0.06	0.18	0.04
Tline-4	Merta	Piparcity	0.06	0.18	0.04
Tline-5	Merta	Jodhpur	0.04	0.12	0.03
Tline-6	Pokran	Piparcity	0.01	0.03	0.02
Tline-7	Piparcity	Jodhpur	0.08	0.24	0.05

Generator Bus Data

Bus no	Bus type	Nodal voltage in p.u	Nodal phase angle in p.u	Active power injected in p.u	Reactive power injected in p.u	Generators reactive power upper limits in p.u	Generators reactive power lower limit in p.u
Giral-1	Slack bus	1.06	0	unknown	unknown	5	−5
Merta-2	Generator PV bus	1.00	To be calculated	0.4	To be calculated	3	−3

Load Bus Data

Bus no	Bus type	Nodal voltage in p.u	Nodal phase angle in p.u	Active power drawn in p.u	Reactive power drawn in p.u
Merta-2	LoadPQ bus	To be calculate	To be calculate	0.20	0.10
Sangariya-3	LoadPQ bus	To be calculate	To be calculate	0.45	0.15
Piparcity-4	LoadPQ bus	To be calculate	To be calculate	0.40	0.05
Jodhpur-5	LoadPQ bus	To be calculate	To be calculate	0.60	0.1

TCSC's DATA

Total no of TCSC = 1.

Connected between: Sangariya-3 to Pokran-6.
Capacitive reactance of TCSC = 9.375e-3.
Inductive reactance of TCSC = 1.625e-3.
Initial Firing angle = 145 degree.
Firing angle lower limit = 90 degree.
Firing angle upper limit = 180 degree.
Active power to be controlled = 0.21 p.u

General Parameters:

Maximum Iteration = 100,
Tolerance = 1e − 12.

Computer Program for Power Flow Control Using Tcsc (thyristor- Controlled Series Capacitor):

To solve the power flow equation of the above network, which contains one TCSC to control the required amount of active power between POKRAN and PIPARCITY, a program is written in MATLAB code.

Summary of Output:

ACTIVE POWER SENT FROM EACH BUS:

GIRAL	GIRAL	MERTA	MERTA	MERTA	POKRAN	PCITY
MERTA	SANGA	SANGA	PCITY	JODHPUR	PCITY	JODHPUR
88.7927	42.6632	25.5906	26.7005	54.1990	21.0060	7.1711

ACTIVE POWER RECEIVED AT EACH BUS:

MERTA	SANGA	SANGA	PCITY	JODHPUR	PCITY	JODHPUR
-86.2091	-40.8924	-25.1076	-26.1804	-52.9147	-20.9535	-7.0853

ACTIVE POWER LOSS ON TRANSMISSION LINE:

GIRAL	GIRAL	MERTA	MERTA	MERTA	POKRAN	PCITY
MERTA	SANGA	SANGA	PCITY	JODHPUR	PCITY	JODHPUR
2.5836	1.7708	0.4830	0.5201	1.2843	0.0524	0.0858

REACTIVE POWER SENT FROM EACH BUS:

GIRAL	GIRAL	MERTA	MERTA	MERTA	POKRAN	PCITY	
MERTA	SANGA	SANGA	PCITY	JODHPUR	PCITY	JODHPUR	
							74.3796
16.9521	-2.6660	-1.5331	5.6513	2.5142	0.4444		

REACTIVE POWER RECEIVED AT EACH BUS::::

MERTA	SANGA	SANGA	PCITY	JODHPUR	PCITY	JODHPUR
-73.1461	-17.3287	-0.0832	-1.0971	-4.9491	-4.3160	-5.0509

REACTIVE POWER LOSS ON TRANSMISSION LINE:

GIRAL	GIRAL	MERTA	MERTA	MERTA	POKRAN	PCITY
MERTA	SANGA	SANGA	PCITY	JODHPUR	PCITY	JODHPUR
1.2335	-0.3766	-2.7493	-2.6302	0.7021	-1.8018	-4.6065

NET ACTIVE POWER INJECTED/OUTAGE AT GENERATOR/LOAD BUS:

GIRAL	MERTA	MERTA	SANAG	PCITY	JODHPUR	POKRAN
131.4559	40.2810	-20.0000	-66.0000	-39.9629	-60.0000	21.0060

NET REACTIVE POWER INJECTED/OUTAGE AT GENERATOR/LOAD BUS:

GIRAL	MERTA	MERTA	SANAG	PCITY	JODHPUR	POKRAN		
							91.3317	-
61.6940	-10.0000	-17.4119	-4.9687	-10.0000	2.5142			

TOTALGENERATION_ACTIVEPOWER = 171.7369
TOTALGENERATION_REACTIVEPOWER = 29.6377
TOTAL_ACTIVE_LOAD = -165
TOTAL_REACTIVE_LOAD = -40
TOTAL_ACTIVEPOWER_LOSS = 6.7369
TOTAL_REACTIVEPOWER_LOSS = -10.3623
ACTIVE_POWER_INJECTED_IN_TCSC = 21.0000
REACTIVE_POWER_INJECTED_IN_TCSC =2.4119
FINAL_FIRINGANGLE = 148.4675
TOTAL_REACTANCE = -0.0216
TOTAL_ITERATION = 8

Result-1:

Figure 7.5 is modified and reproduced in Fig. 7.3, in which one TCSC is connected between Sangariya and Pokran to control 21 MW of power flow from Sangariya to Piparcity. The power flow solution is obtained in 8 iterations to a power mismatch tolerance of 1e-12. The power flow results are shown in figure.

Since the TCSC cannot generate active power, there is an increase in active power flow from GIRAL bus to SANGARIYA bus (i.e., from 42.00 MW to 42.66 MW). At the same time, there is an increase of active power flow from MERTA bus to SANGARIYA bus (i.e., from 24.56 MW to 25.59 MW). In total, there is an increase of active power flow from SANGARIYA to PIPARCITY (i.e., from 19.39 MW to 21 MW).

It should be remarked that transmission line from SANGARIYA to PIPARCITY is series compensated by the use of TCSC-1 and there is an increase of active power flow from 19.38 MW to 21 MW, which is just under 8% active power increase. Thus TCSC with firing angle control provides a good series compensation in the transmission line for controlling the active.

Power Flow Solution of 7-Bus Network:

The network shown in Fig consists of two TCSC for controlling the power flow. One TCSC is connected in between Sangariya and Pokran. Other is connected in between Piparcity and Boranada. The MATLAB code which is written for one TCSC is now modified with additional data for TCSC-2 and executed. The power flows vary in the transmission line and it is different from the case used for one TCSC.

Data for 7-bus Network in Which Tcsc-1 and Tcsc-2 is Connected:

Transmission Line Data

Transmission line	Fm	To	Resistance in p.u	Reactance in p.u	Susceptance in p.u
Tline-1	Giral	Merta	0.02	0.06	0.06
Tline-2	Giral	Sangariya	0.08	0.24	0.05
Tline-3	Merta	Sangariya	0.06	0.18	0.04
Tline-4	Merta	Piparcity	0.06	0.18	0.04
Tline-5	Merta	Jodhpur	0.04	0.12	0.03
Tline-6	Pokran	Piparcity	0.01	0.03	0.02
Tline-7	Boranada	Jodhpur	0.08	0.24	0.05

Generator Bus Data

Bus no	Bus type	Nodal voltage in p.u	Nodal phase angle in p.u	Active power injected in p.u	Reactive power injected in p.u	Generators reactive power upper limits in p.u	Generators reactive power lower limit in p.u
Giral-1	Slack bus	1.06	0	unknown	Unknown	5	−5
Merta-2	Generator PV bus	1.00	To be calculate	0.4	To be calculated	3	−3

Load Bus Data

Bus no	Bus type	Nodal voltage in p.u	Nodal phase angle in p.u	Active power drawn in p.u	Reactive power drawn in p.u
Merta-2	LoadPQ bus	To be calculated	To be calculated	0.20	0.10
Sangariya-3	LoadPQ bus	To be calculated	To be calculated	0.45	0.15
Piparcity-4	LoadPQ bus	To be calculated	To be calculated	0.40	0.05
Jodhpur-5	LoadPQ bus	To be calculated	To be calculated	0.60	0.1

TCSC's DATA

Total no of TCSC = 2.

Data for TCSC-1

Connected between: From Sangariya-3 to Pokran-6.
Capacitive reactance of TCSC -1 = 9.375e-3.
Inductive reactance of TCSC-1 = 1.625e-3.
Initial Firing angle = 145 degree.
Firing angle lower limit = 90 degree.
Firing angle upper limit = 180 degree.
Active power to be controlled = 0.21 p.u

Data for TCSC-2

Connected between: From Piparcity-4 to Boranada-7.
Capacitive reactance of TCSC -2 = 9.375e-3.
Inductive reactance of TCSC-2 = 1.625e-3.
Initial Firing angle = 145 degree.
Firing angle lower limit = 90 degree.
Firing angle upper limit = 180 degree.
Active power to be controlled = 0.22 p.u

General Parameters:
Maximum Iteration = 100,
Tolerance = 1e − 12.

Computer Program for Power Flow Control Using Tcsc (thyristor- Controlled Series Capacitor):

To solve the power flow equation of above network which contains one TCSC to control the required amount of active power between POKRAN and PIPARCITY, a program is written in MATLAB code.

Summary of Outputs:

SUMMARY OF OUTPUTS:

ACTIVE POWER SENT FROM EACH BUS:

GIRAL MERTA	GIRAL SANGA	MERTA SANGA	MERTA PCITY	MERTA JODHPUR	POKRAN PCITY	BNADA JODHPUR
89.2434	42.7349	25.4970	43.1498	38.3180	21.0033	22.9577

ACTIVE POWER RECEIVED AT EACH BUS:

MERTA	SANGA	SANGA	PCITY	JODHPUR	PCITY	JODHPUR
-86.6484	-40.9829	-25.0142	-41.8647	-37.6175	-20.9519	-22.3825

ACTIVE POWER LOSS ON TRANSMISSION LINE:

GIRAL MERTA	GIRAL SANGA	MERTA SANGA	MERTA PCITY	MERTA JODHPUR	POKRAN PCITY	BNADA JODHPUR
2.5951	1.7520	0.4828	1.2851	0.7005	0.0514	0.5752

REACTIVE POWER SENT FROM EACH BUS:

GIRAL MERTA	GIRAL SANGA	MERTA SANGA	MERTA PCITY	MERTA JODHPUR	POKRAN PCITY	BNADA JODHPUR
74.2474	16.0179	-3.8013	-7.0318	10.0700	-0.1455	-4.4705

REACTIVE POWER RECEIVED AT EACH BUS:

MERTA	SANGA	SANGA	PCITY	JODHPUR	PCITY	JODHPUR
-72.9821	-16.4735	1.0360	6.4911	-11.0271	-1.6681	1.0271

REACTIVE POWER LOSS ON TRANSMISSION LINE:

GIRAL MERTA	GIRAL SANGA	MERTA SANGA	MERTA PCITY	MERTA JODHPUR	POKRAN PCITY	BNADA JODHPUR
1.2653	-0.4556	-2.7653	-0.5408	-0.9571	-1.8136	-3.4434

NET ACTIVE POWER INJECTED/OUTAGE AT GENERATOR/LOAD BUS:

GIRAL	MERTA	MERTA	SANGA	PCITY	JODHPUR	POKRAN	BNADA
131.9784	40.3164	-20.0000	-65.9971	-62.8167	-60.0000	21.0033	22.9577

NET REACTIVE POWER INJECTED/OUTAGE AT GENERATOR/LOAD BUS:

GIRAL	MERTA	MERTA	SANGA	PCITY	JODHPUR	POKRAN	BNADA
90.2652	-63.7452	-10.0000	-15.4376	4.8229	-10.0000	-0.1455	-4.4705

TOTALGENERATION_ACTIVE POWER = 172.2948
TOTALGENERATION_REACTIVE POWER = 31.3430
TOTAL_ACTIVE_LOAD = -165
TOTAL_REACTIVE_LOAD = -40
TOTAL_ACTIVEPOWER_LOSS = 7.2948
TOTAL_REACTIVEPOWER_LOSS = -8.6570

Result-2

From Fig. 7.5, we observe that TCSC-1 provides series compensation in the transmission line between SANGARIYA to PIPARCITY bus. This is because active power flow from SANGARIYA to PIPARCITY is increased from 19.39 MW (shown in Fig. 6.5) to 21 MW. At the same time the active power flow from GIRAL to SANGARIYA is increased from 42.00 MW to 42.73 MW and from MERTA to

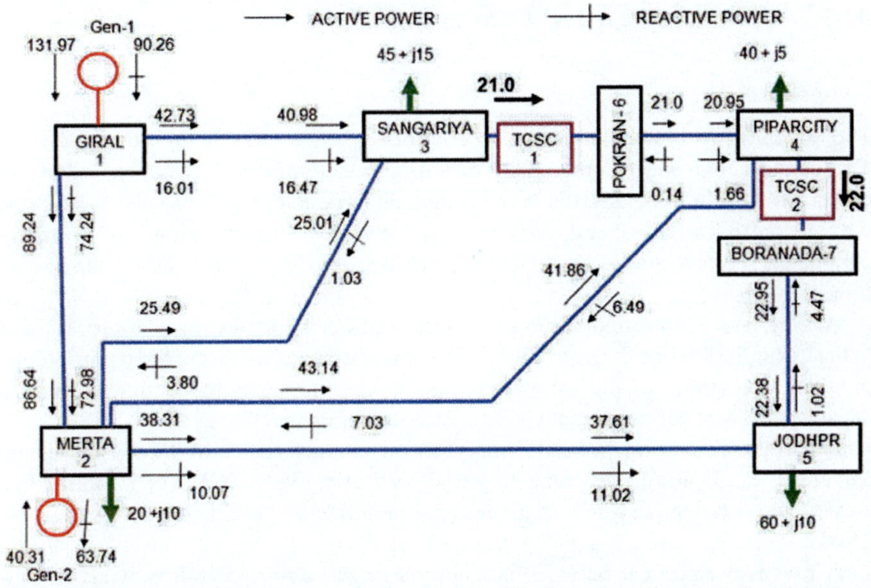

Fig. 5 Active and reactive power flow shown at each bus for 7-bus network where two TCSC is used to control the power flow

SANGARIYA is increased from 24.56 MW to 25.49 MW as TCSC doesn't generate any active power (Fig. 5).

Similarly from Fig. 7.5, we observe that TCSC-2 provides series compensation in the transmission line between PIPARCITY to JODHPUR. This is because active power flow from PIPARCITY to JODHPUR is increased from 6.63 MW to 22 MW. At the same time, the active power flow from MERTA to PIPARCITY is increased from 27.81 MW to 27.81 MW and MERTA to JODHPUR is decreased from 54.75 MW to 38.31 MW as TCSC does not generate any active power.

Also the power mismatch equation is satisfied at each bus, after using two TCSC. Thus from the analysis, it is very clear that both TCSC provides effective series compensation in the two different transmission lines and specified amount of active power is controlled.

Conclusion:

From the Result-1 and Result-2, we reached a conclusion that how effectively Thyristor-Controlled Series capacitor (TCSC) can control the active power flow between buses. With the use of TCSC, a specified amount of power can be transferred from one bus to other as TCSC does not consume or generate active power.

4 Conclusion and Further Scope of Work

Summary of Work

The objective of the whole work is to control the power flow in the power system. This can be achieved by knowing the various parameters which are concerned in the power flow in the transmission line I. For each parameter, we have discussed how we can control the active and reactive power flow in the transmission line. For each parameter, we have applied in MATLAB application software to analyze the power flow.

As Thyristor-Controlled Series Capacitor (TCSC), is a series compensator is used in the transmission line to control the active power flow, so we discussed the principle of series compensation and simulation block is used to verify the truthiness, how a capacitor is an effective element to decrease the series reactance.

In further discussion, we have analyzed the characteristics of TCSC. By using MATLAB application, we came to know how effectively TCSC can control the current and active power flow in the transmission line by varying the firing angle of TCSC.

A practical electrical network is having a large number of buses. Thus in this work, a 5-bus imaginary network is considered for finding the power flow solution. Newton–Raphson method is used to solve this network. By using MATLAB code for this network, power flow between each bus is determined. Power flow solution of this network satisfies the power mismatch equation at each bus.

Jacobian matrix is determined for 6-bus network in which one TCSC is used in between two buses. Again power flow explanation is determined for this network and we came to know that it also satisfies the power mismatch equations at each bus. At the same time, a specified amount of active power is allowed to flow between two buses by using the TCSC. In the next section again, two TCSC is used to analyze the power flow in different transmission lines. Here also MATLAB code is used for finding the power flow solution of the given electrical network.

Conclusion

From the execution of MATLAB code, we reached at a conclusion that Thyristor-Controlled Series Capacitor is one of the fast acting power electronic controllers which can provide current and power flow control in the transmission line by varying its firing angle. Thus Thyristor-Controlled Series Capacitor (TCSC) can be used as a series capacitor to decrease the overall transmission line reactance. Depending on the development of power transfer preferred at that time, without affecting other system-performance criteria, series compensation can be varied by Thyristor-Controlled Series Capacitor (TCSC). Thus, Thyristor-Controlled Series Capacitor (TCSC) is one of the important FACTS controller, which increases the overall power transfer capacity in the transmission line.

Further Scope of Work

Works on this topic never end with unlimited applications. It can be applied to damping of the power swings from local and interconnected area oscillations, voltage regulation of local network, reduction of short-circuit current, etc. Various research works are going on control interaction between multiple TCSC. Also, SVC-TCSC can be combined and used within power systems to enhance inter-area stability. In this work MATLAB code is used for 5-bus network for power flow solution. MATLAB code can also be used for designing a physical 20-bus network to determine the exact power flow from one bus to another. Though in this discussion Mi Power software is used for 5-bus systems, it is capable of being used in a large bus network. Further study can be done on modern power system of TCSC on Fault Component Distance Protection and Impact of TCSC on the Protection of Transmission Lines. Thus TCSC can be used in many fields of the power system.

References

1. Saddat H (2002) Power system analysis (MATLAB concept). Tata McGraw-Hill, Edition
2. Rashid MH (2004) Power electronics (concept of Flexible AC Transmission Systems), 3rd edn. Pearson Education
3. Narain G (2001) Hingorani/Laszlo Gyugyi, Understanding FACTS, 1st edn. IEEE press
4. Nagrath/DP Kothari IJ (2003) Power system engineering (concept of series and shunt compensation). Tata McGraw-Hill
5. C.L. Wadhwa, "Electrical Power Systems (compensation in power system)", *New Age International, 3rd Edition,2004*
6. P.Moore and P. Ashmole, "Flexible ac transmission systems : part 4-advanced FACTS controller," *Power Engineering Journal*, April 1998,pp.95–100
7. E Acha/VG Agelidis/T J E Miller, "Power Electronic Control in Electrical Systems", *Newnes Power Engineering Series,Ist Indian Edition,2006*
8. Fuerte-Esquivel CR, Acha E, Ambriz-Perez H (Feb.) A thyristor controlled series compensator model for the power flow solution of practical power networks. Power Systems, IEEE Transactions on Power Dlivery 15(1):58–64
9. Gama, C.; Tenorio, R.; "Improvements for power systems performance: modeling, analysis and benefits of TCSCs" ,Power Engineering Society Winter Meeting, 2000. IEEE Volume 2, 23–27 Jan. 2000 Page(s):1462 - 1467 vol.2
10. Xie Da; Niu Hui; Chen Chen; Wu Jishun, "An algorithm to control the power flow in large systems based on TCSC " ,Power System Technology, 1998. Proceedings. POWERCON '98. *Volume 1, 18–21 Aug. 1998 Page(s):344 - 348 vol.1*
11. Geng Juncheng; Tong Luyuan; Ge Jun; Wang Zhonghong; "Mathematicalmodel for describing characteristics of TCSC", *Power System Technology, 2002. Proceedings. PowerCon 2002. Volume 3, 13–17 Oct. 2002 Page(s):1498 - 1502*
12. Abdel-Moamen, M.A.; Narayana Prasad Padhy, "Power flow control and transmission loss minimization model with TCSC for practical power networks", *Power Engineering Society General Meeting, 2003,Volume 2, 13–17 July 2003*
13. R. Mohan Mathur, Rajiv K. Varma; " Thyristor-Based FACTS Controllers For Electrical Transmission Systems", *A John Wiley & Sons, Inc. Publication,2002*
14. Matsuki, J.; Ikeda, K.; Abe, M.; "Investigations of a thyristor-controlled series capacitor", Industrial Electronics, Control, and Instrumentation, 1996., Proceedings of the 1996 IEEE IECON 22nd International Conference, *Volume 2, 5–10 Aug. 1996 Page(s):683 – 688*

15. Persson, J.; Rouco, L.; Soder, L "Linear analysis with two linear models of a thyristor-controlled series capacitor", *Power Tech Conference Proceedings, 2003 IEEE Bologna, Volume 3,23–26 June 2003 Page(s):8pp*
16. IEEE Power Engineering Society (1996) FACTS Application. IEEE press, New York
17. William D. Stevenson, Jr. "Elements of Power System Analysis", *McGraw-Hill Series in Electrical Engineering, Fourth Edition,1982*
18. Persson, J.; Rouco, L.; Soder, L ,"Linear analysis with two linear models of a thyristor-controlled series capacitor", *Power Tech Conference Proceedings, 2003 IEEE Bologna*
19. Rudra Pratap Singh, "Intoduction to MATLAB",*Ist Edition,2004*
20. Chulin G, Luyuan T, Zhonghang W (2002) Stability Control of TCSC between interconnected Power networks. In: Power system technology, 2002 proceedings, vol 3, pp 1943–1946
21. Chapman SJ (2004) MATLAB programming for engineers, 3rd edn. Thomsan Learning
22. Ally A, Rigby BS (2004) Member, IEEE, An investigation into the impact of a thyristor controlle series capacitor-based closed loop power flow controller under fault conditions. IEEE AFRICON, pp 675–681
23. Tan X, Tong L (1998) Characteristics and firing angle control of thyristor controlled series compensation installations. In: IEEE Conference, pp 672–676
24. Billinton R, Fotuhi-Firuzaba M, Faried SO (1999) Power system reliability enhancement using a thyristor controlled series capacitor. IEEE Trans Power Syst.14(1)
25. Xueqiang Z, Chen C (1998) Circuit Analysis of a thyristor controlled series compensation. In: IEEE paper, pp 1067–1072

Chapter 37
Stockwell Transform and Hilbert Transform Based Hybrid Algorithm for Recognition of Power Quality Disturbances

Ramesh Aseri and Ashwani Kumar Joshi

1 Introduction

Presently the electrical Power quality has become a major concern for both utilities and customers because of the loads on the customer part get affected and might get damaged. The utilities have opted the new open-access and competitive market power policy. Now, the electricity consumers are in a unique position to demand a higher quality of service. [2, 3]. The utilities and all other power supplier have to certify a high quality service to remain competitive and Other issue which are responsible for low power quality can be considered as: Voltage sag Voltage swell; Voltage flicker; Voltage Interruption; Waveform Distortion; Voltage Fluctuation and Power Frequency Variations A brief overview of the commonly occurred PQ disturbances, their symptoms, possible causes and consequences are described in the Table 1.

2 Proposed Methodology

The proposed methodology based on the combined features of Stockwell transform and Hilbert transform for the recognition or detection of single stage and complex PQ disturbances is detailed below step by step [4–7].

- First generate the voltage signal with PQ disturbance in the MATLAB using mathematical relations.
- Decompose the voltage signal with PQ disturbance using the stockwell transform to obtain the S-matrix.

R. Aseri · A. K. Joshi (✉)
Department of Electrical Engineering, Jodhpur Institute of Engineeringand Technology, Jodhpur, India
e-mail: ashwanijoshi27@gmail.com

M. Shorif Uddin et al. (eds.), *Intelligent Energy Management Technologies*, Algorithms for Intelligent Systems,
https://doi.org/10.1007/978-981-15-8820-4_37

Table 1 Power Quality Standards

Standards	Phenomena
IEEE St. 519-1992	Harmonic control in electrical power system
IEEE St. 1159-1995	Monitoring electric power quality
IEEE St. 1531-2003	Application and specification of harmonic filter
IEEE St. 1573-2003	Electronic power subsystem
IEEE St. 1566-2005	Adjustable speed drives rated 375 KW larger
IEEE St. 995-1987	Efficiency determination of alternating current adjustable speed drives

Table 2 Power Quality Disturbance and Their Quantification

Type of Disturbance	Categories	Subcategory	Typical Spectral Content	Typical Duration	Typical Voltage Magnitude
Transient	Impulsive	Nanosecond Microsecond Millisecond	5 ns rise 1 μs rise 0.1 ms rise	< 50 ns 50 ns to 1 ms > 1 ms	–
	Oscillatory	LF MF HF	<5 kHz 5–500 kHz 0.5–5 MHz	0.3–50 ms 20 μs 5 μs	0–4 pu 0–8 pu 0–4 pu
Short-Duration Variations	Instantaneous	Sag (dip) Swells	–	0.5–30 cycles 0.5–30 cycles	0.1–0.9 pu 1.1–1.8 pu
	Momentary	Interruption Sag Swells	–	0.5 cycles-3 s 30 cycles-3 s 30 cycles-3 s	< 0.1 pu 0.1–0.9 pu 1.1–1.4 pu
	Temporary	Interruption Sag Swells	–	3 s to 1 min. 3 s to 1 min. 3 s to 1 min.	< 0.1 pu 0.1–0.9 pu 1.1–1.2 pu
Long-Duration Variations	Interruption Sustained	–	–	> 1 min.	0.0 pu
	Under voltages Over voltages	–	–	> 1 min. > 1 min.	0.8–0.9 pu 1.1–1.2 pu
Voltage Unbalance	–	–	–	Steady state	0.5–2%
Waveform Distortion	D.C. Offset	–	–	Steady State	1–0.1%
	Harmonics Inter harmonics Notching Noise	–	0–100th 0–6 kHz – Broadband	Steady State Steady State Steady State Steady State	0–20% 0–2% 0–1%
Voltage Fluctuations	–	–	<25 Hz	Intermittent	0.1–7%
Power Frequency Variations	–	–		< 10 s	–

Table 3 Different Type of Disturbance, Their Symptoms, Possible Causes and Consequences

Disturbance Type	Symptom(s)	Possible Cause(s)	Consequence(s)
Interruption	Complete loss of supply (exceeding 1 min)	• Weather, storms, lightning and heavy winds • Accidents and excavation • Planned maintenance • Line faults, blown fuse	Affects all equipment
Overvoltage	Long-term increase in supply voltage (>+ 6%)	• Light system loading • Poor regulation	Affects most equipment without internal backup facilities
Under voltage	Long-term lowering of the supply voltage (<–10%)	• Heavy, peak network loading • Lack of VAR support • Poor power factor	
Momentary Interruption	Short-term power loss (200 ms-few seconds)	• Circuit breaker tripping • Fault clearing • Bus transfer	
Voltage Swell	Medium-term (> 200 ms) duration coupled with an increasing amplitude (+10 to +30%)	• Circuit Capacitance • Switching out large loads • Load rejection • Phase fault	Protection tripping or possible damage to tabulation and windings
Voltage Transient	Short-duration (sub-cycle) impulse voltage/current spikes, large amplitude high voltage gradients	• Lightning/capacitive switching • Low fault current trip protection • Non-linear switching loads • Transmitted noise through the supply system	Control resetting and major damage to sensitive electronic components and insulation
Current Harmonics	Steady-state periodic waveforms which deform the supply signal	• Increased use of non-linear circuit elements • High frequency switches, TVs, computers and fluorescent lighting • Power factor correction capacitor • Negligent users unaware of signal pollution generated by equipment	Overheating of transformers and motor drives, increased power- loss and control command interference

(continued)

Table 3 (continued)

Disturbance Type	Symptom(s)	Possible Cause(s)	Consequence(s)
EMC (Electromagnetic Compatibility) and EMI Effects	Radiated and conducted interference, spurious signals which proliferate on the supply	• Incorrect wiring • Common disturbances between supply and earth • Series disturbance between supply and neutral • Generated by unshielded electrical equipment • Unknown effects regarding human health matters	Interference with control signals, spurious noise and induced voltages
Flicker	Series of systematic voltage fluctuations	• Variable frequency voltage variation, fluorescent lights • Erratic loads, reactive power variation	Irritating light flicker and control reset
Voltage Unbalance	3-phase load interruptions, negative sequence and unsymmetrical voltages	• Unbalanced poly-phase loads, e.g. capacitor banks and motors	Interruption of 3-phase operations

- Obtain the sum absolute values of the absolute values of S-matrix.
- Decompose the voltage signal with PQ disturbance using the Hilbert transform.
- Obtain the absolute values of the output of Hilbert transform.
- Obtain power quality index by multiplying sum absolute values of the output of S-matrix and absolute values of the output of the Hilbert transform. This power quality index clearly detects and localizes the PQ disturbances.
- Plot the voltage signal and power quality index.

3 Recognition of Single Stage Power Quality Disturbances: Simulation Results

Single stage PQ disturbances considered in the proposed study include pure sine wave, voltage swell, voltage sag, momentary interruption, oscillatory transient, impulsive transient, harmonics, notch and spike [8–12]. Analysis of above mentioned PQ disturbances using the proposed Stockwell transform and Hilbert Transform based algorithm is introduced in this chapter. Detailed analysis of the proposed or planned PQ index to detect the various single stage PQ disturbances is provided in the following subsection.

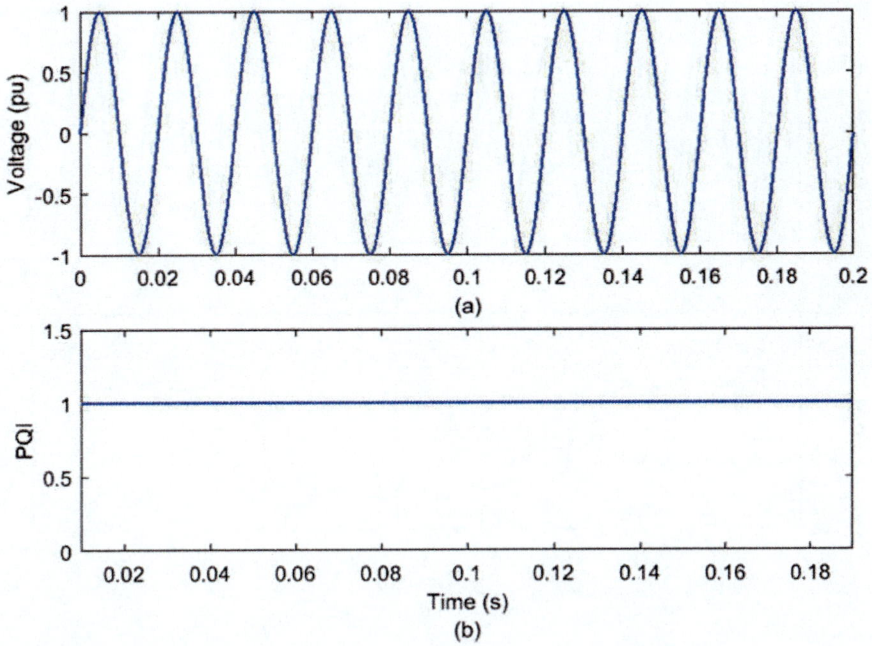

Fig. 1 Recognition of pure sinusoidal waveform of voltage signal **a** voltage signal without PQ disturbance **b** proposed PQ index based on Stockwell transform and Hilbert transform

3.1 Pure Sinusoidal Voltage Signal

See Fig. 1.

3.2 Voltage Sag

See Fig. 2.

3.3 Voltage Swell

See Fig. 3.

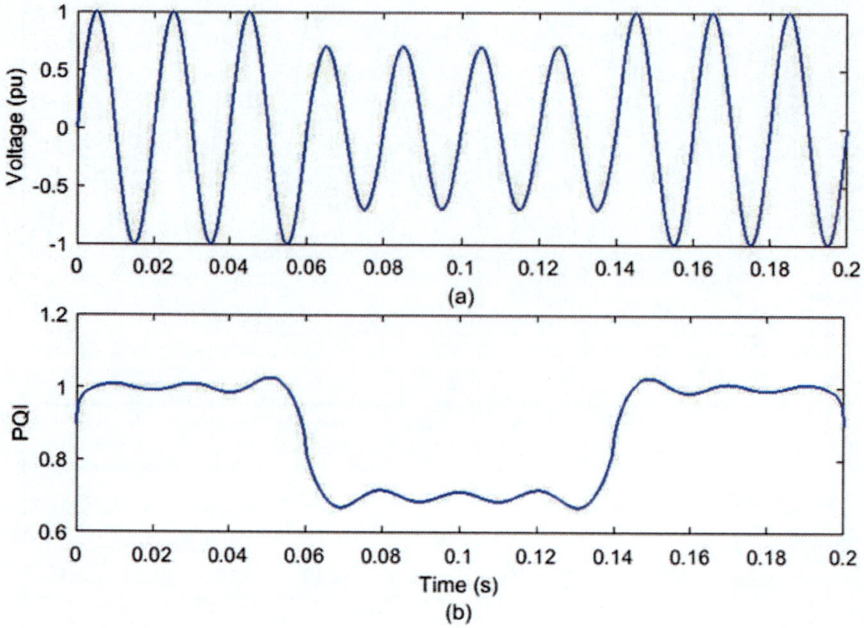

Fig. 2 Recognition of voltage sag **a** voltage signal with PQ disturbance **b** proposed PQ index based on Stockwell transform and Hilbert transform

3.4 Momentary Interruption

See Fig. 4.

3.5 Harmonics

See Fig. 5.

3.6 Flicker

See Fig. 6.

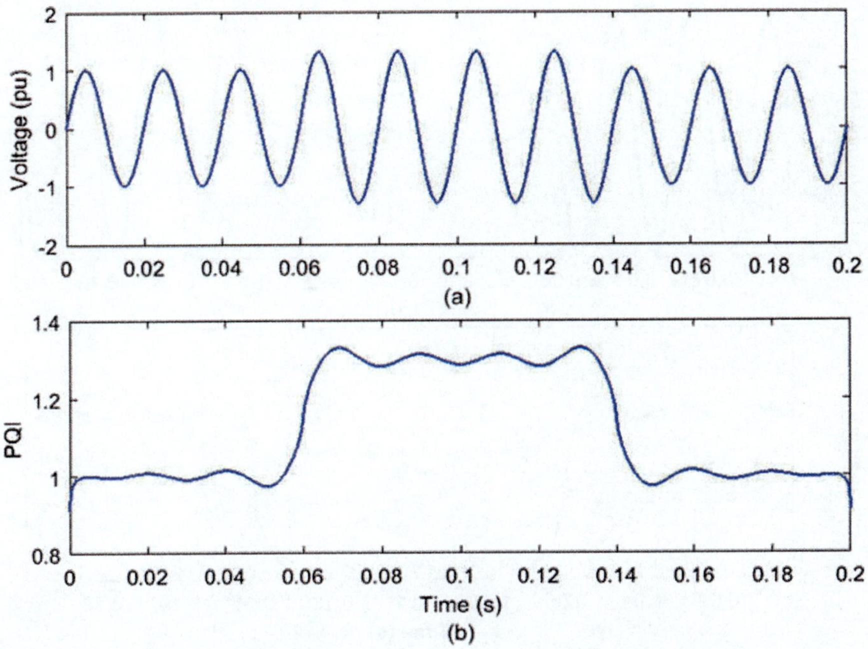

Fig. 3 Recognition of voltage swell **a** voltage signal with PQ disturbance **b** proposed PQ index based on Stockwell transform and Hilbert transform

3.7 Notch

See Fig. 7.

3.8 Spike

See Fig. 8.

3.9 Oscillatory Transient

See Fig. 9.

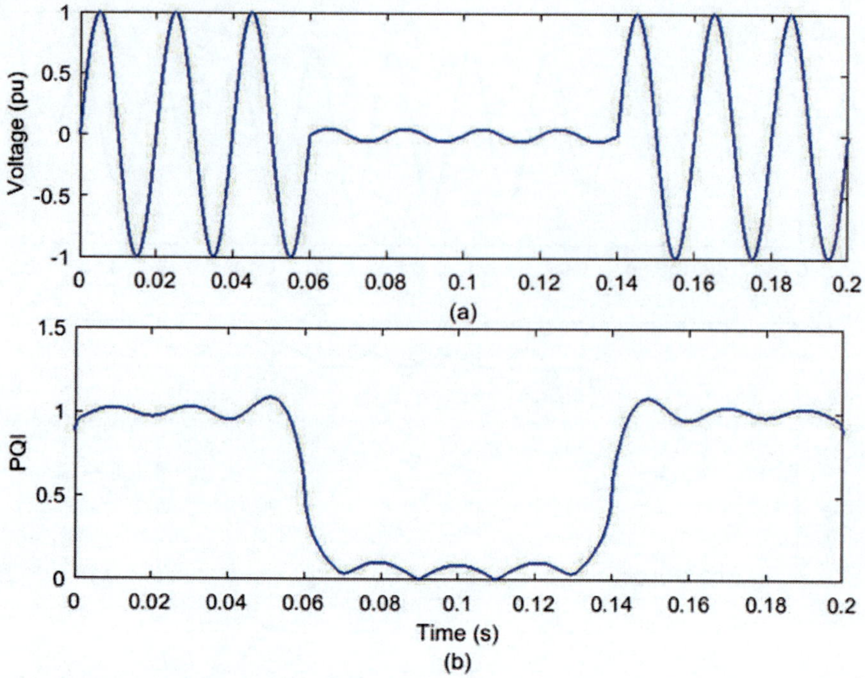

Fig. 4 Recognition of momentary interruption **a** voltage signal with PQ disturbance **b** proposed PQ index based on Stockwell transform and Hilbert transform

3.10 Impulsive Transient

See Fig. 10.

3.11 Classification of Single Stage PQ Disturbance

See Fig. 11.

The classification of the single stage PQ disturbances has been achieved using the maximum values of the proposed PQ disturbances [13–15]. The maximum value of the proposed PQ index for all the single stage PQ disturbances considered in the proposed study is provided in the Table 4.

Based on the values of proposed PQ index, flowchart for the classification of various single stage PQ disturbances is provided in Fig. 12. PQ disturbances have been classified effectively using a set of if then else rules using the maximum values of the proposed PQ index.

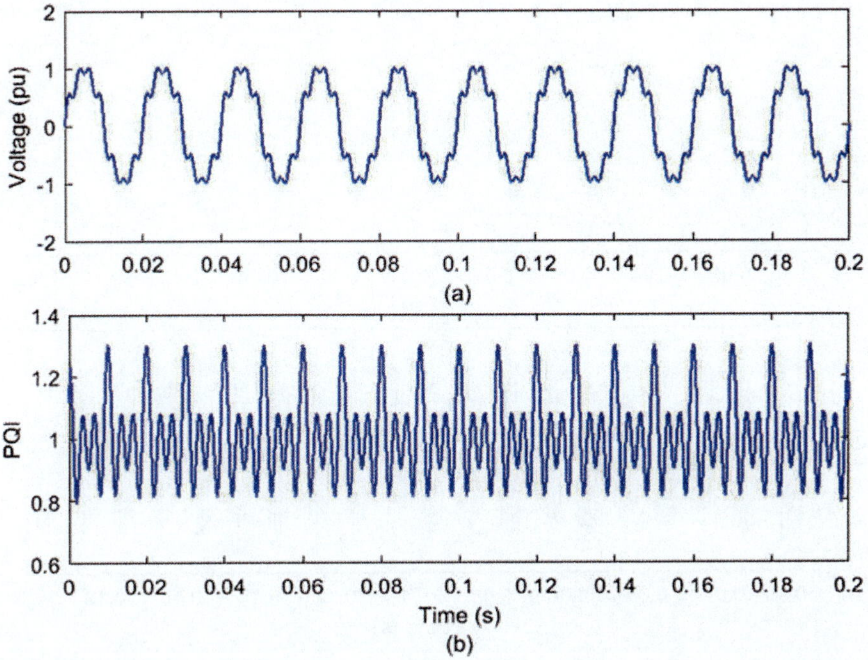

Fig. 5 Recognition of harmonics **a** voltage signal with PQ disturbance **b** proposed PQ index based on Stockwell transform and Hilbert transform

4 Recognition of Complex Power Quality Disturbances: Simulation Results

The simulation results related to detection of complex (multiple) PQ disturbances using the proposed algorithm are described in this point. Complex PQ disturbances considered in the proposed study include voltage sag with harmonics, voltage swell with harmonics, momentary interruption with harmonics, Flicker and Harmonics, Voltage Sag and Oscillatory Transient, Voltage Swell and Oscillatory Transient, Momentary Interruption and Oscillatory Transient, Flicker and Oscillatory Transient, Harmonics and Oscillatory Transient, Voltage Sag and Impulsive Transient, Voltage Swell and Impulsive Transient, Momentary Interruption and Impulsive Transient, Flicker and Impulsive Transient, Harmonics and Impulsive Transient, Voltage Sag and Spike, Voltage Sag and Harmonics and Oscillatory Transient, Flicker and Harmonics and Impulsive Transient, Voltage Sag, Oscillatory Transient, Harmonics and Impulsive Transient. Analysis of above mentioned complex PQ disturbances using the proposed Stockwell transform and Hilbert Transform based algorithm is presented in this chapter [16–20]. Detailed analysis of the proposed PQ index to detect the various complex PQ disturbances is provided in the following subsection.

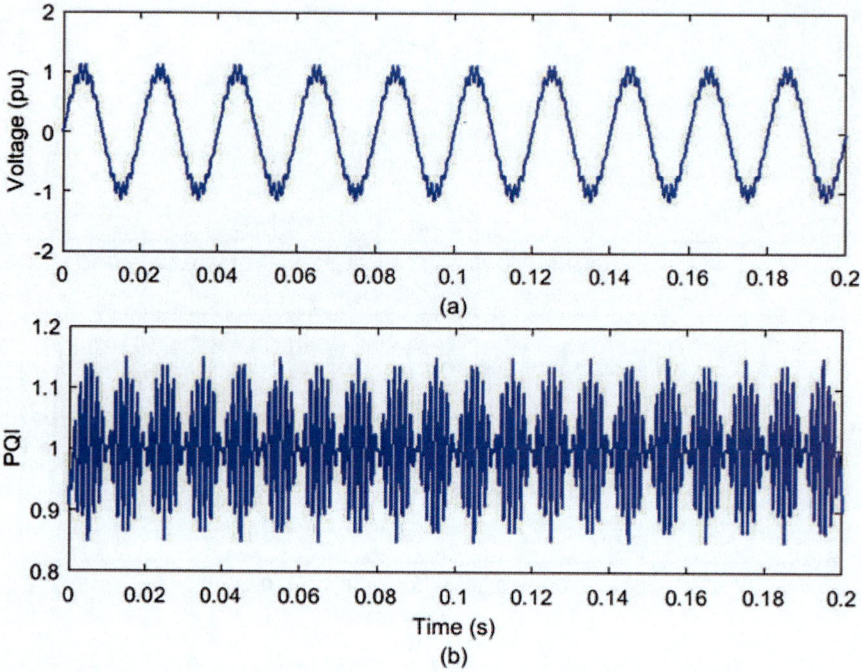

Fig. 6 Recognition of flicker **a** voltage signal with PQ disturbance **b** proposed PQ index based on Stockwell transform and Hilbert transform

4.1 Voltage Sag with Harmonics

See Fig. 11.

4.2 Momentary Interruption and Harmonics

See Fig. 12.

4.3 Flicker and Harmonics

See Fig. 13.

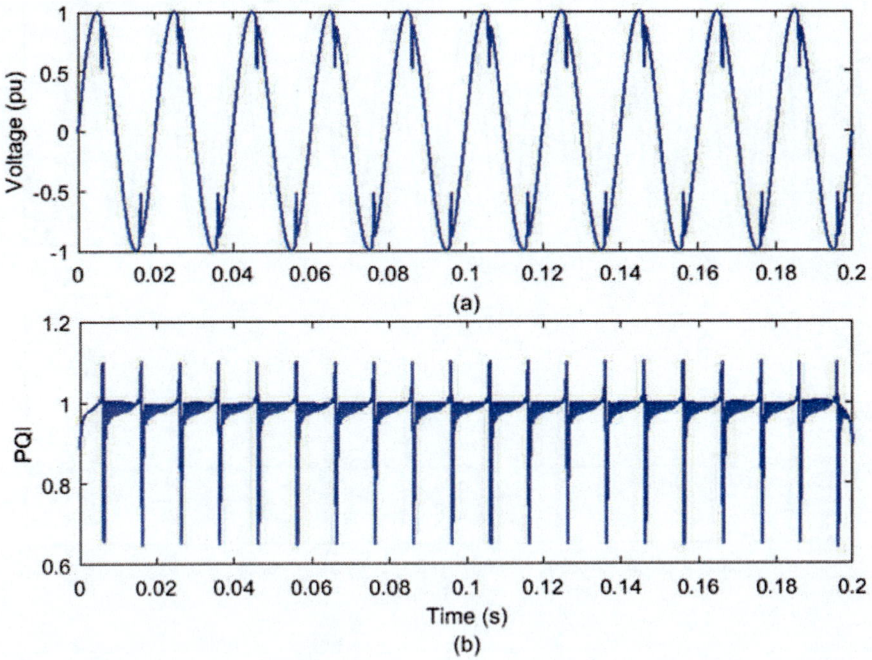

Fig. 7 Recognition of notch **a** voltage signal with PQ disturbance **b** proposed PQ index based on Stockwell transform and Hilbert transform

4.4 Voltage Swell and Oscillatory Transient

See Fig. 14.

4.5 Momentary Interruption and Oscillatory Transient

See Fig. 15.

4.6 Flicker and Oscillatory Transient

See Fig. 16.

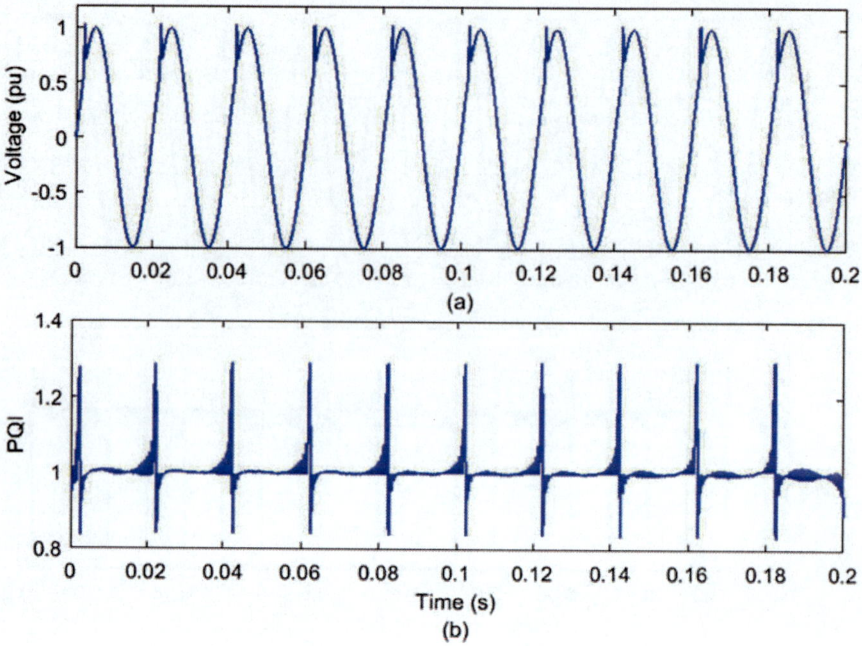

Fig. 8 Recognition of spike **a** voltage signal with PQ disturbance **b** proposed PQ index based on Stockwell transform and Hilbert transform

4.7 Harmonics and Oscillatory Transient

See Fig. 17.

4.8 Voltage Sag and Impulsive Transient

See Fig. 18.

4.9 Voltage Swell and Impulsive Transient

See Fig. 19.

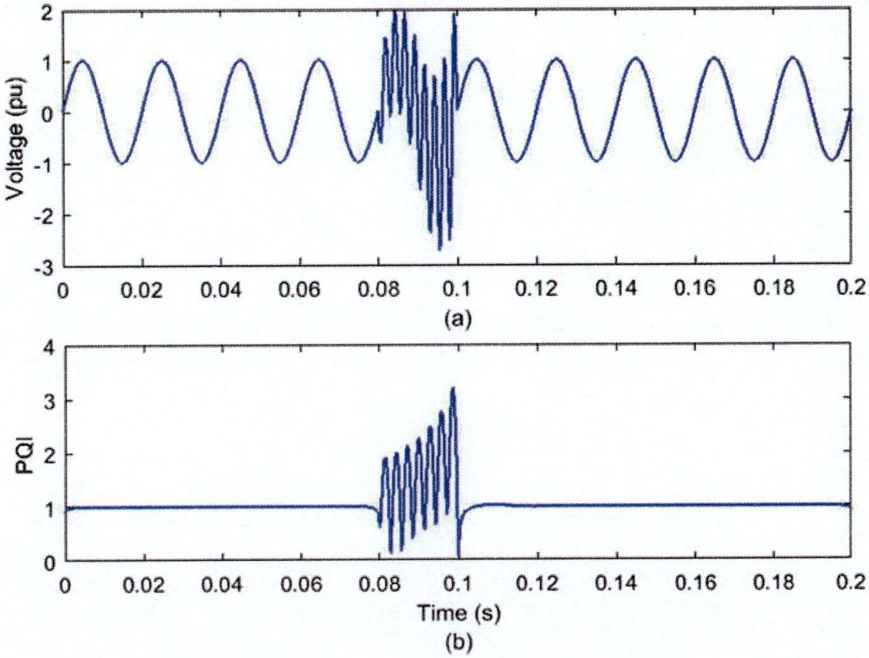

Fig. 9 Recognition of oscillatory transient **a** voltage signal with PQ disturbance **b** proposed PQ index based on Stockwell transform and Hilbert transform

4.9.1 Classification of Complex Power Quality Disturbances

The classification of complex stage PQ disturbances has been achieved using the maximum values of the proposed PQ disturbances [16–20]. The maximum value of the proposed PQ index for all the complex PQ disturbances considered in the proposed study is provided in the Table 3. Based on the values of the proposed PQ index, flowchart for the classification of various complex PQ disturbances is provided in Fig. 4. PQ disturbances have been classified effectively using a set of if then else rules using the maximum values of the proposed PQ index.

5 Conclusions

Detection of Single Stage Power Quality Disturbances

An algorithm based on the combined features of Stockwell transform and Hilbert transform has been proposed for the recognition of single stage PQ disturbances. Voltage signal with single stage PQ disturbance has been decomposed using the Stockwell transform to obtain the S-matrix. Sum absolute values of the absolute

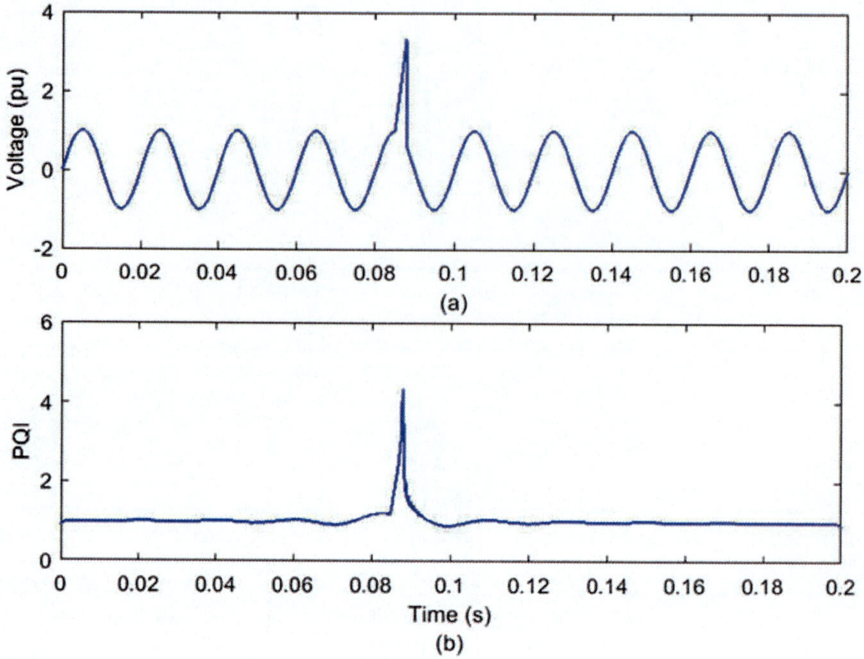

Fig. 10 Recognition of impulsive transient **a** voltage signal with PQ disturbance **b** proposed PQ index based on Stockwell transform and Hilbert transform

values of S-matrix are obtained. Further, the voltage signal with PQ disturbance is also decomposed using the Hilbert transform and absolute values of the output of Hilbert transform is obtained. Proposed PQ index is obtained by multiplying sum absolute values of the output of S-matrix and absolute values of output of the Hilbert transform. This PQ index clearly detects and localizes the PQ disturbances. Classification of the PQ disturbances has been achieved using the maximum values of proposed PQ index. Investigated PQ disturbances include voltage sag, voltage swell, momentary interruption, oscillatory transient, impulsive transient, harmonics, notch and spike. This has been concluded that proposed approach based on combined features of the Stockwell Transform and Hilbert Transform is found to be effective in the recognition of the single stage PQ disturbances. This algorithm is more effective compared to the algorithm based on the discrete wavelet transform.

Detection of Complex Power Quality Disturbances

Proposed algorithm based on the combined features of Stockwell transform and Hilbert transform has also been tested for recognition of complex PQ disturbances. Voltage signal with complex PQ disturbance has been decomposed using the Stockwell transform to obtain S-matrix. Sum absolute values of the absolute values of S-matrix are obtained. Further, the voltage signal with complex PQ disturbance is also decomposed using the Hilbert transform and absolute values of output of Hilbert

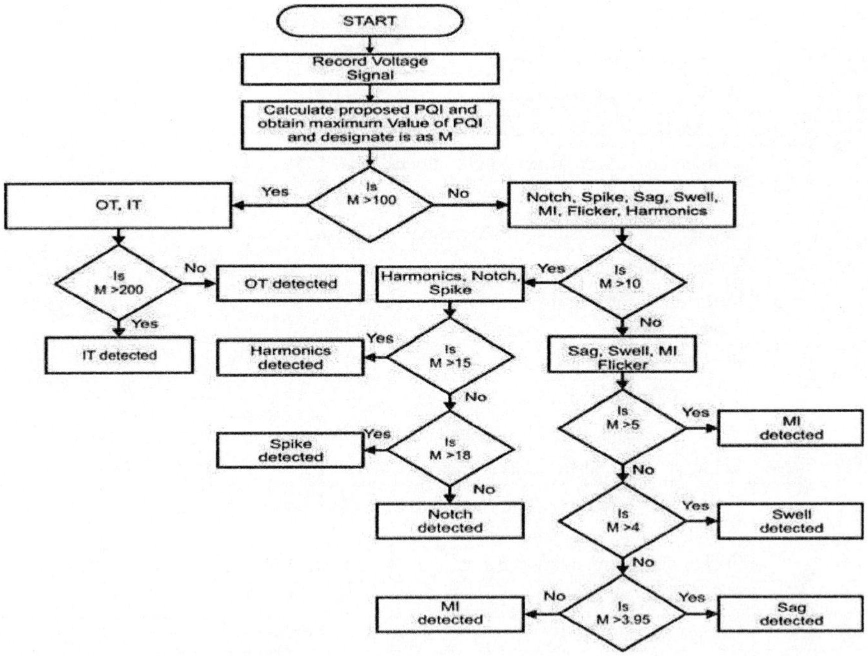

Fig. 11 Flowchart for classification of single stage PQ disturbances

Table 4 Maximum values of Proposed PQ Index For Single Stage PQ Disturbances

Sr. No.	Operational Event	Maximum value of PQI Magnitude
1	Pure sine wave	3.9895
2	Voltage sag	3.99
3	Voltage swell	4.0437
4	Momentary Interruption	3.8951
5	Harmonics	11.2380
6	Flicker	8.2938
7	Notch	15.5470
8	Spike	19.3988
9	Oscillatory Transient	148.0551
10	Impulsive Transient	304.4628

transform is obtained. Proposed PQ index is obtained by multiplying sum absolute values of output of S-matrix and absolute values of output of the Hilbert transform. This PQ index clearly detects and localizes the complex PQ disturbances. Classification of complex PQ disturbances has been achieved using the maximum values of proposed PQ index. Investigated complex PQ disturbances include voltage sag with

Table 5 Maximum Values Of Proposed Power Quality Index For Complex PQ Disturbances

Sr. No.	Complex PQ Disturbance	Maximum value of PQI Magnitude
1	Voltage Sag with Harmonics	11.2121
2	Voltage Swell and Harmonics	13.2071
3	Momentary Interruption and Harmonics	09.1619
4	Flicker and Harmonics	12.2506
5	Voltage Sag and Oscillatory Transient	142.6547
6	Voltage Swell and Oscillatory Transient	148.6776
7	Momentary Interruption and Oscillatory Transient	138.9654
8	Flicker and Oscillatory Transient	145.7373
9	Harmonics and Oscillatory Transient	137.9313
10	Voltage Sag and Impulsive Transient	304.8193
11	Voltage Swell and Impulsive Transient	310.1465
12	Momentary Interruption and Impulsive Transient	300.7539
13	Flicker and Impulsive Transient	285.6506
14	Harmonics and Impulsive Transient	289.0105
15	Voltage Sag and Spike	19.3822
16	Voltage Sag, Harmonics and Oscillatory Transient	135.5474
17	Flicker, Harmonics and Impulsive Transient	271.0759
18	Voltage Sag, Oscillatory Transient, Harmonics and Impulsive Transient	161.7251

harmonics, voltage swell with harmonics, momentary interruption with harmonics, Flicker and Harmonics, Voltage Sag and Oscillatory Transient, Voltage Swell and Oscillatory Transient, Momentary Interruption and Oscillatory Transient, Flicker and Oscillatory Transient, Harmonics and Oscillatory Transient, Voltage Sag and Impulsive Transient, Voltage Swell and Impulsive Transient, Momentary Interruption and Impulsive Transient, Flicker and Impulsive Transient, Harmonics and Impulsive Transient, Voltage Sag and Spike, Voltage Sag and Harmonics and Oscillatory Transient, Flicker and Harmonics and Impulsive Transient, Voltage Sag, Oscillatory Transient, Harmonics and Impulsive Transient. This has been concluded that proposed approach based on combined features of the Stockwell Transform and Hilbert Transform is found to be effective in the recognition of the complex PQ disturbances. This algorithm is more effective compared to the algorithm based on the discrete wavelet transform.

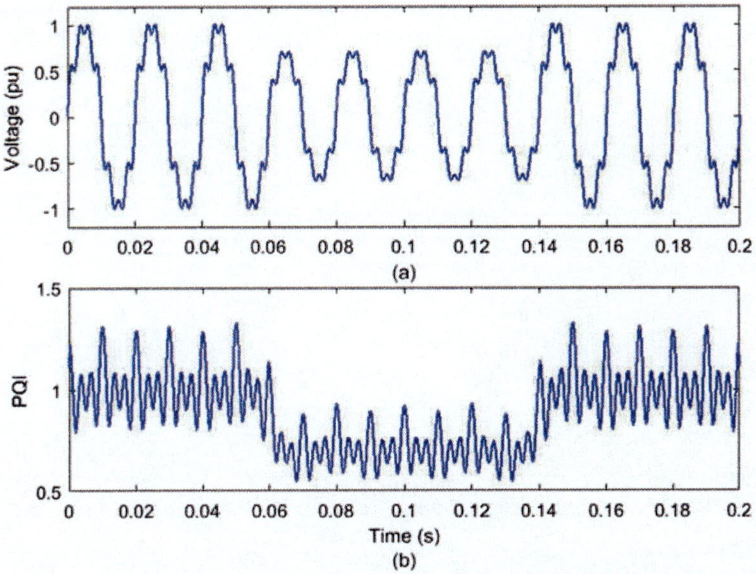

Fig. 12 Recognition of simultaneous occurrence of voltage sag and harmonics **a** voltage signal with PQ disturbance **b** proposed PQ index based on Stockwell transform and Hilbert transform

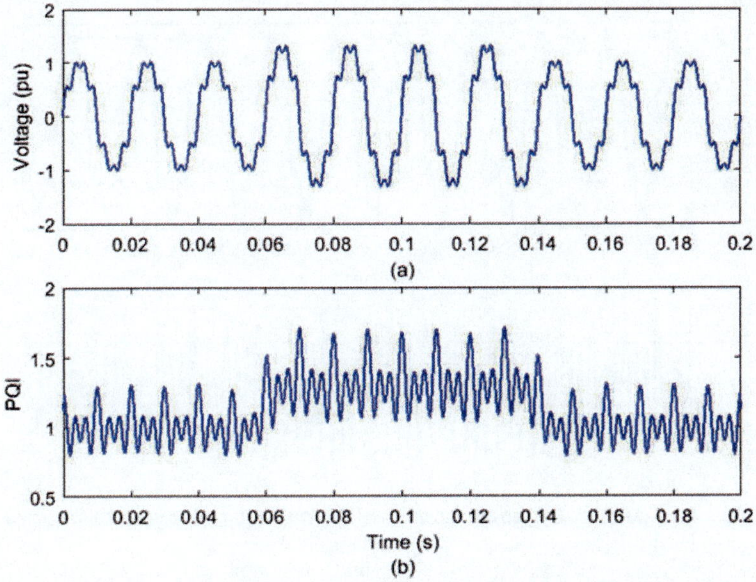

Fig. 13 Recognition of simultaneous occurrence of voltage swell and harmonics **a** voltage signal with PQ disturbance **b** proposed PQ index based on Stockwell transform and Hilbert transform

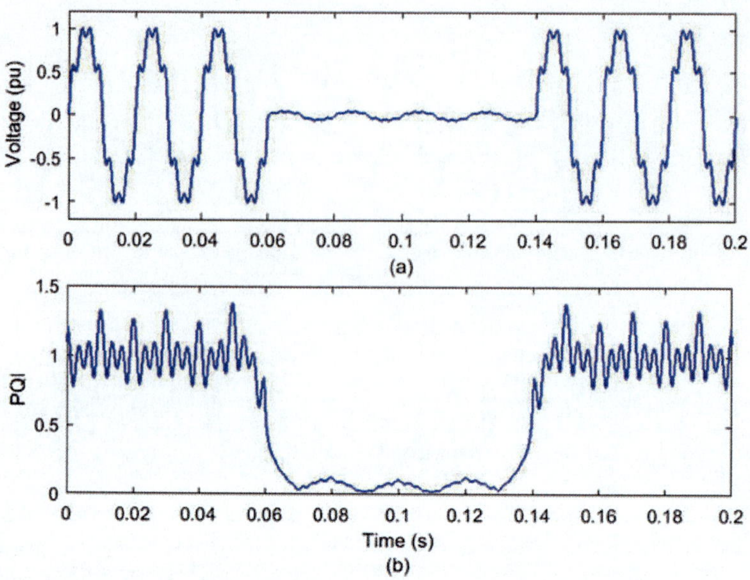

Fig. 14 Recognition of simultaneous occurrence of momentary interruption and harmonics **a** voltage signal with PQ disturbance **b** proposed PQ index based on Stockwell transform and Hilbert transform

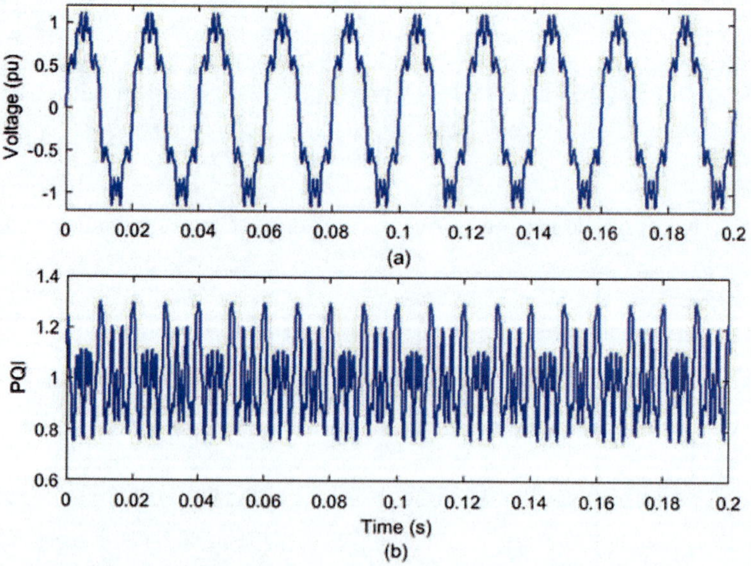

Fig. 15 Recognition of simultaneous occurrence of flicker and harmonics **a** voltage signal with PQ disturbance **b** proposed PQ index based on Stockwell transform and Hilbert transform

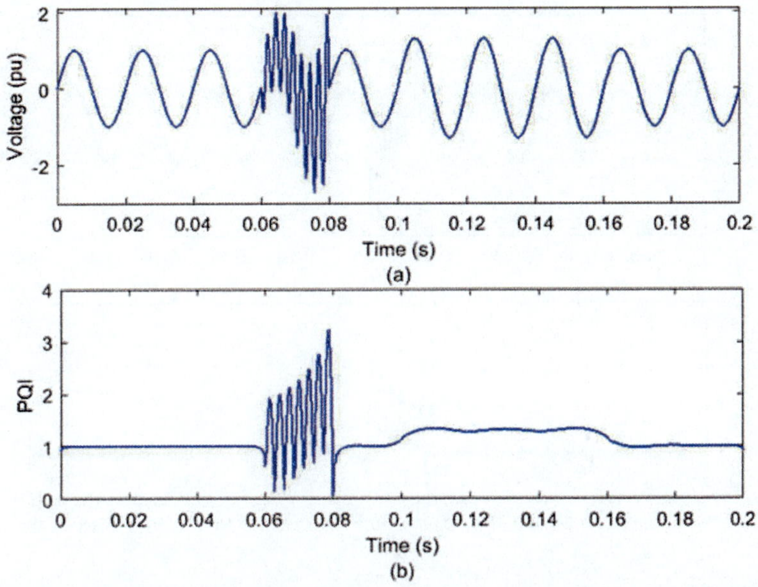

Fig. 16 Recognition of simultaneous occurrence of voltage swell and oscillatory transient **a** voltage signal with PQ disturbance **b** proposed PQ index based on Stockwell transform and Hilbert transform

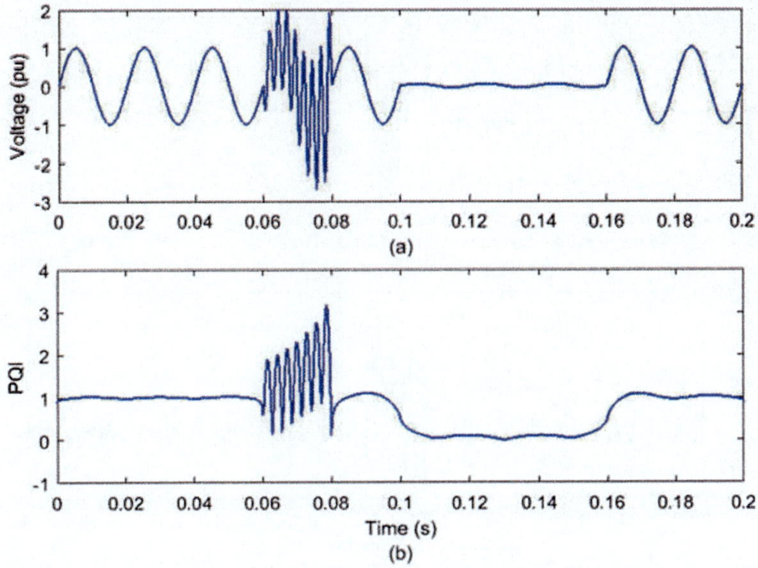

Fig. 17 Recognition of simultaneous occurrence of momentary interruption and oscillatory transient **a** voltage signal with PQ disturbance **b** proposed PQ index based on Stockwell transform and Hilbert transform

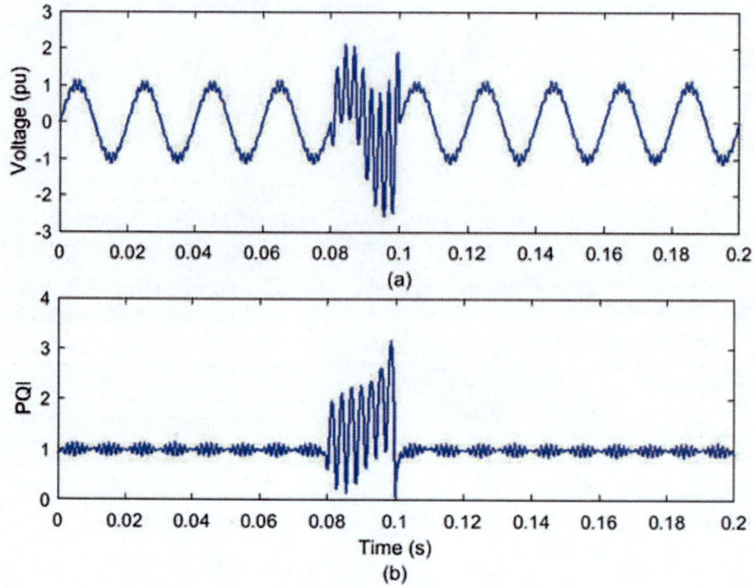

Fig. 18 Recognition of simultaneous occurrence of flicker and oscillatory transient **a** voltage signal with PQ disturbance **b** proposed PQ index based on Stockwell transform and Hilbert transform

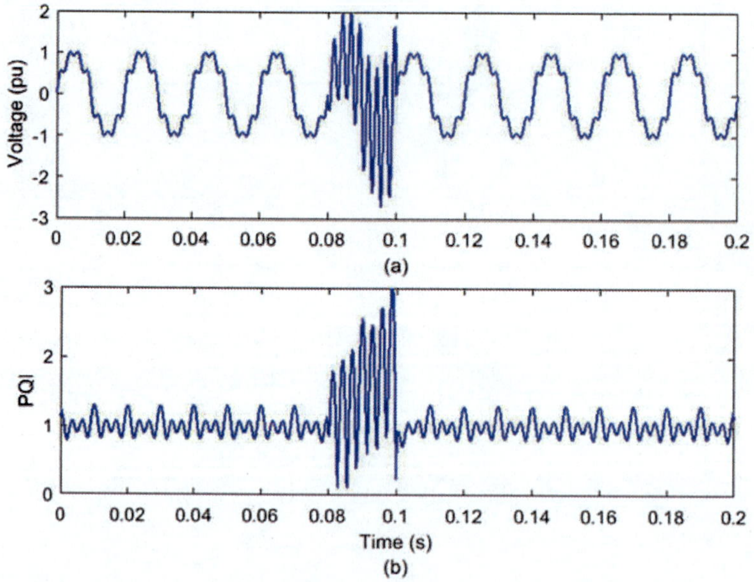

Fig. 19 Recognition of simultaneous occurrence of harmonics and oscillatory transient **a** voltage signal with PQ disturbance **b** proposed PQ index based on Stockwell transform and Hilbert transform

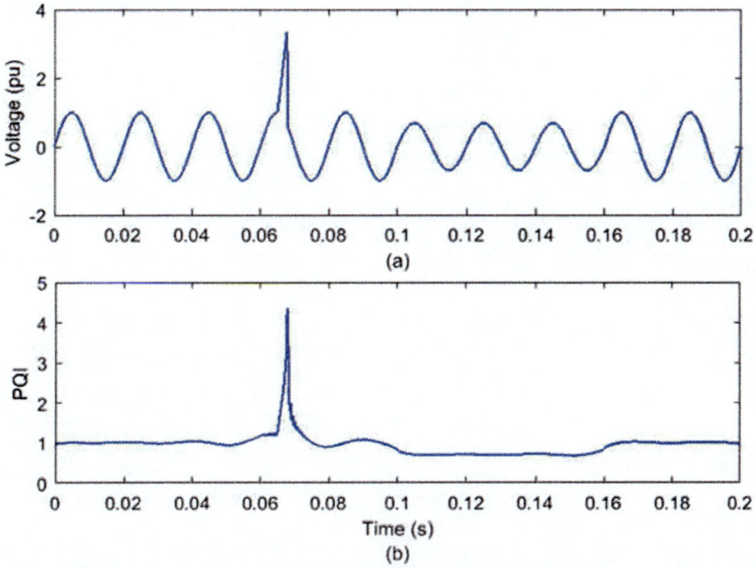

Fig. 20 Recognition of simultaneous occurrence of voltage sag and impulsive transient **a** voltage signal with PQ disturbance **b** proposed PQ index based on Stockwell transform and Hilbert transform

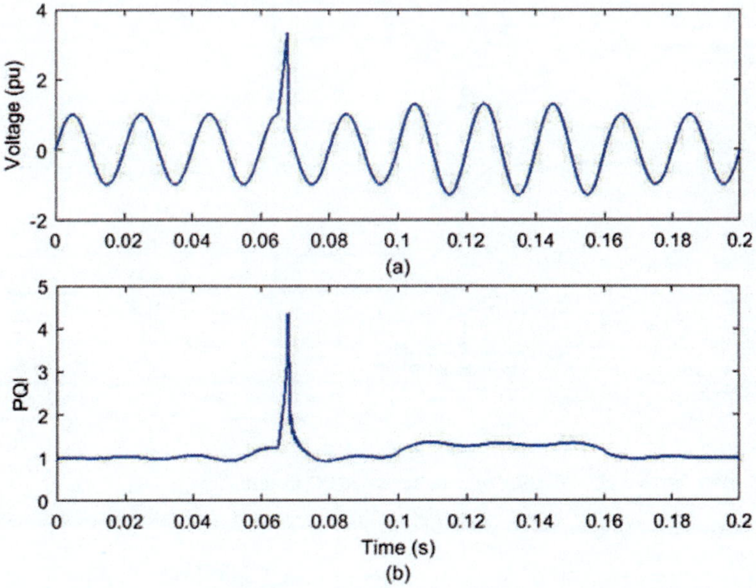

Fig. 21 Recognition of simultaneous occurrence of voltage swell and impulsive transient **a** voltage signal with PQ disturbance **b** proposed PQ index based on Stockwell transform and Hilbert transform

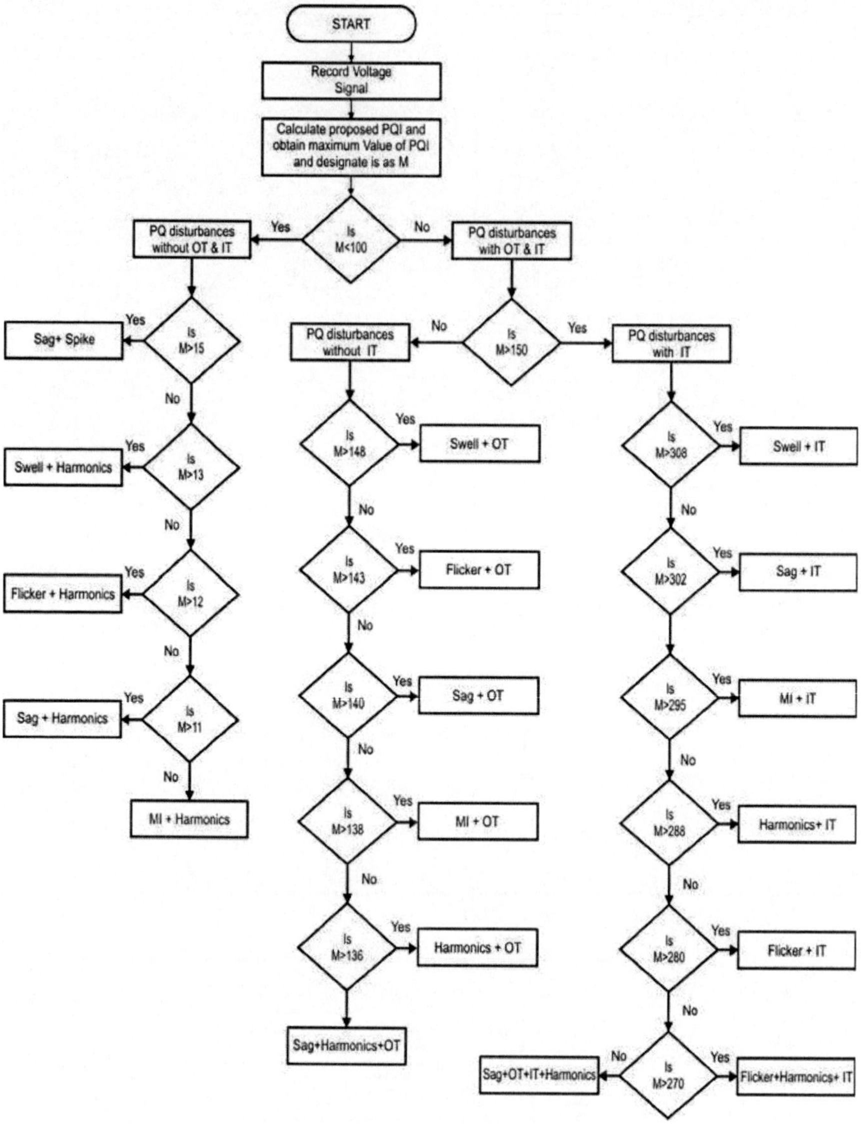

Fig. 22 Flowchart for classification of complex stage PQ disturbances

Appendix A

STOCKWELL TRANSFORM PROGRAMMING

```
function ST=stran(h)
% COMPUTE S-TRANSFORM WITHOUT FOR LOOPS
[~,N]=size(h); % h is a 1xN one-dimensional series
nhaf=fix(N/2);
odvn=1;
if nhaf*2==N;
   odvn=0;
end
f=[0:nhaf -nhaf+1-odvn:-1]/N;
Hft=fft(h);
%Compute all frequency domain Gaussians as one matrix
invfk=[1./f(2:nhaf+1)]';
W=2*pi*repmat(f,nhaf,1).*repmat(invfk,1,N);
G=exp((-W.^2)/2); %Gaussian in freq domain
% End of frequency domain Gaussian computation
% Compute Toeplitz matrix with the shifted fft(h)
HW=toeplitz(Hft(1:nhaf+1)',Hft);
% Exclude the first row, corresponding to zero frequency
HW=[HW(2:nhaf+1,:)];
% Compute Stockwell Transform
ST=ifft(HW.*G,[],2); %Compute voice
%Add the zero freq row
st0=mean(h)*ones(1,N);
ST=[st0;ST];
end
```

References

1. Lima MAA, Cerqueira AS, Courya DV, Duqueb CA (2012) A novel method for power quality multiple disturbance decomposition based on Independent Component Analysis. International Journal of Electrical Power and Energy Systems, vol. 42, pp 593–604
2. Saurabh Kamble and Ishita Dupare, "Detection of power quality disturbances using Wavelet Transform and artificial neural network," International Conference on Magnetics, Machines and Drives, 2014
3. Rahul Dubey, S. R.Samantaray, B. Chitti Babu and S. Nandha Kumar, "Detection of power quality disturbances in presence of DFIG wind farm using Wavelet Transform based energy function," IEEE International Conference, 2011

4. Norman C.F.Tse, John Y.C.Chan, Wing-Hong Lau and Loi Lei Lai, "Hybrid Wavelet and Hilbert Transform with frequency-shifting decomposition for power quality analysis," IEEE Transactions on Instrumentation and Measurement, vol. 61, no. 12, pp-3225–3233, Dec 2012
5. Mahela OP, Shaik AG, Gupta N (2015) A critical review of detection and classification of power quality events. Renew Sustain Energy Rev 41:495–505
6. Tse NCF, Chan JYC, Lau W-H, Lai LL (2012) Hybrid wavelet and hilbert transform with frequency-shifting decomposition for power quality analysis. Instrumentation and Measurement, IEEE Transactions on 61(12):3225–3233
7. Stockwell RG, Mansinha L, Lowe RP (1996) Localization of the complex spectrum: the s transform. Signal Processing, IEEE Transactions on 44(4):998–1001
8. Eristi H, Yildirim O, Eristi B, Demir Y (2014) Automatic recognition system of underlying causes of power quality disturbances based on s-transform and extreme learning machine. International Journal of Electrical Power Energy Systems 61:553–562
9. Biswal M, Dash PK (2013) Detection and characterization of multiple power quality disturbances with a fast s-transform and decision tree based classifier. Digit Signal Proc 23(4):1071–1083
10. Kaddah SS, Abo-Al-Ez KM, Megahed TF, Osman MG (2016) Probabilistic power quality indices for electric grids with increased penetration level of wind power generation. International Journal of Electrical Power Energy Systems 77:50–58
11. Ray PK, Mohanty SR, Kishor N (2011) Disturbance detection in grid-connected distributed generation system using wavelet and s-transform. Electr Power Syst Res 81(3):805–819
12. C.-T. Hsu, R. Korimara, and T.-J. Cheng, "Power quality analysis for the distribution systems with a wind power generation system," Computers Electrical Engineering, pp. –, 2015
13. E.A. Cano Plataa and H.E. Tacca, "Power load identification," Journal of the Franklin Institute, vol. 342, pp 97–113, AUG2004
14. Akhbardeha Alireza, Junnilaa Sakari, Koivuluomaa Mikko, Koivistoinenb Teemu (Oct 2007) Vaino¨ Turjanmaab, Tiit Koobib and Alpo Varria, "Towards a heart disease diagnosing system based on force sensitive chair's measurement, biorthogonal wavelets and neural networks". Eng Appl Artif Intell 20:493–502
15. Moravej Zahra (2015) Jamal Dehghani Ashkezari and Mohammad Pazoki, "An effective combined method for symmetrical faults identification during power swing". Electrical Power and Energy Systems 64:24–34
16. Bhim Singh Sunil Kumar Dube and Sabha Raj Arya, "An improved control algorithm of DSTATCOM for power quality improvement," Electrical Power and Energy Systems, vol. 64,pp 493–504, AUGUST 2015
17. Sundarabalan CK, Selvi K (2015) Compensation of voltage disturbances using PEMFC supported Dynamic Voltage Restorer. Electrical Power and Energy Systems 71:77–92
18. Om Prakash Mahela and Abdul Gafoor Shaik, "Power quality improvement in distribution network using DSTATCOM with battery energy storage system," Electrical Power and Energy Systems, vol. 83, pp 229–240, APRIL 2016
19. C.K. Sundarabalan and K. Selvi, "PEM fuel cell supported distribution static compensator for power quality enhancement in three-phase four-wire distribution system," international journal of hydrogen energy, vol. 39, pp 19051–19066, OCT. 2014
20. N. Srinivasa Raoa, Dr. A. Selwin Mich Priyadharsonb and Dr. J. Praveen, "Simulation of Artificial Intelligent Controller based DVR for Power Quality Improvement," Procedia Computer Science, vol.47, pp 153 – 167, july 2015

Chapter 38
Detection and Classification of Transmission Line Faults Using Combined Features of Stockwell Transform, Hilbert Transform, and Wigner Distribution Function

Tanmay Bhati and Harish Kumar Khyani

1 Introduction

Introduction mainly focuses on the detection and classification of transmission line faults using combined features of the Stockwell transform, Hilbert transform, and Wigner Distribution features. Timely detection of fault is main aspect of any kind of High-Tension lines weather it is of transmission system or distribution system. Detection and classification subjects are to be studied in details to avoid repetitive and dummy trappings of system. Each time when a system exhibits a fault it creates jerk on equipment's and affects stability as well. All this factor is necessary to design a precise relay system looking to the study parameters [1].

1.1 Types of Faults

Following are the faults generally take place in power system:

- **Over current**: It occurs mainly due to short-circuit or leakage or due to corona effect and sometimes due to overload on the supply system.
- **Under volt**age: It mainly occurs either on the failure of alternator's field or short circuits because of more voltage drop in machines and lines.
- **Unbalance**: It occurs either on two phases or breaking of one of the conductors or grounding of one or on a short circuit of two phases. In such type of condition,

T. Bhati (✉) · H. K. Khyani
Department of Electrical Engineering, Jodhpur Institute of Engineering & Technology, Jodhpur, India
e-mail: tanmaybhati@gmail.com

M. Shorif Uddin et al. (eds.), *Intelligent Energy Management Technologies*, Algorithms for Intelligent Systems,
https://doi.org/10.1007/978-981-15-8820-4_38

different current flow through different phases and fault is known as an unbalanced fault.

- **Reversed power**: This fault occurs only in interconnected systems. On failure of the field, a generator starts working as a motor and takes power instead of delivering the power means the flow of power is reversed. Similarly in case of feeders which are connected in parallel, if some fault occurs on any one of the feeders, then the fault is fed from both the ends, means again direction of flow of power is reversed in case of the faulty feeder.
- **Surges**: Whenever a severe fault occurs or lightning takes place in the neighboring circuits, then in the lines, some short-lived waves of very high voltages and currents are setup. It may be considered as a high voltage of high frequency and such fault is known as the surge.
- **Short-circuit currents**

When the current gets least resistive path or non-resistive path a very heavy current flow through this least resistive path or non-resistive path. This phenomenon is known as short-circuiting. The heavy current is known as short-circuit current. This short-circuit current is very harmful to the equipment. It is necessary to disconnect the faulty section to save the equipment from complete interruption of supply and damage for which protective gear such as fuses, circuit of suitable capacity are required. The fault in three-phase power system may occur:

- Single phase to earth,
- Phase to phase,
- Two phases to earth,
- Phase to phase and at the same time from the third phase to earth,
- All the three phases to earth,
- All the three phases short-circuited.

The first four types of fault are called an unsymmetrical fault. While the latter two types of fault are called a symmetrical fault. Although the symmetrical fault gives the more severe duty on the circuit breaker and the unsymmetrical faults are more prevalent. The calculations of a symmetrical short circuit are very important for the purpose of determination of circuit breaker ratings.

Various Signal Processing Techniques like wavelet transform (WT), artificial neural network (ANN), Fuzzy logic (FL) etc. are used for the identification of faults and its classification for the protection system [2]. In this context, a comprehensive survey is performed to review conceptually, intelligent fault detection and classification techniques that are worked upon for diagnosis of fault on transmission lines [3]. A simplified fault locating approach which can be used for multi-terminal transmission lines using unsynchronized measurements is proposed by the author in [4]. Procedures for data synchronization are deployed for spotting/identifying faulted phase before location of fault location has been initiated. Fault location algorithm will not depend on variations in the fault resistance and variations source impedance. Location algorithm and faulted phase is effectively tested for all types of faults and various high fault resistances have been used. Result states that newly proposed multi-terminal fault location algorithm is not only fast and accurate but also immune to power system transients. In [5] authors proposed a method supported by WT fort identification, classification and location of faults on power transmission lines. A Global Positioning System clock is being used to synchronize sampling of voltage and current signals at both the ends of the transmission line. The detailed coefficients of current signals of both the ends are employed or utilized to calculate fault indices. These resultant fault indices are compared with threshold values for the detection and classification of the faults. In order to locate the fault (or to identify the location of fault), Artificial Neural Networks are being employed, which uses the approximate decompositions of the voltages and currents of the local end [6]. Accurate estimation of fault location on transmission line is required for inspection of line, maintenance of line, and rectification of fault incident on line. Hence, for protection relay functioning, the identification of fault location is really very important subjects in power system [7]. Various mathematical and signal processing techniques have been used for recognition and classification of faults on transmission. Yadav and Swetapadma [8] discussed a fault detection and classification algorithm which effectively protects the transmission line against the faults is proposed. Algorithm is based on the combination of Wavelet transform (WT) and linear discriminant analysis (LDA).

2 Proposed Test System and Algorithm for Recognition of Faults on Transmission Line

In this section, the proposed test transmission line used for the study of detection and classification of transmission line faults is detailed. The methodology based on combined features of Stockwell transform, Hilbert transform, and Wigner Distribution Function has been illustrated for implementation using current signals. A brief description of the Stockwell transform, Hilbert transform, and Wigner Distribution Function is also presented.

2.1 Stockwell's Transform

There are some different methods of achieving the S-transform. At this point, the relationship between S-transform and STFT is introduced and the type of S-transform is derived from the "phase correction" of the CWT. The short-time Fourier transform of a signal, i.e., h(t) is defined by means of the following relation:

$$STFT(\tau, f) = \int_{-\infty}^{+\infty} h(t)g(\tau - t)e^{-j2\pi ft}dt \quad (1)$$

where τ and f denote the time of spectral localization and Fourier frequency, respectively, and window function is denoted by $g(t)$. This S-transform may be derived with the help of the above equation by replacing $g(t)$ i.e. the window function by means of the Gaussian function which is shown below.

$$g(t) = \frac{|f|}{\sqrt{2\pi}} e^{-\frac{f^2 t^2}{2}} \quad (2)$$

Then the S-transform is defined as

$$S(\tau, f) = \int_{-\infty}^{+\infty} h(t) \frac{|f|}{\sqrt{2\pi}} e^{\frac{f^2(\tau - t)^2}{2}} e^{-j2\pi f t} dt \quad (3)$$

Hence, the S-transform is a special case of STFT with a Gaussian window function. If the window of S-transform is wider in time domain, S-transform can provide better frequency resolution for lower frequency. While the window is narrower, it can provide better time resolution for higher frequency. The output of the S-transform is a matrix known as S-matrix. The information related to the frequency and amplitude of the signal can be derived from the S-matrix.

2.2 *Hilbert Transform*

The main objective of the Hilbert Transform is to express or demonstrate a substitute/alternate process to present spectral analysis tools intended to provide the time–frequency–energy depiction of time series data. Also, the method attempts to describe non-stationary data locally. In order to calculate instantaneous frequencies as well as amplitudes and explain the signal more locally, Hilbert transform was used rather than a wavelet-based transform or Fourier transform. The below given equation explains the Hilbert transform, which can be explained/written for any function, i.e., x(t) of LP class. PV denotes Cauchy's principal value integral.

$$H[x(t)] = \frac{1}{\pi} PV \int_{-\infty}^{\infty} \frac{x(\tau)}{t - \tau} d\tau \quad (4)$$

Not all functions give "good" Hilbert transforms, meaning those which produce physical instantaneous frequencies. For example, functions with nonzero means will give negative frequency contributions using the Hilbert transform. Therefore, the signals which can be analyzed using the Hilbert transform must be restricted so that their calculated instantaneous frequency functions have a physical connotation. Subsequently, it describes the empirical mode decomposition. Essentially, it is an algorithm which decomposes nearly every signal to a finite/limited set of functions, which contain "good quality" Hilbert transforms that create physically meaningful instantaneous frequencies.

2.3 Wigner Distribution Function

The Wigner distribution function (WDF) is utilized in signal processing to transform in time–frequency analysis. The Wigner distribution function was initially proposed by Eugene Wigner in physics to account for quantum corrections toward classical statistical mechanics in the year 1932, and it is of much significance in quantum mechanics in phase space (Wigner quasi-probability distribution is also called the Wigner function or the Wigner–Ville distribution). The shared algebraic structure between time–frequency conjugate pairs and position–momentum is specified, it too beneficially serves in signal processing, as a change/transform in time–frequency analysis, which is the main subject of this article. In comparison with a short-time Fourier transform such as the Gabor transform, the Wigner distribution function which will be able to furnish superior lucidity in some cases. The Gabor transform is given by the following mathematical relation:

$$G_x(t, f) = \int_{-\infty}^{\infty} e^{-\pi(\tau-t)^2} e^{-j2\pi f\tau} x(\tau) d\tau \tag{5}$$

The Wigner distribution function is given as

$$W_x(t, f) = \int_{-\infty}^{\infty} x\left(t + \frac{\tau}{2}\right) x^*\left(t - \frac{\tau}{2}\right) e^{-j2\pi f\tau} d\tau \tag{6}$$

2.4 Proposed Transmission Line Used as Test System

The proposed test transmission line is connected between two buses of the complex power system network. The large area utility on both ends of the transmission line is represented by generators 1 and 2 (G1 and G2). The test system of transmission line used for the study is illustrated in Fig. 1. Total line length is taken as 230 km. The line is modeled in four L1, L2, L3, and L4 each having a length of 57.5 km. Fault is created at points F1, F2, and F3. The fault location points F1 and F3 are used to investigate the effect of fault location whereas fault location point F2 situated at the middle of the line is used in all other cases. The measurement of the voltage

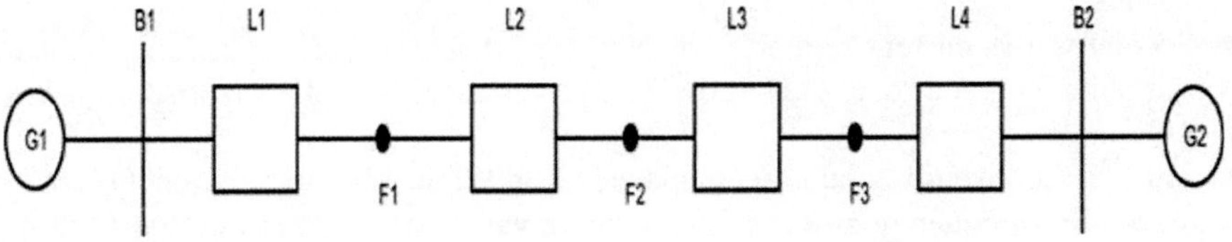

Fig. 1 Test transmission line

Table 1 Technical parameters of test system [5]

Particulars of parameters	Values of parameters
Voltage rating of system	500 kV
Frequency	60 Hz
Source resistance (ohm)	17.77
Source inductance	0.1218 H
Positive sequence resistance of line	0.01273 /km
Zero sequence resistance of line	0.3864 /km
Positive sequence inductance of line	0.9337e-3 H/km
Zero sequence inductance of line	4.1264e-3 H/km
Positive sequence capacitance of line	12.74e-9 F/km
Zero sequence capacitance of line	7.751e-9 F/km

and current is carried out using the three-phase VI measurement block of Simulink (Table 1).

2.5 Proposed Methodology

The methodology proposed for detection and classification of transmission line faults to design an effective transmission line protection scheme is designed for the current-based protection scheme used the Stockwell Transform, Hilbert Transform, and Wigner Distribution Function for processing of the signals.

2.5.1 Protection Scheme Using Current Signal

Current signal based algorithm is proposed for detection and classification of faults on the transmission line to design an effective transmission line protection scheme consisting of the design with the following steps detailed below.

- To start with (Initially), the current signal is decomposed by means of Stockwell transform with a sampling frequency of 3.2 kHz to gain S-matrix. Further, S-matrix is converted to a matrix having the absolute values corresponding to each element of S-matrix. Median of this new matrix is calculated and designated as ST-index.
- Secondly, the current signal is decomposed using Hilbert transform with a sampling frequency of 3.2 kHz. Absolute values output so obtained are designated as H-index.
- Thirdly, the current signal is decomposed using Wigner distribution function with a sampling frequency of 3.2 kHz. Absolute values obtained at the first level of decomposition are designated as WDF-index.

- Finally, element by element product of ST-index, H-index, and DWF-index is obtained and designated as Fault index. A threshold value of 5000 is selected to distinguish the faulty phase from the healthy phase as well as the faulty event from the healthy event.
- The proposed algorithm has been implemented under MATLAB/Simulink environment.

3 Detection and Classification of Transmission Line Faults Using Current-Based Fault Index

Simulation results related to detection and classification of transmission line faults using combined features of Stockwell Transform, Hilbert Transform, and Wigner Distribution Function based approach are detailed in this section. The results related to case studies such as reverse power flow, effects of fault location, switching of inductive load, and switching of capacitive load have been described in this section.

3.1 Healthy Condition

It is observed from Fig. 2b that the proposed ST-index has zero values throughout the range. Figure 2c depicts that the value of the proposed H-index is constant over the entire time range. It is observed from Fig. 2d that the value of WDF-index is zero except at the central location where it is non-zero representing the frequency distribution of fundamental frequency. The proposed fault index has zero values over entire time range as depicted in Fig. 2e. These plots are taken as reference plots for detection of faulty events.

3.2 Line to Ground Fault

This can be observed from Fig. 3a current in Phase-A increases after the faulty event. It is observed from Fig. 3b that the proposed ST-index has zero values throughout the range corresponding to Phases-B and C whereas these values corresponding to Phase-A are high indicating that there is a fault on this phase. Figure 3c indicates that the value of proposed H-index is constant over the entire time range for the Phases-B and C whereas it has high values for Phase-A indicating the faulty phase.

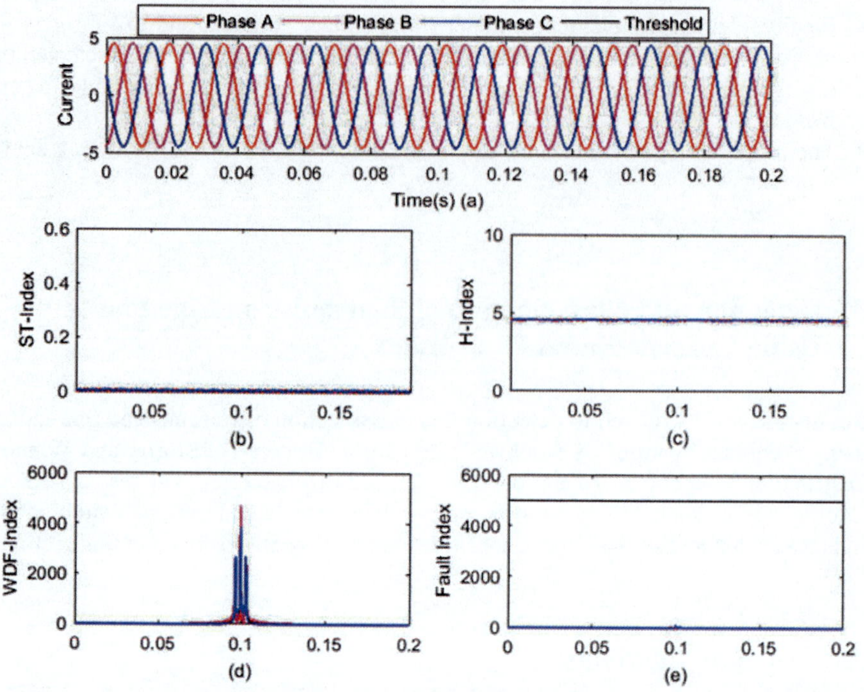

Fig. 2 Detection of transmission line faults using hybrid algorithm based on combined features of Stockwell transform, Hilbert transform, and Wigner distribution function results for healthy condition **a** current waveform **b** ST-index **c** H-index **d** WDF-index **e** proposed fault index

Figure 3d indicates that the value of proposed WDF-index has very high values corresponding to Phase-A near the central location of time whereas these values are low for Phases-B and C. It is depicted from Fig. 3e that proposed fault index corresponding to Phase-A has higher values above the threshold indicating the presence of a fault. The values of the fault index corresponding to Phases-B and C are below the threshold and approximately equal to zero indicating the healthy phases. Hence, LG fault on Phase-A has been successfully detected using the proposed algorithm based on Stockwell Transform, Hilbert Transform, and Wigner Distribution Function.

Similarly, in the rest of the section the results are shown through the figure from which we can analyze the result of different fault conditions.

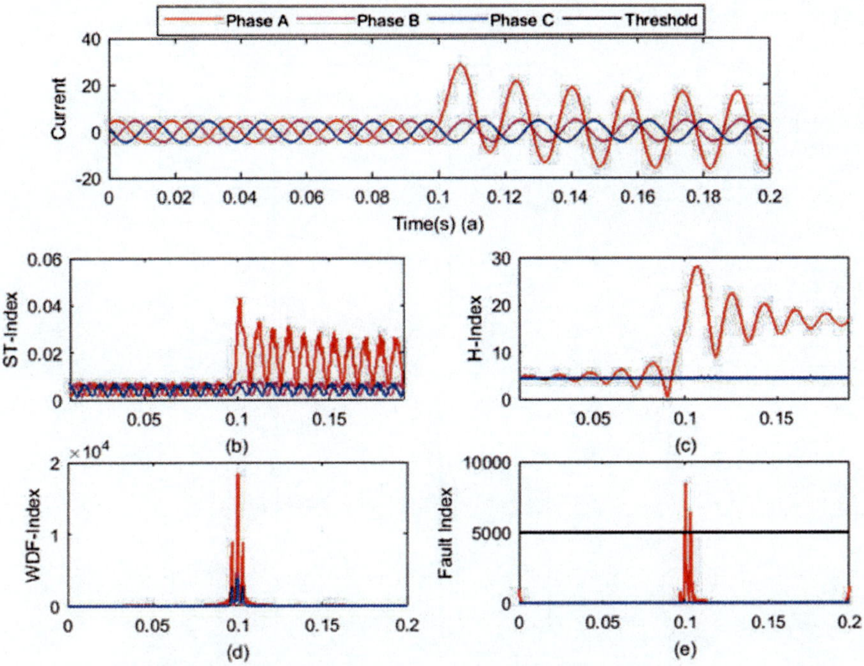

Fig. 3 Detection of transmission line faults using hybrid algorithm based on combined features of Stockwell transform, Hilbert transform, and Wigner distribution function results for line to ground fault **a** current waveform **b** ST-index **c** H-index **d** WDF-index **e** proposed fault index

3.3 Double Line Fault and Double Line to Ground Fault (LLG)

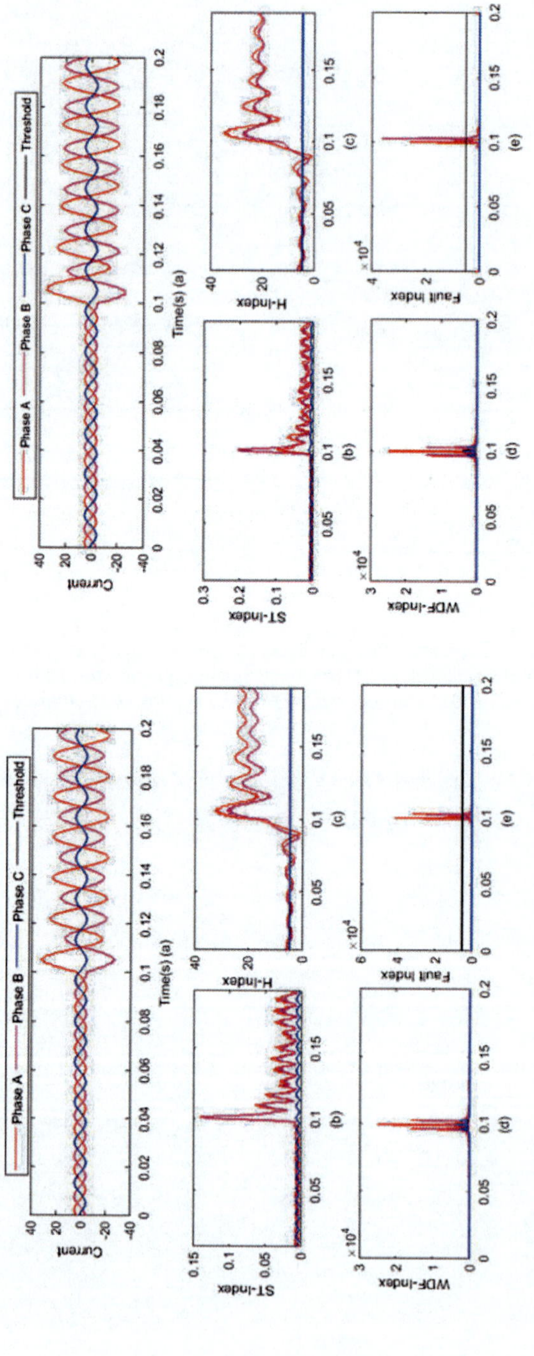

Fig. 4 Detection of transmission line faults using hybrid algorithm based on combined features of Stockwell transform, Hilbert transform, and Wigner distribution function results for double line fault and double line to ground fault **a** current waveform, **b** ST-index, **c** H-index, **d** WDF-index, **e** proposed fault index

3.4 Three Phase to Ground Fault

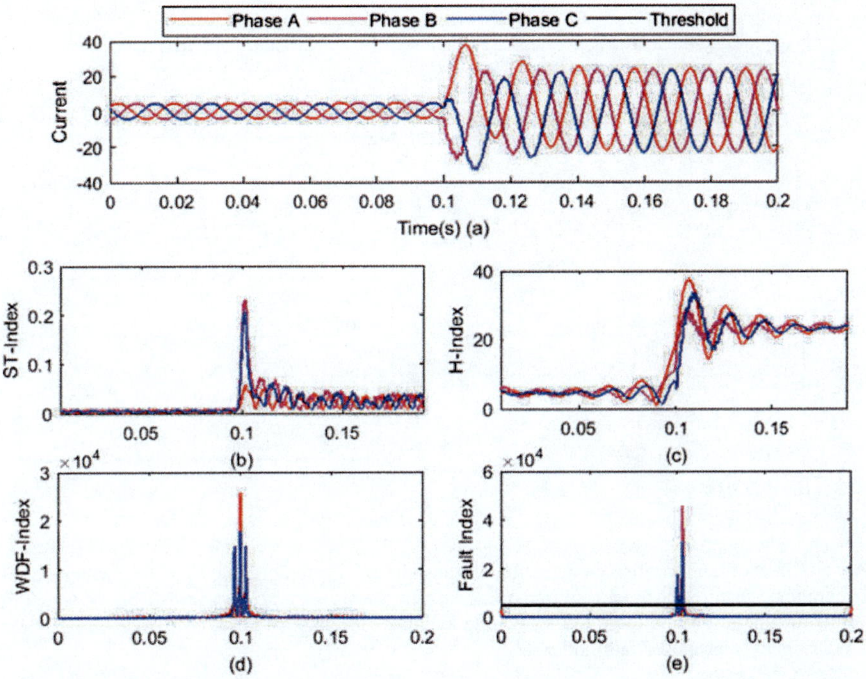

Fig. 5 Detection of transmission line faults using hybrid algorithm based on combined features of Stockwell transform, Hilbert transform, and Wigner distribution function results for three phase to ground fault **a** current waveform, **b** ST-index, **c** H-index, **d** WDF-index, **e** proposed fault index

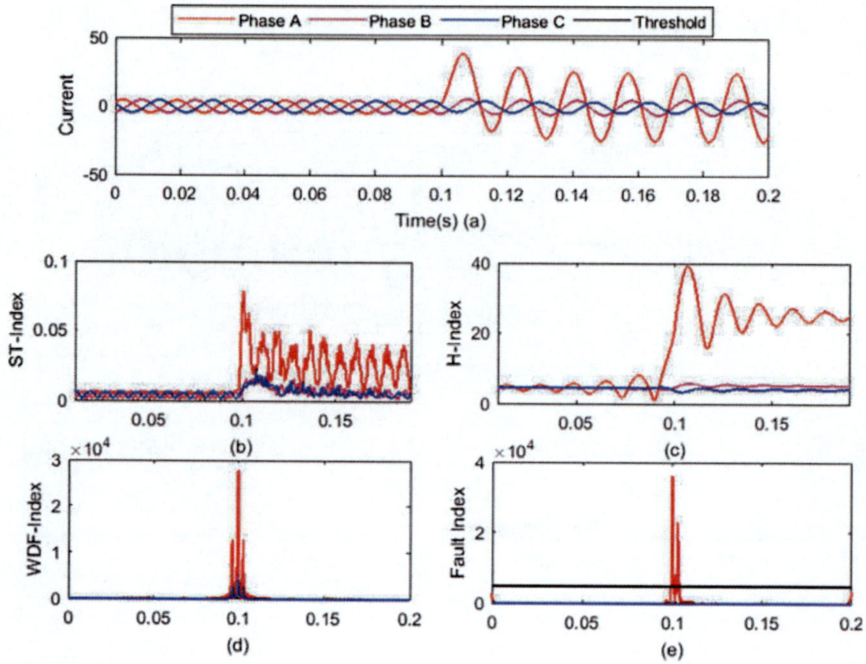

Fig. 6 Effect of fault location on detection of transmission line faults using hybrid algorithm based on combined features of Stockwell transform, Hilbert transform, and Wigner distribution function results for line to ground fault at 25% of line length **a** current waveform, **b** ST-index **c** H-index, **d** WDF-index, **e** proposed fault index

3.5 *Effect of Fault Location*

To generalize the proposed algorithm, it is tested at different locations of fault on the transmission line. All the faults have been tested at 25 and 75% of line length from the sending end of the line. However, the results of LG fault are shown as 25 and 75% of line length in the following sections. The results have been discussed for all the phases at 50% of line length in the above sections.

3.5.1 Line to Ground Fault at 25% Line Length

This can be observed from Fig. 6a that current in Phase-A increases after the faulty event. It is observed from Fig. 6b that the proposed ST-index has zero values throughout the range corresponding to Phases-B and C whereas these values corresponding to Phase-A are high indicating that there is a fault on this phase. Figure 6c indicates that the value of proposed H-index is constant over the entire time range for the Phases-B and C whereas it has high values for the Phase-A indicating the faulty phase. Figure 6d indicates that the value of proposed WDF-index has very high values corresponding to Phase-A near the central location of time whereas these values are low for Phases-B and C. It is depicted from Fig. 6e that proposed fault index corresponding to Phase-A has higher values above the threshold indicating the presence of fault. The values of fault index corresponding to Phases-B and C are below the threshold and approximately equal to zero indicating the healthy phases. Hence, LG fault on Phase-A at a line length of 25% from sending end of the transmission line has been successfully detected using the proposed algorithm based on Stockwell Transform, Hilbert Transform, and Wigner Distribution Function.

3.5.2 Line to Ground Fault at 75% Line Length

This can be observed from Fig. 7a that current in Phase-A increases after the faulty event. It is observed from Fig. 7b that the proposed ST-index has zero values throughout the range corresponding to Phases-B and C whereas these values corresponding to Phase-A are high indicating that there is a fault on this phase. Figure 7c

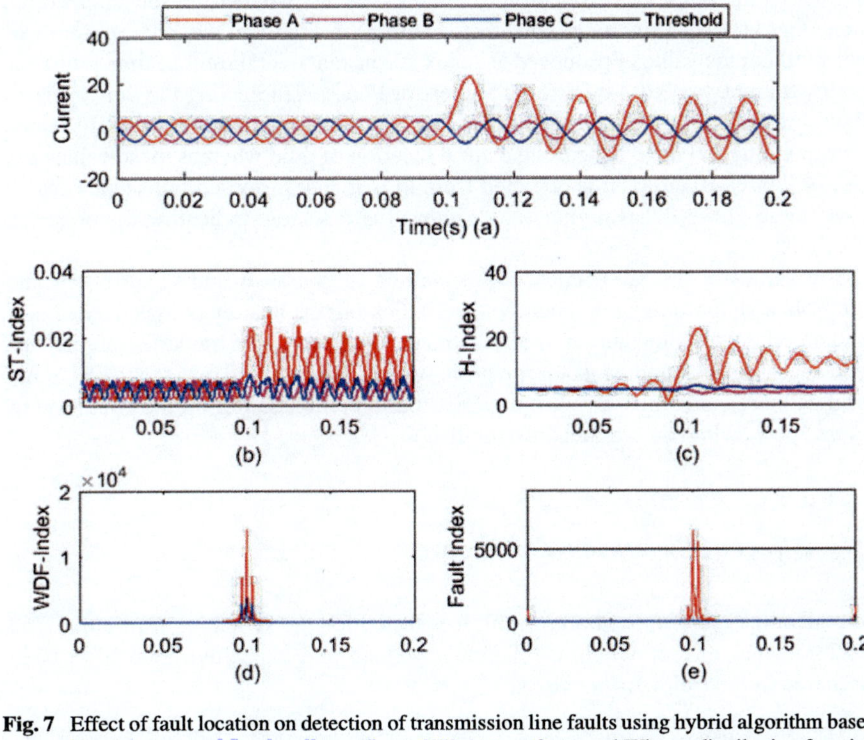

Fig. 7 Effect of fault location on detection of transmission line faults using hybrid algorithm based on combined features of Stockwell transform, Hilbert transform and Wigner distribution function results for line to ground fault at 75% of line length **a** current waveform, **b** ST-index, **c** H-index, **d** WDF-index, **e** proposed fault index

indicates that the value of proposed H-index is constant over the entire time range for Phases-B and C whereas it has high values for Phase-A indicating the faulty phase. Figure 7d indicates that the value of proposed WDF-index has very high values corresponding to Phase-A near the central location of time whereas these values are low for Phases-B and C. It is depicted from Fig. 7e that proposed fault index corresponding to Phase-A has higher values above the threshold indicating the presence of fault. The values of fault index corresponding to Phases-B and C are below the threshold and approximately equal to zero indicating the healthy phases. Hence, LG fault on Phase-A at a line length of 75% from sending end of the transmission line has been successfully detected using the proposed algorithm based on Stockwell Transform, Hilbert Transform, and Wigner Distribution Function.

3.6 Effect of Reverse Power Flow

This can be observed from Fig. 8a that current in Phase-A increases after the faulty event. It is observed from Fig. 8b that the proposed ST-index has zero values throughout the range corresponding to Phases-B and C whereas these values corresponding to Phase-A are high indicating that there is a fault on this phase. Figure 8c indicates that the value of proposed H-index is constant over the entire time range for Phases-B and C whereas it has high values for Phase-A indicating the faulty phase. Figure 8d indicates that the value of proposed WDF-index has very high values corresponding to Phase-A near the central location of time whereas these values are low for Phases-B and C. It is depicted from Fig. 8e that proposed fault index corresponding to Phase-A has higher values above the threshold indicating the presence of a fault.

The values of the fault index corresponding to phases-B and C are below the threshold and approximately equal to zero indicating the healthy phases. Hence, LG fault on Phase-A at line length of 50% from sending end of the transmission line has been successfully detected using the proposed algorithm based on Stockwell Transform, Hilbert Transform, and Wigner Distribution Function during the condition of reverse power flow on the transmission line.

3.7 Effect of Switching Transients

To generalize the proposed algorithm, it is tested for the switching transients. The results related to the switching of inductive load and capacitive load have been discussed in the following sections.

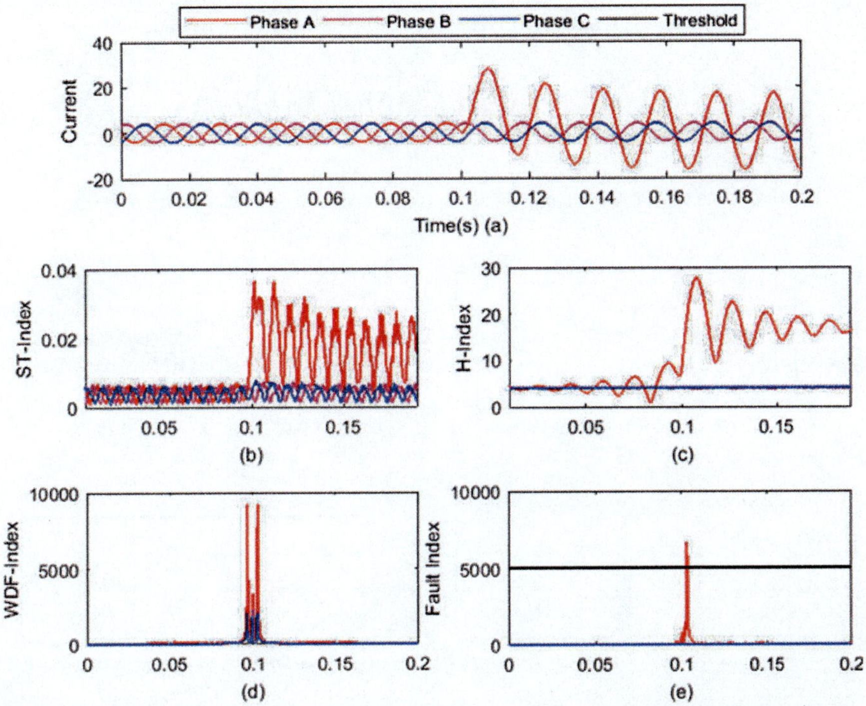

Fig. 8 Effect of reverse power flow on detection of transmission line faults using hybrid algorithm based on combined features of Stockwell transform, Hilbert transform and Wigner distribution function results for line to ground fault at 50% of line length **a** current waveform, **b** ST-index, **c** H-index, **d** WDF-index, **e** proposed fault index

3.7.1 Switching of Capacitive Load

This can be observed from Fig. 9a that current contains switching transients due to switching the capacitive load. It is observed from Fig. 9b that the proposed ST-index has zero values before switching on the capacitive load and finite values at the time of switching the capacitor. Figure 9c indicates that the value of the proposed H-index is constant before switching on the capacitive load and finite values at the time of switching the capacitor. Figure 9d indicates that the value of proposed WDF-index has high values corresponding to all phases near the central location of time. It is depicted from Fig. 9e that proposed fault index corresponding to all phases has higher values but below the threshold indicating that the event is switching and not the faulty. Hence, switching of the capacitive load has been differentiated from the faulty events using the proposed algorithm based on Stockwell Transform, Hilbert Transform, and Wigner Distribution Function.

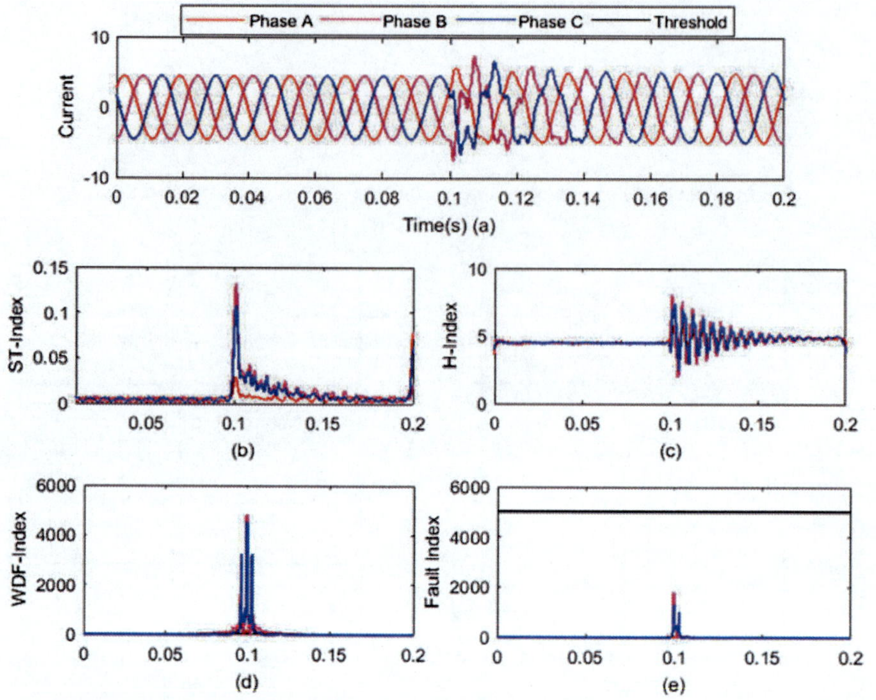

Fig. 9 Simulation results using hybrid algorithm based on combined features of Stockwell transform, Hilbert transform, and Wigner distribution function for the effect of capacitive load switching **a** current waveform, **b** ST-index, **c** H-index, **d** WDF-index, **e** proposed fault index

3.8 Switching of Inductive Load

This can be observed from Fig. 10a that current in all phases increases due to increased load after switching the inductive load. It is observed from Fig. 10b that the proposed ST-index has zero values before switching on the inductive load and finite values at the time of switching the load. Figure 10c indicates that the value of the proposed H-index is constant before switching on the inductive load and finite values at the time of switching the load. Figure 10d indicates that the value of the proposed WDF-index has high values corresponding to all phases near the central location of time. It is depicted from Fig. 10e that proposed fault index corresponding to all phases has higher values but below the threshold indicating that the event is switching and not the faulty. Hence, switching of the inductive load has been differentiated from the faulty events using the proposed algorithm based on Stockwell Transform, Hilbert Transform, and Wigner Distribution Function.

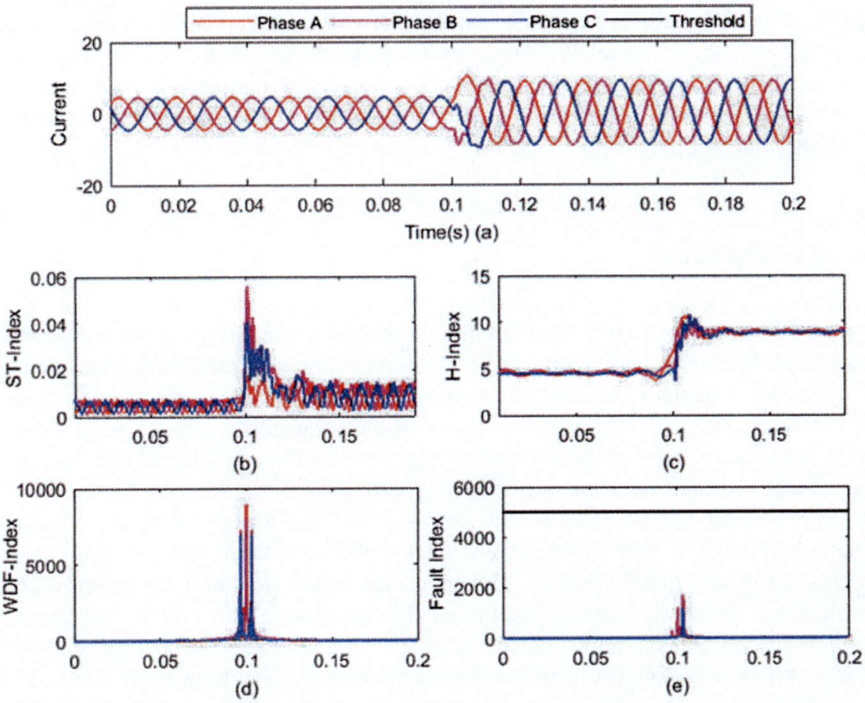

Fig. 10 Simulation results using hybrid algorithm based on combined features of Stockwell transform, Hilbert Transform and Wigner Distribution Function for the effect of inductive load **a** current waveform **b** ST-index **c** H-index **d** WDF-index **e** proposed fault index

3.9 Classification of Faults

The peak values of the proposed fault index based on the current-based features are provided in Table 2. It can be observed that if values of fault index are below 5000 for all the phases then there is no fault on the system and this indicates the healthy condition. If there are only values of fault index is above 5000 corresponding to one

Table 2 Peak values of current-based fault index

Sr. No.	Type of fault	Peak values of fault index		
		Phase-A	Phase-B	Phase-C
1	Healthy condition	0	0	0
2	Line to ground fault	8000	100	100
3	Double line fault	4×10^4	3×10^4	100
4	Double line to ground fault	2.5×10^4	3.5×10^4	100
5	Three-phase fault involving ground	4.5×10^4	3×10^4	4×10^4

phase, then this indicates the presence of LG fault. If the fault index corresponding to two phases is above the threshold then this indicates the presence of LL or LLG fault. These can be discriminated from each other by using certain ground fault index. If the fault index corresponding to all the three phases is above the threshold then this indicates the presence of LLLG fault.

4 Conclusions

In this paper, an accurate protection system against various types of power system faults like line to ground, double line, double line to ground, and three-phase faults including ground three phase to ground have been developed. An algorithm based on combined features of Stockwell transform, Hilbert transform, and Wigner distribution function has been proposed for the detection and classification of transmission line faults to design effective protection scheme.

The method is tested on a test transmission line for different cases which includes change in line length, reverse power flow condition and switching transients using MATLAB/Simulink. The advantage of this method is its effectiveness for different cases where many traditional algorithms have failed. From the simulation results, it has been found that the normal value of fault index in healthy condition stays zero. In case of faulty event the value of fault index varies between 6.121e+03 as minimum value and 45.36e+03 as maximum value in case of current-based analysis when threshold is 5000. In case of voltage-based analysis, the value of fault index varies between 1.290e+03 as minimum value and 9.101e+3 as maximum value when threshold is 1000. Hence, there exist an accurate difference between a healthy and a faulty condition.

References

1. Yang L, Xiao H, Zidong W, Dong-Hua X (2015) Fault Detection and Diagnosis for a Class of Nonlinear Systems with Decentralized Event-triggered Transmissions. IFAC-Pap Line (Elsevier) 48–21:1134–1139
2. Durga Prasad Ch, Srinivasu N (2015) Fault detection in transmission lines using instantaneous power with ED based fault index. Proceedia Technol 21:132–138
3. Ferreira VH, Zanghi R, Fortes MZ, Sotelo GG, Silva RBM, Souza JCS, Guimarães CHC, Gomes Jr.Fsjjglds S (2016) A survey on intelligent system application to fault diagnosis in electric power system transmission lines Elect. Power Syst Res 136:135–153
4. Hussain S, Osman AH (2016) Fault location scheme for multi-terminal transmission lines using unsynchronized measurements. Int J Elect Power Energy Syst 78:277–284
5. Gafoor Shaik A, Pulipaka RRV (2015) A new wavelet based fault detection, classification and location in transmission lines. Int J Elect Power Energy Syst 64:35–40
6. Dehghani M, Khooban MH, Niknam T (2016) Fast fault detection and classification based on a combination of wavelet singular entropy theory and fuzzy logic in distribution lines in the presence of distributed generations. Int J Elect Power Energy Syst 78:455–462

7. Morais AP, Cardoso Júnior G, Mariotto L, Marchesan G (2016) Fault location scheme for multi-terminal transmission lines using unsynchronized measurements. Int. J. Elect. Power Energy Syst 78:277–284
8. Swetapadma A, Yadav A (2016) Directional relaying using support vector machine for double circuit transmission lines including cross-country and inter-circuit faults. Int J Electr Power Energy Syst 81:254–264

Author Index

A
Agarwal, Kusum Lata, 251, 335, 343, 357
Agarwal, Nitesh, 227
Akter, Morium, 95
Ali, Md. Hayder, 53, 65
Ali, Mohammad Hanif, 53, 65
Aseri, Ramesh, 425

B
Bhandari, Sanjay, 191
Bhansali, Pratik, 303
Bhati, Tanmay, 449

C
Chaudhary, Laxmi, 159
Choudhary, Arjun, 9
Choudhary, Ranjana, 397

D
Dave, Sushma, 101
Dhanraj, 251
Dhariwal, Saraswati Chand, 209

E
Eva, Mahmuda Najnin, 95

F
Fahmida Islam, Sk., 95
Ferdous, Jannatul, 95

G
Gehlot, Kishore Singh, 409
Gupta, Utkarsh, 1, 139

J
Jangid, Aisha, 159
Jha, Aashish Kumar, 111
Jindal, Aditya, 335
Joshi, Ashwani Kumar, 425
Joshi, Yogesh, 381

K
Kanwar, Khamma, 285
Kapoor, Gaurav, 73, 119
Karia, Megha C., 25
Khyani, Harish, 295
Khyani, Harish Kumar, 199, 219, 449
Kulshrestha, Abhijit, 171
Kumar, Amit, 181
Kumar, Nitesh, 181
Kushwaha, Ajay, 267

L
Lehri, Divam, 9

M
Mangal, Rajat, 209
Mathur, Akleshwar, 219
Mathur, Akshat, 101
Mathur, Ashish, 171
Mathur, Geetika, 171
Mehta, Sandip, 303

M. Shorif Uddin et al. (eds.), *Intelligent Energy Management Technologies*,
Algorithms for Intelligent Systems,
https://doi.org/10.1007/978-981-15-8820-4

Mehta, Vinit, 319, 381
Mohanty, A. M., 181

N
Namjoshi, Sadanand A., 45
Nehiwal, Jayesh, 295

P
Panchauli, Dheeraj, 1
Pandya, Bhavik J., 25
Panwar, Kapil, 335
Patel, Shrawan Ram, 397
Patra, Prashanjeet, 181
Purohit, Rajendra, 227

R
Ram Patel, Shrawan, 295
Ram, Shrawan, 409

S
Saini, Pawan Kumar, 133
Sangani, Kamlesh B., 25
Shalini, 133
Sharma, Komal, 159
Sharma, Nidhi, 111
Sharma, Santosh Kumar, 139
Sharma, Shubham, 343
Sharma, Yatendra Mohan, 133
Shorif, Sumaita Binte, 95
Shringi, Surbhi, 139
Singhal, Amit, 85, 149
Singh, Bablu Kumar, 191, 277
Singh, Chandershekhar, 267, 295
Singh, Samit Kumar, 45
Singh, Sudeep Kumar, 181
Swain, Suvam Sourav, 181

T
Tandon, Ankit, 85, 149

U
Uddin, Mohammad Shorif, 95

V
Vajpai, Jayashri, 199, 285, 319
Vishnoi, Shyam Lal, 357
Vyas, Supriya, 277

Y
Yadav, D. K., 1